Ultrafilters across Mathematics

CONTEMPORARY MATHEMATICS

530

Ultrafilters across Mathematics

International Congress
ULTRAMATH 2008: Applications of Ultrafilters and Ultraproducts in Mathematics
June 1–7, 2008
Pisa, Italy

Vitaly Bergelson
Andreas Blass
Mauro Di Nasso
Renling Jin
Editors

American Mathematical Society
Providence, Rhode Island

2000 *Mathematics Subject Classification.* Primary 03C20, 03E05, 03H05, 05C55, 28E15.

Library of Congress Cataloging-in-Publication Data

ULTRAMATH 2008: Applications of Ultrafilters and Ultraproducts in Mathematics (2008 : Pisa, Italy)

Ultrafilters across mathematics : international congress, ULTRAMATH 2008, Applications of Ultrafilters and Ultraproducts in Mathematics, June 1–7, 2008, Pisa, Italy / Vitaly Bergelson ... [et al.], editors.

p. cm. — (Contemporary mathematics ; v. 530)

Includes bibliographical references.

ISBN 978-0-8218-4833-3 (alk. paper)

1. Ultrafilters (Mathematics)—Congresses. 2. Combinatorial set theory—Congresses. 3. Ramsey theory—Congresses. I. Bergelson, V. (Vitaly), 1950– II. Title.

QA9.7.I58 2010
511.3—dc22

2010025146

∞ The paper used in this book is acid-free and falls within the guidelines
established to ensure permanence and durability.
Visit the AMS home page at `http://www.ams.org/`

10 9 8 7 6 5 4 3 2 1 15 14 13 12 11 10

Contents

Preface

The project of this volume originated from the International Conference:

ULTRAMATH 2008: *Applications of Ultrafilters and Ultraproducts in Mathematics*

held in Pisa (Italy), June 1-7, 2008. ULTRAMATH aimed to present recent results in the whole spectrum of mathematics which are grounded on the use of ultrafilters and ultraproducts. One of its goals was to disseminate information about the various ultrafilter techniques and their usefulness in dealing with open problems. Another goal of the ULTRAMATH was to bring together researchers with different backgrounds, and to facilitate their collaboration and interactions, especially on topics which bridge different areas of mathematics. About 100 mathematicians participated, and the program consisted of 12 invited talks and 32 contributed talks (a complete list is given at the end of this preface).

The intention of this volume is to present the state-of-the-art of applications of ultrafilter and ultraproduct techniques.

To this end, the editors chose experts on various aspects of ultrafilters, ultrapowers, and their applications, and invited them to contribute research expositions or surveys on those aspects. All papers included in this volume are the final versions and they have all been individually refereed.

Besides two general surveys on ultrafilters in set theory and on the ultraproduct construction, this volume includes papers that cover the following areas of applications: additive and combinatorial number theory, nonstandard methods in stochastic differential equations, measure theory, dynamics, Ramsey theory, algebra in the space of ultrafilters, and large cardinals.

The papers in this volume are intended to be accessible and interesting for students and researchers who are not experts on ultrafilters and ultraproducts. Greater prominence has been given to results that can be formulated and presented in non-specialist terms and be in principle understandable by any mathematician, and to those results that connect different areas of mathematics, revealing new facets of known important topics.

Acknowledgements. We are grateful to the following programs and organizations for the financial support that made "ULTRAMATH 2008" possible:

• Research program P.R.I.N. 2007 "Teoria dei Modelli, Teoria degli Insiemi e Applicazioni";

• Centro di Ricerca Matematica "E. De Giorgi", Pisa;
• Istituto Nazionale di Alta Matematica I.N.D.A.M. - G.N.S.A.G.A.;
• School of Graduate Studies "Galileo Galilei", Pisa;
• Università di Pisa;
• Journal "Logic and Analysis", Springer.

We are also grateful to the A.M.S. for agreeing to publish the volume. A special thanks to Christine M. Thivierge, the American Mathematical Society's Associate Editor for Conference Proceedings, for her assistance and patience.

Vitaly Bergelson
Andreas Blass
Mauro Di Nasso
Renling Jin

ULTRAMATH 2008

Organizer: M. Di Nasso; Scientific committee: V. Bergelson, A. Blass, R. Jin; WEB: `http://www.dm.unipi.it/~ultramath`.

List of invited talks. Vieri Benci, *Nonstandard Methods in Analysis*; Vitaly Bergelson, *Ultrafilters and Ergodic Ramsey Theory*; Andreas Blass, *Ultrafilters and Set Theory*; Nigel Cutland, *Stochastic Navier-Stokes Equations: Ideas and Results Using Nonstandard Analysis*; David H. Fremlin, *Ultrafilters in Measure Theory*; Neil Hindman, *Algebra in the Space of Ultrafilters and Ramsey Theory*; Renling Jin, *Ultrapower of* $\mathbb{N}$ *and Density Problems*; Marius Junge, *Ultraproducts in Functional Analysis*; Melvyn B. Nathanson, *Problems in Additive Number Theory with an Ultrafilter Flavor*; Itay Neeman, *Ultrafilters, Determinacy, and Large Cardinals*; Jaroslav Nešetřil, *Some Topics on Non-Ramsey Connection of Combinatorics and Ultrafilters*; Boban Velickovic, *Set Theoretic Aspects of the Space of Ultrafilters* $\beta\mathbb{N}$.

List of contributed talks. Mathias Beiglböck, *A Variant of the Hales-Jewett Theorem*; Piotr Borodulin-Nadzieja, *On Pseudo-intersections and Condensers*; Abdelmadjid Boudaoud, *On the Terms of Unlimited Rank of Lucas Sequences*; Ulf Clotz, *Topology from a Remote Point of View*; Daryl Cooper, *Real Projective Structures on Manifolds and the Hyper-reals*; Abhijit Dasgupta, *Logicless Nonstandard Analysis: an Axiom System*; Marco Forti, *Counting Infinite Point Sets*; Stefano Francaviglia, *Asymptotic Cones of Metric Spaces and Related Topics*; Arthur D. Grainger, *Remarks on Ultrafilters on the Collection of Finite Subsets of an Infinite Set*; Gonsalo Gutierres, *Ultrafilters, Closure Operators and the Axiom of Choice*; Frederik Herzberg, *A Definable Framework for Nonstandard Analysis*; Dirk Hofmann, *Some Aspects of Ultrafilter Convergence in Topology*; Karel Hrbàček, *Relative Set Theory*; Peter Krautzberger, *Idempotent Ultrafilters and Finer Topologies on* $\beta\mathbb{N}$; Leonard Kwuida, *Prime Ideal Theorem for Weakly Dicomplemented Lattices*; Sonia L'Innocente, *Ultraproducts and Lie Algebras: Some Possible Interactions*; Ya-Qing Li, *Defining New Generalized Functions by Nonstandard Discrete Functions and Difference Quotients*; Paolo Lipparini, *Variations on Properties of Ultrafilters Related to Compactness of Topological Spaces and to Model-Theoretical Principles*; Daniar Mushtari, *Ultraproducts and Contiguity*; Richard O'Donovan, *Analysis Using Relative Infinitesimals*; Serpil Pehlivan, *Weak Ideal Convergence in Banach Spaces*; Aleksandar Perovic, *A Remark on Ultrapower Cardinality and the Continuum Problem*; Igor Protasov, *Dynamical Equivalence on* G^*; Yves Raynaud, *Ultraproducts and Characterization of Certain Classes of Classical Banach Lattices*; Ireneusz Recław, *Ideal Limits of Sequences of Continuous Functions and a Game*; David Ross, *Automatic Continuity of Nonstandard Measures*; Denis I. Saveliev, *On Hindman Sets*; Christian Serpé, *Nonstandard Methods in Algebraic Geometry*; João Paulo Teixeira, *Local-in-Time Existence and Regularity of the* n*-Dimensional Navier-Stokes Equations via Discretizations*; M. Akbari Tootkaboni, *Comfort Order on Locally Compact Topological Spaces*; Boaz Tsaban, *Super Ultrafilters and Ramsey Theory*; Hans Vernaeve, *Algebras of Generalized Functions and Nonstandard Analysis.*

Contemporary Mathematics
Volume **530**, 2010

An elementary approach to Stochastic Differential Equations using the infinitesimals.

Vieri Benci, Stefano Galatolo, and Marco Ghimenti

Abstract. The aim of this paper is to evidence two points relative to NSA:

- In most applications of NSA to analysis, only elementary facts and techniques of nonstandard calculus seems to be necessary.
- The advantages of a theory which includes infinitesimals rely more on the possibility of making new models rather than in proving techniques.

These two points will be illustrated by the theory of Brownian motion which can be considered as a *classical model* to test the power of the infinitesimal approach. Starting from a naive idea of Brownian motion, we deduce the Fokker-Plank equation in a simple and rigorous way. It is possible to keep every things to a simple level since all the theory of stochastic differential equations is treated as a hyperfinite theory and it is not translated in a *standard model.*

The only standard object is the final one: the Fokker-Plank equation.

1. Introduction

One of the fields of Mathematics where the ultrafilters play a crucial role is Nonstandard Analysis and α-theory which is a variant of it (see [**5**]). In this paper we present an application of α-theory to stochastic differential equations.

Suppose that x is a physical quantity whose evolution is governed by a deterministic force which has small random fluctuations; such a phenomenon can be described by the following equation

$$\dot{x} = f(x) + h(x)\xi(t) \tag{1.1}$$

where $\dot{x} = \frac{dx}{dt}$, and ξ is a "white noise". Intuitively, a *white noise* is the derivative of a *Brownian motion*, which is a continuous function which is not differentiable in any point.

There is no function ξ which has such a property, actually the mathematical object which models ξ is a distribution. Thus Equation (1.1) makes sense if it lives in the world of distributions.

On the other hand the kind of problems which an applied mathematician asks are of the following type. Suppose that $x(0) = 0$ and that $\xi(t)$ is a random noise of

2010 *Mathematics Subject Classification.* 03H10, 26E35, 35Q84.

which only the statistical properties are known. What is the probability distribution $P(t,x)$ of x at the time t?

This question can be formalized by the theory of stochastic differential equation and eq. (1.1) takes the form

$$dx = f(x)dt + h(x)dw. \tag{1.2}$$

thus, the white noise dw is regarded as the "differential" of a Wiener process w. In this case, both $x(t)$ and $w(t)$ are modelled, not by distributions, but by stochastic processes.

By the Ito theory, the above question can be solved rigorously: the probability distribution can be determined solving the Fokker-Plank equation:

$$\frac{\partial P}{\partial t} = \frac{1}{2}\frac{\partial^2}{\partial x^2}\left(h(x)^2 P\right) + \frac{\partial}{\partial x}\left(f(x)P\right) \tag{1.3}$$

Eq. (1.1) (or (1.2)) and (1.3) are very relevant in applications of Mathematics and the practitioners of mathematics such as engineers, physicists, economists, etc. make a large use of it. However the mathematics used in these equations is rather involved and many of them are not able to control it.

Usually people think of some intuitively simpler model. For example, $\xi(t)$ is considered as a force which acts at discrete instants of time t_i; it is supposed that the difference of two successive times $dt = t_{i+1} - t_i$ is infinitesimal and its strength is infinite; namely

$$\xi(t) = \pm\frac{1}{\sqrt{dt}} \tag{1.4}$$

The sign of this force is determined by a fair coin tossing. Clearly eq. (1.4) makes no sense and the gap between the rigorous mathematical description and the intuitive model is quite large.

The main purpose of this paper is to reduce this gap. We will use infinite and infinitesimal numbers in such a way that eqs. (1.4) and (1.1) make sense and, in this framework, we will deduce eq.(1.3) rigorously. Our proof is relatively simple and very close to intuition.

The use of infinite and infinitesimal numbers naturally leads to Nonstandard Analysis (*NSA*). Actually, some mathematicians have described the stochastic differential equation by Nonstandard Analysis (cf. e.g. [**2**], [**8**], [**1**], [**11**] and references therein). However the machinery of NSA is too complicated for practitioners of mathematics even if its ideas are simpler.

In this paper we will not use NSA but α-theory which is a variant of it; α-theory has been introduced in [**5**] (see also [**4**] and [**3**]) with the purpose to provide a simpler approach to NSA. In fact, in the quoted paper it has been proved that a particular model of NSA can be deduced by the axioms of α-theory (we refer also to [**6**] and to [**7**] for the reader interested to investigate in these questions).

The main difference between α-theory and the usual Nonstandard Analysis consists in the fact that it does not need to distinguish two mathematical universes, (the standard universe and the nonstandard one). α-theory just postulates the "existence" of an infinite integer number called α and it provides the rules necessary to deal with the mathematical objects which can be constructed by its introduction. For example, α-theory allows to define functions such as "$\sin(\alpha t)$" and to manage with it.

In particular, using this theory, it is possible to define the "grid functions" which are functions defined for times t_i belonging to a set $\mathbb{H}$ which models the axis of time. Using the notion of grid function, we are able to give a sense to (1.1) and (1.4) and to deduce eq. (1.3) rigorously.

Our approach presents the following peculiarities:

- we will rewrite eq. (1.1) as a "grid" differential equation:

$$\frac{\Delta x}{\Delta t} = f(x) + \xi(t) \tag{1.5}$$

  where $\frac{\Delta x}{\Delta t}$ denotes the grid derivative (see Def. 3.1). From this equation, it easy to recover both a distribution equation and a stochastic equation, and, at the same time, eq. (1.5) has a very intuitive meaning.
- when eq. (1.5) is considered from the stochastic point of view, the noise ξ is regarded as a *grid function* belonging to the space of all possible noises $\mathcal{R}$. If ξ is regarded as a random variable, the probability on the sample space $\mathcal{R}$ can be defined in a naive way, namely every noise has the same probability. This is the basic idea of the Loeb measure ([**9**]) which is an important tool in the applications of NSA , but we do not need to use it. Actually we do not need to introduce any kind of measure.

Concluding, this paper wants to show the following two points:

(a) α-theory provides an entry to the use of infinite and infinitesimal numbers (and to nonstandard methods) that is easier for outsider then the usual ones.

(b) By directly working with infinitesimal and skipping the step of making connections with the traditional approaches to problems, one get simple new models of natural phenomena.

2. The Alpha-Calculus

2.1. Basic notions of Alpha-Theory. In this section, we will expose the basic facts of α-theory and the basic tools which will be used in the paper in a elementary and self contained way.

α-theory is based on the existence of a new mathematical object, namely α which is added to the other entities of the mathematical universe. We may think of α as a new "*ideal*" natural number added to $\mathbb{N}$, in a similar way as the imaginary unit i can be seen as a new ideal number added to the real numbers $\mathbb{R}$. Before going to the axioms for α, we remark that *all* usual principles of mathematics are implicitly assumed. Informally, we can say that, by adopting α-theory, one can construct sets and functions according to the "usual" practice of mathematics, with no restrictions whatsoever. A precise definition of what we mean by "usual principles of mathematics" (i.e. of our underlying set theory) is given in [**5**].

Like the use of the imaginary entity i is governed by simple properties like $i^2 = -1$ and the usual rules for the product and sum, the use of α is governed by the following five axioms.

In the following, by *sequence* we mean any function defined on the set of natural numbers. In particular we give no restriction on the set of its values.

α1. Extension Axiom.
Every sequence φ can be uniquely extended to $\mathbb{N} \cup \{\alpha\}$. The corresponding value at infinity will be denoted by $\varphi(\alpha)$ and called the value of φ at the point α or

more simply "α-value". If two sequences φ, ψ are different at all points, then $\varphi(\alpha) \neq \psi(\alpha)$.

We remark that if $\varphi : \mathbb{N} \to A$, then in general $\varphi(\alpha) \notin A$. The "difference preserving" condition given above can be rephrased as follows: "If two sequences are different at all n then they must be different at the point "α" as well". It is a non-triviality condition, that will allow plenty of values at infinity. Moreover remark that the α-value of a sequence should not be confused with its limit. In fact, the α-value differs from a limit even for this first axiom; in fact different sequences might have the same limit.

The next axiom gives a natural coherence property with respect to compositions. If $g : A \to B$ and $h : B \to C$, denote by $h \circ g : A \to C$ the composition of h and g, i.e. $(h \circ g)(x) = h(g(x))$.

α2. Composition Axiom.
If φ and ψ are sequences and if f is any function such that compositions $f \circ \varphi$ and $f \circ \psi$ make sense, then

$$\varphi(\alpha) = \psi(\alpha) \Rightarrow (f \circ \varphi)(\alpha) = (f \circ \psi)(\alpha)$$

So, if two sequences takes the same value at infinity, by composing them with any function we again get sequences with the same α-value.

α3. Real Number Axiom.
Let r be a real number. If $c_r : n \mapsto r$ is the constant sequence with value r, then $c_r(\alpha) = r$; if $1_{\mathbb{N}} : n \mapsto n$ is the immersion of $\mathbb{N}$ in $\mathbb{R}$, then $1_{\mathbb{N}}(\alpha) = \alpha \notin \mathbb{R}$.

We remark that, here, the real numbers need to be considered as atoms namely mathematical objects which are not sets. This axiom simply says that, for real numbers, the notions of constant sequence is preserved at infinity. The latter condition says that the ideal number α is actually a *new* number. Thus the immersion $1_{\mathbb{N}}$ provides a first example of sequence $\varphi : \mathbb{N} \to \mathbb{R}$ such that $\varphi(\alpha) \notin \mathbb{R}$.

α4. Internal Set Axiom.
If ψ is a sequence of sets, then also $\psi(\alpha)$ is a set and

$$\psi(\alpha) = \{\varphi(\alpha) : \varphi(n) \in \psi(n) \text{ for all } n\}.$$

Thus, the membership relation is preserved at infinity. That is, if $\varphi(n) \in \psi(n)$ for all n, then $\varphi(\alpha) \in \psi(\alpha)$. Besides, all elements of $\psi(\alpha)$ are obtained in this way. That is, they all are values at infinity of sequences which are pointwise members of ψ. The sets considered above will be called *Internal sets.*

α5. Pair Axiom.
If $\vartheta(n) = \{\varphi(n), \psi(n)\}$ for all n, then $\vartheta(\alpha) = \{\varphi(\alpha), \psi(\alpha)\}$.

Thus, if the sequence ξ is such that either $\xi(n) = \varphi(n)$ or $\xi(n) = \psi(n)$ for all n, then either $\xi(\alpha) = \varphi(\alpha)$ or $\xi(\alpha) = \psi(\alpha)$ at infinity as well. As a straight consequence of the last two axioms, any constant sequence with value a finite set of natural numbers, or a finite set of finite sets of natural numbers etc., takes the same value at infinity as well. We remark that this is not true for arbitrary constant sequences.

We remark that the above five axioms are given somewhat "informally". Precise indications for a rigorous formulation as sentences of a suitable first-order language are given in [**5**]. Also, we refer to [**5**] for the proofs of the propositions below, but we suggest the reader to try them by himself to get acquainted with α-theory.

DEFINITION 2.1. If A is a set, the $*$-transform of A is defined as follows:

$$A^* = \{\varphi(\alpha) : \varphi \in A^{\mathbb{N}}\}.$$

If ψ is a sequence such that $\psi(n) = A$ *for all* n then by the Internal Set Axiom, we have that $\psi(\alpha) = A^*$. Then constant set-valued sequences behave differently than real valued sequences (cf. the Real Number Axiom).

DEFINITION 2.2. The set of the *hyperreal numbers* is the $*$-transform of the set of the real numbers:

$$\mathbb{R}^* = \{\varphi(\alpha) : \varphi \in \mathbb{R}^{\mathbb{N}}\}.$$

In other words, the hyperreal numbers are the α-values assumed by real sequences. With obvious notation, for instance we will write $\sin\frac{2}{\alpha}$ to mean the hyperreal number obtained as the values at infinity of the sequence $\left\{\sin\frac{2}{n}\right\}_{n\in\mathbb{N}}$.

The sum and product operation are naturally transported on the hyperreal set. Moreover we have the following:

PROPOSITION 2.3. *The hyperreal number system* $\langle\mathbb{R}^*; +, \cdot, 0, 1, <\rangle$ *is an ordered field.*

Besides the considered sets of hyper-numbers, another fundamental notion in nonstandard analysis is the following.

DEFINITION 2.4. A set $\Gamma \subset A^*$ is called *hyperfinite* if

$$\Gamma = \{\varphi(\alpha) : \varphi(n) \in A_n\}$$

where $A_n \subset A$ is a sequence of finite sets. Given a hyperfinite set Γ, we define its cardinality $|\Gamma|$ as follows:

$$|\Gamma| = \psi(\alpha) \in \mathbb{N}^*$$

where $\psi(n) = |A_n|$ is the cardinality of the finite set A_n.

In general hyperfinite sets are infinite; their importance relies in the fact that they retain all "elementary" properties of finite sets. Applications of hyperfinite sets will be given in subsequent sections. For example the following holds

PROPOSITION 2.5. *Every nonempty hyperfinite subset of* $\mathbb{R}^*$ *has a greatest and a smallest element.*

A very important example of a hyperfinite set which we will use in this paper is the *hyperfinite grid* $\mathbb{H}$. The *hyperfinite grid* $\mathbb{H}_\alpha$ is defined as the α-value of the set

$$\mathbb{H}_n = \left\{\frac{k}{n} : k \in \mathbb{Z},\ -\frac{n^2}{2} \leq k < \frac{n^2}{2}\right\};$$

namely,

$$\mathbb{H}_\alpha := \left\{ \frac{k}{\alpha} : k \in \mathbb{Z}^*,\ -\frac{\alpha^2}{2} \le k < \frac{\alpha^2}{2} \right\}$$

In the following, for short, usually we will write $\mathbb{H}$ instead of $\mathbb{H}_\alpha$. Clearly $\mathbb{H}$ is a hyperfinite set with $|\mathbb{H}| = \alpha^2$. Given $a, b \in \mathbb{H}$, we set

$$\begin{aligned} [a,b]_{\mathbb{H}} &= \{x \in \mathbb{H} : a \le k \le b\} \\ [a,b)_{\mathbb{H}} &= \{x \in \mathbb{H} : a \le k < b\} \end{aligned}$$

If we identify the functions with their graphs, f^* is defined by Definition 2.1 and it is not difficult to prove the following

PROPOSITION 2.6. *Let $f : A \to B$ be a function. Then its star-transform f^* is a function $f^* : A^* \to B^*$ and, for every sequence $\varphi : \mathbb{N} \to A$,*

$$f^*(\varphi(\alpha)) = (f \circ \varphi)(\alpha)$$

Moreover, f^* is 1-1 (or onto) iff f is 1-1 (or onto, respectively).

In the special case that $f : \mathbb{N} \to B$, Proposition 2.6 gives $f^*(\alpha) = f(\alpha)$. When confusion is unlikely, we will omit the symbol "$*$" and "f^*" will be denoted by "f".

Let $f_n : A \to B$ be a sequence of functions; then identifying the functions with their graphs f_α is well defined by axiom (α3) and we have that

$$f_\alpha : A^* \to B^*$$

is a function defined by

$$f_\alpha(\varphi(\alpha)) = \psi(\alpha)$$

where $\psi(n) := f_n(\varphi(n))$ is a sequence in B.

DEFINITION 2.7. A function

$$f : A^* \to B^*$$

is called internal if it is the graph of an internal set, namely if there is a sequence of functions $f_n : A \to B$ such that

$$f = f_\alpha$$

2.2. Infinitesimally small and infinitely large numbers. A fundamental feature of α-calculus is that the intuitive notions of "infinitesimally small" number and "infinitely large" number can be formalized as actual objects of the hyperreal line. This gives many possibilities to simplify proofs and statements in calculus theory.

DEFINITION 2.8. A hyperreal number $\xi \in \mathbb{R}^*$ is *bounded* or *finite* if its absolute value $|\xi| < r$ for some $r \in \mathbb{R}$. We say that ξ is *unbounded* or *infinite* if it is not bounded. ξ is *infinitesimal* if $|\xi| < r$ for all positive $r \in \mathbb{R}$.

Clearly, the inverse of an infinite number is infinitesimal and vice versa, i.e. the inverse of a (nonzero) infinitesimal number is infinite. An example of an infinitesimal is given by ⓔ $:= 1/\alpha$, the α -value of the sequence $\{1/n\}$.

From now on, the symbol ⓔ will always denote $1/\alpha$.

All infinitesimal and all real numbers are bounded. However there are finite hyperreals that are neither infinitesimal nor real, for example $5 +$ ⓔ and $7 + \sin\alpha$.

DEFINITION 2.9. We say that two hyperreal numbers ξ and η are *infinitesimally close* if $\xi - \eta$ is infinitesimal. In this case we write $\xi \sim \eta$.

It is easily seen that $\sim$ is an equivalence relation.

On the other hand (as it is intuitive) each bounded hyperreal is infinitely close to some real. The following indeed comes from the completeness of the real line.

THEOREM 2.10 (Shadow Theorem). *Every bounded hyperreal number ξ is infinitesimally close to a unique real number r, called the shadow of ξ. Symbolically $r = sh(\xi)$.*

The notion of a shadow is extended to every hyperreal number, by setting $sh(\xi) = +\infty$ if ξ is positive unbounded, and $sh(\xi) = -\infty$ if ξ is negative unbounded.

DEFINITION 2.11. Given two hyperreal numbers ξ and $\zeta \in \mathbb{R}^* \setminus \{0\}$, we say that they have the same order if ξ/ζ and ζ/ξ are bounded numbers and we will write

$$\xi \approx \zeta$$

(notice the difference between "$\sim$" and "$\approx$" since these symbols will be largely used in the rest of this paper). We say that ξ has a larger order than ζ if ξ/ζ is an infinite number and we will write

$$\xi \gg \zeta$$

We say that ξ has a smaller order than ζ if ξ/ζ is an infinitesimal number and we will write

$$\xi \ll \zeta$$

2.3. Some notions of infinitesimal calculus. Now we see how all this machinery can be used to build a rigorous "infinitesimal" calculus. We present how the definition of limit can be given in our setting.

DEFINITION 2.12. We say that $\lim_{x \to x_0} f(x) = l$ if $f^*(\xi) \sim l$ for all $\xi \sim x_0$ $(\xi \neq x_0)$.

With the definition of limit all the elementary calculus can be reconstructed, but the features of our method allow one to avoid the use of limits and work with *real* infinitesimal and infinite numbers. Let us see some examples: the definitions of continuity and derivative. We remark that the theory given by these definitions is equivalent to the standard calculus and all the known results (as for example the Lagrange's or Fermat's theorems) apply.

DEFINITION 2.13. A real function $f : A \to \mathbb{R}$ is *continuous* at $x_0 \in A$ if for every $\xi \in A^*$, $\xi \sim x_0 \Rightarrow f^*(\xi) \sim f^*(x_0)$.

Let f be any real function defined on a neighborhood of x_0.

DEFINITION 2.14. We say that f has derivative at x_0 if there exists $f'(x_0) \in \mathbb{R}$ such that for all infinitesimals $\varepsilon \neq 0$,

$$\frac{f^*(x_0 + \varepsilon) - f^*(x_0)}{\varepsilon} \sim f'(x_0)$$

Equivalently, f has *derivative* $f'(x_0)$ at x_0 if for every infinitesimal ε there is an infinitesimal δ such that $f^*(x_0 + \varepsilon) = f(x_0) + f'(x_0)\varepsilon + \delta\varepsilon$.

As said before all the classical results of calculus hold in this framework. An example which will be used in the following is the Taylor formula (with infinitesimal remainder).

THEOREM 2.15. *If $f \in C^{n+1}(\mathbb{R})$ then for each infinitesimal ε there is an infinitesimal η such that*

$$f^*(x+\varepsilon) = \sum_{k\leq n} \frac{f^{(k)}(x)\varepsilon^k}{k!} + \eta\varepsilon^n.$$

Now we introduce a concept of integral. This concept is more general than the Riemann integral and will allow us to integrate noises and stochastic equations. Intuitively this integral is just an infinite sum of hyperreal numbers. This sum will be done on a hyperfinite set.

DEFINITION 2.16. If $\Gamma = \chi(\alpha)$ is a hyperfinite set of hyperreal numbers, then its *hyperfinite sum:*

$$\sum_{x\in\Gamma} x = Sum_\chi(\alpha)$$

is defined as the value at infinity of the sequence of finite sums

$$Sum_\chi(n) = \sum_{x\in\chi(n)} x.$$

It is easily checked that this definition does not depend on the choice of the sequence $\{\chi(n)\}$, but only on its value at infinity Γ. Using this definition, we define the α-*integral.*

DEFINITION 2.17. Let $f: A \to \mathbb{R}$ be *any* function, where $A \subseteq \mathbb{R}$. Its α-*integral* on A, denoted by $\int_A f(x)\,\Delta x$, is the number in $\mathbb{R}\cup\{\pm\infty\}$ defined as the shadow of the following hyperfinite sum:

$$\int_A f(x)\,\Delta x \;=\; sh\left(ⓔ\cdot\sum_{\xi\in\mathbb{H}\cap A^*} f^*(\xi)\right)$$

Notice that

$$\int_A f(x)\,\Delta x \;=\; sh\,(S_A(\alpha)) \text{ where } S_A(n) = \frac{1}{n}\cdot\sum_{x\in\mathbb{H}(n)\cap A} f^*(x)$$

Of course, if $A = [a,b]$ is a closed interval, we adopt the usual notation $\int_a^b f(x)\,\Delta x$.

The α-integral $\int_a^b f(x)\,\Delta x$ is defined for every function. In fact, while the sequence

$$S_a^b(n) = \frac{1}{n}\cdot\sum_{x\in\mathbb{H}(n)\cap(a,b)} f^*(x)$$

may not have a limit in the classic sense, its α-value $S_a^b(\alpha)$ is always defined. If the function f is Riemann integrable then $\lim_{n\to\infty} S_a^b(n)$ exists and coincides with the α-integral (notice that if a real sequence $\{\varphi(n)\}$ has "classic" limit $l \in \mathbb{R}\cup\{\pm\infty\}$, then it must be $sh\,(\varphi(\alpha)) = l$). Thus the α-integral actually generalizes the Riemann integral.

3. Grid functions

A grid function is a function whose argument ranges on a hyperfinite "grid" whose elements are the (hypernatural) multiples of $\frac{1}{\alpha}$. Since the grid is hyperfinite these functions are easy to handle and from many points of view they behave similarly to functions on finite sets. We will see that this simple kind of function is flexible enough to contain elements representing distributions. This flexibility will allow us to obtain in a simple way a kind of stochastic calculus (see, e.g. the Ito's formula, Th. 3.4).

3.1. Basic notions. An internal function

$$\xi : \mathbb{H} \to \mathbb{R}^*$$

is called a *grid function.*

DEFINITION 3.1. Given a grid function $\xi : \mathbb{H} \to \mathbb{R}^*$, we define its grid derivative $\frac{\Delta\xi}{\Delta t}$ as

$$\frac{\Delta\xi}{\Delta t}(t) = \frac{\xi(t + ⓔ) - \xi(t)}{ⓔ};$$

The grid integral of ξ is defined as

$$\mathbb{I}[\xi] = ⓔ \sum_{t\in\mathbb{H}} \xi(t);$$

if $\Gamma \subset \mathbb{H}$ is a hyperfinite set we define $\mathbb{I}_\Gamma[\xi]$, its grid integral in Γ, as

$$\mathbb{I}_\Gamma[\xi] = ⓔ \sum_{t\in\Gamma} \xi(t)$$

Most of the properties of the usual derivative hold also for the grid derivative, for example we have that, if ξ and ζ are continuous functions, with finite grid derivative,

$$\begin{aligned} \frac{\Delta(\xi\zeta)}{\Delta t} &= \frac{\xi(t+ⓔ)\zeta(t+ⓔ) - \xi(t)\zeta(t)}{ⓔ} = \\ &= \frac{\xi(t+ⓔ)\zeta(t+ⓔ) - \xi(t+ⓔ)\zeta(t) + \xi(t+ⓔ)\zeta(t) - \xi(t)\zeta(t)}{ⓔ} = \\ &= \frac{\Delta\xi}{\Delta t}(t)\cdot\zeta(t) + \xi(t+ⓔ)\cdot\frac{\Delta\zeta}{\Delta t}(t) \sim \frac{\Delta\xi}{\Delta t}\cdot\zeta + \xi\cdot\frac{\Delta\zeta}{\Delta t}. \end{aligned}$$

These notions can be easily extended to functions of more variables; for example if

$$\rho(t,x) : \mathbb{H}\times\mathbb{H} \to \mathbb{R}^*$$

we set

$$\begin{aligned} \frac{\Delta\rho}{\Delta t}(t,x) &= \frac{\rho(t+ⓔ,x) - \rho(t,x)}{ⓔ} \\ \frac{\Delta\rho}{\Delta x}(t,x) &= \frac{\rho(t,x+ⓔ) - \rho(t,x)}{ⓔ} \end{aligned}$$

and if $\Gamma \subset \mathbb{H}^2$ is a hyperfinite set we define its grid integral $\mathbb{I}_\Gamma[\rho]$ as

$$\mathbb{I}_\Gamma[\rho] = ⓔ^2 \sum_{(t,x)\in\Gamma} \rho(t,x).$$

It is clear that the derivative of a grid function ξ is a grid function. Moreover, if ξ is a grid function, then the *grid integral function* $x \mapsto \mathbb{I}_{[a,x)}[\xi]$ is a grid function. We have the following relation between the grid-derivative and the grid-integral:

THEOREM 3.2. *If ξ is a grid function, then*

$$\begin{aligned} \mathbb{I}_{[x,y)}\left[\frac{\Delta\xi}{\Delta x}\right] &= \xi(y) - \xi(x) \\ \frac{\Delta}{\Delta x}\mathbb{I}_{[a,x)}[\xi] &= \xi(x) \end{aligned}$$

PROOF. Obviously we have

$$\begin{aligned} \mathbb{I}_{[x,y)}\left[\frac{\Delta\xi}{\Delta x}\right] &= \text{Ⓔ}\sum \frac{\xi(x+\text{Ⓔ}) - \xi(x)}{\text{Ⓔ}} + \frac{\xi(x+2\cdot\text{Ⓔ}) - \xi(x+\text{Ⓔ})}{\text{Ⓔ}} + \ldots \\ &\quad + \frac{\xi(y) - \xi(y-\text{Ⓔ})}{\text{Ⓔ}} = \\ &= \xi(y) - \xi(x). \end{aligned}$$

Furthermore

$$\begin{aligned} \frac{\Delta}{\Delta x}\mathbb{I}_{[a,x)}[\xi] &= \frac{\mathbb{I}_{[a,x+\text{Ⓔ})}[\xi] - \mathbb{I}_{[a,x+\text{Ⓔ})}[\xi]}{\text{Ⓔ}} = \\ &= \sum_{t\in[a,x+\text{Ⓔ})} \xi(t) - \sum_{t\in[a,x)} \xi(t) = \xi(x). \end{aligned}$$

□

DEFINITION 3.3. A grid function ξ is called integrable in $[a,b]$ if $\mathbb{I}_{[a,b]}[\xi]$ is finite; in this case, we set

$$\int_a^b \xi(s)\,\Delta s := sh\left(\mathbb{I}_{[a,b)}[\xi]\right) = sh\left(\text{Ⓔ}\sum_{t\in\mathbb{H}\cap[a,b)} \xi(t)\right)$$

ξ is called absolutely integrable in $[a,b]$ if $\mathbb{I}_{[a,b)}[|\xi|]$ is finite. $\int_a^b \xi(s)\,\Delta s$ will be called the α-integral of ξ.

Of course, this integral is strictly related to the α-integral given in Def. 2.17. In fact, to every real function

$$f : [a,b] \to \mathbb{R}$$

it is possible to associate its natural extension

$$f^* : [a,b]^* \to \mathbb{R}^*$$

and a grid function

$$\tilde{f} : [a,b]_{\mathbb{H}} \to \mathbb{R}^* \tag{3.1}$$

obtained as the restriction of f^* to $[a,b]_{\mathbb{H}}$. When no ambiguity is possible we will denote f^* and $\tilde{f}$ with the same symbol.

The α-integral of f coincides with the α-integral of $\tilde{f}$ given by Def. 3.3.

3.2. The Ito formula. We show the power of the grid functions approach by stating in a very simple way a proposition which is, in some sense, a variant of the Ito's formula. As in the standard approach this formula will be the main tool in the study of grid stochastic equations.

THEOREM 3.4 (Nonstandard Ito's Formula). *Let $\varphi \in C_0^3(\mathbb{R}^2)$ and $x(t)$ be a grid function such that*

$$\left|\frac{\Delta x}{\Delta t}(t)\right| \le \eta\alpha^{2/3}, \tag{3.2}$$

where $\eta \sim 0$.

Then

$$\frac{\Delta}{\Delta t}\varphi(t,x(t)) \sim \varphi_t(t,x(t)) + \varphi_x(t,x(t))\frac{\Delta x}{\Delta t}(t) + \frac{\textcircled{\varepsilon}}{2}\varphi_{xx}(t,x(t))\cdot\left(\frac{\Delta x}{\Delta t}(t)\right)^2.$$

Here φ_t, φ_x and φ_{xx} denote the usual partial derivative of φ.

PROOF. By definition of grid derivative we have that

$$\begin{aligned}
\frac{\Delta}{\Delta t}\varphi(t,x(t)) &= \frac{\varphi(t+\textcircled{\varepsilon}, x(t+\textcircled{\varepsilon})) - \varphi(t, x(t+\textcircled{\varepsilon}))}{\textcircled{\varepsilon}} \\
&\quad + \frac{\varphi(t, x(t+\textcircled{\varepsilon})) - \varphi(t,x(t))}{\textcircled{\varepsilon}} \\
&\sim \varphi_t(t, x(t+\textcircled{\varepsilon})) + \frac{\varphi(t, x(t+\textcircled{\varepsilon})) - \varphi(t,x(t))}{\textcircled{\varepsilon}} \\
&\sim \varphi_t(t, x(t)) + \frac{\varphi(t, x(t+\textcircled{\varepsilon})) - \varphi(t,x(t))}{\textcircled{\varepsilon}}
\end{aligned}$$

But

$$\varphi(t, x(t+\textcircled{\varepsilon})) = \varphi\left(t, x(t) + \textcircled{\varepsilon}\frac{\Delta x}{\Delta t}(t)\right),$$

and $\left|\textcircled{\varepsilon}\frac{\Delta x}{\Delta t}(t)\right| \le \eta\alpha^{2/3}\cdot\textcircled{\varepsilon} = \eta\cdot\textcircled{\varepsilon}^{1/3}$ is infinitesimal. Then, using the Taylor formula (Theorem 2.15), we have that

$$\begin{aligned}
\varphi\left(t, x(t) + \textcircled{\varepsilon}\frac{\Delta x}{\Delta t}(t)\right) &= \varphi(t,x(t)) + \varphi_x(t,x(t))\cdot\textcircled{\varepsilon}\frac{\Delta x}{\Delta t}(t) \\
&\quad + \frac{1}{2}\varphi_{xx}(t,x(t))\left(\textcircled{\varepsilon}\frac{\Delta x}{\Delta t}(t)\right)^2 \\
&\quad + \frac{1}{3!}\varphi_{xxx}(t,x(t))\left(\textcircled{\varepsilon}\frac{\Delta x}{\Delta t}(t)\right)^3 + \delta\left(\textcircled{\varepsilon}\frac{\Delta x}{\Delta t}(t)\right)^3
\end{aligned}$$

where δ is an infinitesimal; hence

$$\begin{aligned}
\frac{\varphi(t, x(t+\textcircled{\varepsilon})) - \varphi(t,x(t))}{\textcircled{\varepsilon}} &= \varphi_x(t,x(t))\frac{\Delta x}{\Delta t}(t) + \frac{\textcircled{\varepsilon}}{2}\varphi_{xx}(t,x(t))\cdot\left(\frac{\Delta x}{\Delta t}(t)\right)^2 \\
&\quad + \frac{\textcircled{\varepsilon}^2}{6}\varphi_{xxx}\cdot\left(\frac{\Delta x}{\Delta t}(t)\right)^3 + \delta\cdot\textcircled{\varepsilon}^2\left(\frac{\Delta x}{\Delta t}(t)\right)^3
\end{aligned}$$

By the assumption (3.2) the last two terms are infinitesimal and we get the required result. □

3.3. Distributions and grid functions. The grid functions can be considered as a sort of generalization of the usual real functions.

In fact to every real function there corresponds a unique grid functions given by (3.1). In the traditional analysis the most important generalization of the real function is given by the distribution. In this section we will show that the grid functions represent also a generalization of the notion of *distribution.*

First of all we recall some notation: given a set $A \subset \mathbb{R}^N$, $\mathcal{D}(A)$ denotes the space of C^∞ functions with compact support of A. The space of the distributions $\mathcal{D}'(A)$ is the topological dual of $\mathcal{D}(A)$ when $\mathcal{D}(A)$ is equipped with the Schwartz topology.

Actually, $\mathcal{D}'(A)$ can also be constructed without knowing the Schwartz topology by using the notion of grid function. Next, we will show how to do it.

Let $\mathfrak{G}(A)$ denote the set of grid function defined on

$$A_{\mathbb{H}} := A^* \cap \mathbb{H}^N$$

On $\mathfrak{G}(A)$ we define the following equivalence relation:

DEFINITION 3.5. Two grid functions ξ_1, ξ_2 are said to be equivalent if

$$\forall \varphi \in \mathcal{D}, \quad \int (\xi_1 - \xi_2)\, \varphi \Delta s = 0$$

In this case we will write

$$\xi_1 \sim_{\mathcal{D}} \xi_2$$

We may think that two grid functions are equivalent if they are *macroscopically equal.*

Moreover, we set

$$\mathfrak{G}_0(A) = \{\xi \in \mathfrak{G}(A) : \forall \varphi \in \mathcal{D},\ \mathbb{I}_{A_{\mathbb{H}}}[\xi\varphi] \quad is\ finite\}$$

The set of distributions $\mathcal{D}'(A)$ can be defined as follows

$$\mathcal{D}'(A) = \frac{\mathfrak{G}_0(A)}{\sim_{\mathcal{D}}}.$$

Thus a distribution can be considered as an equivalence class T_ξ of some grid function $\xi \in \mathfrak{G}_0(A)$.

T_ξ can be identified with an element of $\mathcal{D}'(A)$ by the following formula:

$$\langle T_\xi, \varphi \rangle = \int_A \xi\varphi\, ds = sh\left(ⓔ \cdot \sum_{t \in A_{\mathbb{H}}} \xi(t)\varphi^*(t) \right), \quad \varphi \in \mathcal{D}. \tag{3.3}$$

To each distribution we can associate a grid function. For example, if $T \in \mathcal{D}'(\mathbb{R})$ we can do in the following way. Since a distribution T has the following representation[1]:

$$T = \sum_{k=0}^{\infty} D^k f_k$$

where f_k are continuous, the grid function ξ corresponding to T is given by

$$\xi(t) = \sum_{k=0}^{\alpha} \frac{\Delta^k}{\Delta t^k} f_k(t).$$

[1] See Rudin, functional analysis, Th. 6.28, pag.169

Let us see some simple examples. The function

$$\delta(t) = \alpha \delta_{0,t}$$

where $\delta_{i,j}$ is the Kronecker symbol corresponds to the Dirac δ. But also the following grid functions

$$\alpha \frac{\delta_{0,t} + \delta_{\text{ⓔ},t}}{2}; \ \sum \delta_{0,t+\frac{k}{\alpha}} \ (k \in \mathbf{Z}); \ \text{etc}$$

correspond to the Dirac δ. The grid function

$$\frac{\Delta \delta}{\Delta t}(t) = \alpha^2 \left(\delta_{0,t} - \delta_{\text{ⓔ},t} \right)$$

corresponds to δ'.

The grid function $\alpha^2 \delta_{0,t}$ is not in $\mathfrak{G}_0(A)$ and hence it does not correspond to any distribution.

4. Stochastic differential equations

4.1. Grid differential equations. A grid ordinary differential equation is a differential equation whose time step ranges on the hyperfinite grid. This fact makes it work as a discrete time object and simplifies many formal aspects.

A grid ordinary differential equation is then an equation of the kind

$$\frac{\Delta x}{\Delta t}(t) = f(t, x(t)), \tag{4.1}$$

where $t \in \mathbb{H}$, $x(t)$ is a grid function and $f : \mathbb{H} \times \mathbb{R}^* \to \mathbb{R}^*$ is an internal function. A grid function $x(t)$ is a solution of the grid equation if the equation is satisfied at each point of the grid.

The following result shows that such an equation has an unique solution. This holds without regularity assumptions on f. Hence, this kind of equation has a solution even if the equation contains a noise term (see section (4.2)).

THEOREM 4.1. *Given an initial time $t_0 \in \mathbb{H}$ and an initial data $x_0 \in \mathbb{R}^*$, the Cauchy problem associated to (4.1), that is*

$$\begin{cases} \frac{\Delta x}{\Delta t}(t) = f(t, x(t)) & t \in \mathbb{H} \\ x(t_0) = x_0 \end{cases} \tag{4.2}$$

admits for $t \geq t_0$ an unique solution $x : \mathbb{H} \to \mathbb{R}^*$.

PROOF. We know that f is an ideal value of a sequence $\{f_n\}_{n \in \mathbb{N}}$. Also, we have that $t_0 = t_{0,\alpha}$, $x_0 = x_{0,\alpha}$ are the ideal values associated to $\{t_{0,n}\}_{n \in \mathbb{N}}$, $\{x_{0,n}\}_{n \in \mathbb{N}}$. For each $n \in \mathbb{N}$ and $m \in \mathbb{Z}$, we can construct by induction a sequence of functions.

$$x_n : \frac{1}{n}\mathbb{Z} \to \mathbb{R}$$

as follows:

$$x_n(t_{0,n}) = x_{0,n} \tag{4.3}$$

$$x_n\left(t_{0,n} + \frac{m+1}{n}\right) = x_n\left(t_{0,n} + \frac{m}{n}\right) + \tag{4.4}$$

$$+\frac{1}{n} f_n\left(\left(t_{0,n} + \frac{m}{n}\right), x_n\left(t_{0,n} + \frac{m}{n}\right)\right). \tag{4.5}$$

Then by definition of internal function we have that, for $t = t_0 + m \cdot \varepsilon$, $m \in \mathbb{Z}^*$

$$x_\alpha(t_0) = x_0 \tag{4.6}$$

$$x_\alpha(t + \varepsilon) = x_\alpha(t) + \varepsilon f_\alpha(t, x_\alpha(t)). \tag{4.7}$$

Thus $x = x_\alpha$ solves (4.2). It is easy to check that this solution is also unique. □

Given $x_0 \sim x_1$, it may happen that

$$x(t, x_0) \nsim x(t, x_1)$$

where $x(t, x_i)$ is the solution of (4.2) with initial data x_i. Sometimes we would like to have

$$x(t, x_0) \sim x(t, x_1) \quad \forall x_1 \sim x_0; \tag{4.8}$$

this can be useful, for example, when we want to consider the standard part of a hyperreal differential equation.

We have the following proposition :

PROPOSITION 4.2. *Consider the following Cauchy problem*

$$\begin{cases} \frac{\Delta x}{\Delta t}(t) = f(t, x(t)) & t \in \mathbb{H}; \\ x(t_0) = x_0, \end{cases} \tag{4.9}$$

and suppose that, there exists L s.t.

$$|f(t, x) - f(t, y)| \le L|x - y|. \tag{4.10}$$

Let x_1 be a bounded initial data for the problem (4.9) and $x(t, x_1)$ be the solution of this problem. Then, if $x_1 \sim x_0$, then for all $0 \le t < T$ we have

$$x(t, x_0) \sim x(t, x_1) \tag{4.11}$$

PROOF. Arguing as in standard analysis, we have that (4.10) guarantees that, for any $T_1 < T$, the solution is bounded. Moreover, in standard analysis, the condition (4.10) guarantees the continuous dependence of the solution from initial data x_0. In our case, as long as the solution is finite, we can proceed in the same way to prove that, choosing an arbitrary $T_1 < T$

$$|x(t, x_0) - x(t, x_1)| \le |x_0 - x_1| e^{LT_1} \tag{4.12}$$

for all $t \in [0, T_1]$. Because $x_0 \sim x_1$ by hypothesis, we have that

$$x(t, x_0) \sim x(t, x_1) \tag{4.13}$$

for all $t \in [0, T_1]$. This assures the proof. □

4.2. Stochastic grid equations and the Fokker-Plank equation. In our approach, a stochastic differential equation consists of a set of grid differential equations. Each differential equation has a *noise* term and gives a trajectory which can be considered as a realization of a process.

Let $\mathcal{R} \subset \mathfrak{G}[0, 1]$ be a hyperfinite set of grid functions and consider the class of Cauchy problems

$$\begin{cases} \frac{\Delta x}{\Delta t}(t) = f(t, x) + h(t, x)\xi, \\ x(0) = x_0, \\ \xi(t) \in \mathcal{R}. \end{cases} \tag{4.14}$$

where

$$f, h : [0, 1]_{\mathbb{H}} \times \mathbb{R}^* \to \mathbb{R}^*$$

We want to study the statistical behavior of the set of solutions of the above Cauchy problems

$$\mathcal{S} = \{x_\xi(t) : \xi \in \mathcal{R}\};$$

More precisely we want to describe the behavior of the density function

$$\rho : [0,1]_{\mathbb{H}} \times \mathbb{H} \to \mathbb{Q}^*$$

defined as follows

$$\rho(t,x) = \frac{|\{x_\xi \in \mathcal{S} : x \le x_\xi(t) < x + \textcircled{\varepsilon}\}|}{\textcircled{\varepsilon}\,|\mathcal{R}|}.$$

We are interested in the case in which $\mathcal{R}$ models a *white noise*; roughly speaking we can define a white noise as the hyperfinite set of all the grid functions with values $\pm\sqrt{\alpha}$. Here is its precise definition:

DEFINITION 4.3. The white noise is the set of grid functions defined by

$$\mathcal{R} = \mathcal{R}_\alpha$$

where

$$\mathcal{R}_n = \{-\sqrt{n}, +\sqrt{n}\}^{[0,1]_{\mathbb{H}_n}}$$

Hence $\mathcal{R}$ is a hyperfinite set with $|\mathcal{R}| = 2^{\alpha+1}$.

REMARK 4.4. We would be tempted to write

$$\mathcal{R}_\alpha = \{-\sqrt{\alpha}, +\sqrt{\alpha}\}^{[0,1]_{\mathbb{H}_\alpha}}$$

but this notation is very ambiguous; in fact $\mathcal{R}_\alpha$ is a set defined by the Internal Set Axiom and it contains only *internal* functions. However the symbol $\{-\sqrt{\alpha}, +\sqrt{\alpha}\}^{[0,1]_{\mathbb{H}_\alpha}}$ usually represents the set of *all* the functions $f : [0,1]_{\mathbb{H}_\alpha} \to \{-\sqrt{\alpha}, +\sqrt{\alpha}\}$.

Now, we can state the main result of this paper:

THEOREM 4.5. *Assume that $\mathcal{R}$ is a white noise and that $f(t,x)$ and $h(t,x)$ are continuous functions. Then the distribution T_ρ relative to the density function ρ is a measure and satisfies the* Fokker-Plank equation

$$\frac{dT_\rho}{dt} + \frac{d}{dx}\left(f(t,x)T_\rho\right) - \frac{1}{2}\frac{d^2}{dx^2}\left(h(t,x)^2 T_\rho\right) = 0. \tag{4.15}$$

$$T_\rho(0,x) = \delta \tag{4.16}$$

in the sense of distribution.

REMARK 4.6. We recall that (4.15) and (4.16) "in the sense of distributions" means that T_ρ satisfies the equation

$$\left\langle \varphi_t + f\varphi_x + \frac{1}{2}h^2\varphi_{xx}, T_\rho \right\rangle + \varphi(x_0) = 0 \tag{4.17}$$

for any $\varphi \in \mathcal{D}([0,1) \times \mathbb{R})$. The duality $\langle\cdot,\cdot\rangle$ is between the space of continuous function and the space of measures. Equation (4.17) can be expressed using the grid function ρ and the α-integral by the following equation:

$$\forall\varphi \in \mathcal{D}([0,1) \times \mathbb{R}), \quad \iint \left(\varphi_t + f\varphi_x + \varphi_{xx}h^2\right)\rho\,\Delta x\,\Delta t + \varphi(0,x_0) = 0 \tag{4.18}$$

Actually, we will prove Th. 4.5 just proving the above equation.

REMARK 4.7. If $f(t,x)$ and $h(t,x)$ are smooth functions, by standard results in PDE, we know that, for $t>0$, the distribution T_ρ coincides with a smooth function $u(t,x)$. Then, for any $t>0$, ρ defines a smooth function u by the formula

$$\forall\varphi \in \mathcal{D}\left((0,1)\times\mathbb{R}\right),\ \iint \rho\varphi\ \Delta x\Delta t = \iint u\varphi\ dx\ dt$$

and u satisfies the Fokker-Plank equation in $(0,1)\times\mathbb{R}$ in the usual sense.

REMARK 4.8. We will see in the proof of Th. (4.5) that if the functions $f(t,x)$ and $h(t,x)$ are not continuous, but only bounded on compact sets, the equation (4.18) still holds. However in this case, equation (4.18) cannot be interpreted so easily. For example, if $f(t,x)$ and $h(t,x)$ are not measurable, there is no simple standard interpretation.

Given $t\in[0,1]_{\mathbb{H}}$ we set

$$\mathcal{R}[0,t)=\mathcal{R}_\alpha[0,t);\ \ \mathcal{R}_n := \left\{-\sqrt{n},+\sqrt{n}\right\}^{[0,t)_{\mathbb{H}_n}}; \tag{4.19}$$

namely, $\mathcal{R}[0,t)$ is the set of the restrictions of the functions of $\mathcal{R}$ to $[0,t)_{\mathbb{H}}$. Moreover, for $\tau\in\mathcal{R}\left[0,s\right)$, we set

$$\mathcal{R}_\tau\left[s,1\right]=\left\{\xi\in\mathcal{R}:\xi\left(t\right)=\tau\left(t\right)\ \text{ for } t<s\right\}$$

So we have the following decomposition:

$$\mathcal{R}=\bigcup_{\tau\in\mathcal{R}[0,s)}\mathcal{R}_\tau\left[s,1\right]. \tag{4.20}$$

We define the mean value of a grid function in the set $[x,y]\cap\mathbb{H}$ as follows:

$$\mathbb{E}_{[x,y)}\left[f\right]=\frac{1}{(y-x)}\mathbb{I}_{[x,y]}\left[f\right]=\frac{\textcircled{\varepsilon}}{(y-x)}\sum_{t\in[x,y)\cap\mathbb{H}}f\left(t\right)$$

In general, if Γ is a hyperfinite set and $\Phi:\Gamma\to\mathbb{R}^*$ is an internal function, the mean value of Φ in Γ is defined as follows:

$$\mathbb{E}_{\xi\in\Gamma}\left[\Phi\right]=\frac{1}{|\Gamma|}\sum_{\xi\in\Gamma}\Phi\left(\xi\right)$$

PROPOSITION 4.9. *If $\mathcal{R}$ is a white noise, then for any $t\in[0,1]_{\mathbb{H}}$, and $\tau\in\mathcal{R}\left[0,t\right]$, we have*

(4.21) *the hyperfinite number $\left|\mathcal{R}_\tau\left[t,1\right]\right|$ does not depend on $\tau\in\mathcal{R}\left[0,t\right]$*

and

$$\mathbb{E}_{\xi\in\mathcal{R}_\tau[t,1]}\left[\xi(t)\right]\ \sim\ 0, \tag{4.22}$$

$$\mathbb{E}_{\xi\in\mathcal{R}_\tau[t,1]}\left[\xi(t)^2\right]\ \sim\ \alpha. \tag{4.23}$$

PROOF. The proof is almost immediate: first of all we have that

$$\left|\mathcal{R}_\tau\left[t,1\right]\right|=2^{\alpha(1-t)+1};$$

moreover

$$\begin{aligned}\mathbb{E}_{\xi\in\mathcal{R}_\tau[t,1]}\left[\xi(t)\right] &= \frac{1}{\left|\mathcal{R}_\tau\left[t,1\right]\right|}\sum_{\xi\in\mathcal{R}_\tau[t,1]}\xi(t)\\ &= \frac{1}{2\left|\mathcal{R}_\tau\left[t,1\right]\right|}\sqrt{\alpha}-\frac{1}{2\left|\mathcal{R}_\tau\left[t,1\right]\right|}\sqrt{\alpha}=0\sim 0\end{aligned}$$

and

$$\mathbb{E}_{\xi\in\mathcal{R}_\tau[t,1]}\left[\xi(t)^2\right] = \frac{1}{|\mathcal{R}_\tau[t,1]|}\sum_{\xi\in\mathcal{R}_\tau[t,1]}\xi(t)^2 = \alpha$$

□

REMARK 4.10. The conclusion of Th. 4.5 holds not only if the "stochastic class" $\mathcal{R}$ is defined by (4.19), but for any class $\mathcal{R}$ which satisfies the properties (4.21), (4.22) and (4.23). For example we can take

$$\mathcal{R} = \mathcal{R}_\alpha;\quad \mathcal{R}_n := \left\{q_1\sqrt{n}, ..., q_k\sqrt{n}\right\}^{[0,1]_{\mathbb{H}_n}};\quad k\in\mathbb{N}$$

with $q_i \in \mathbb{R}^*$,

$$\sum_{i=1}^{k} q_i = 0;\ \sum_{i=1}^{k} q_i^2 = 1.$$

The following two lemmas are a direct consequence of properties (4.21), (4.22) and (4.23).

LEMMA 4.11. *Let $G : [0,1]_{\mathbb{H}} \times \mathbb{R}^* \times \mathbb{R}^* \to \mathbb{R}^*$ be any internal function. Then, for every $t \in [\varepsilon, 1]$*

$$\mathbb{E}_{\xi\in\mathcal{R}}\left[G(t, x_\xi(t), \xi(t))\right] = \mathbb{E}_{\tau\in\mathcal{R}[0,t)}\mathbb{E}_{\xi\in\mathcal{R}_\tau[t,1]}\left[G(t, x_\xi(t), \xi(t))\right]$$

PROOF. By (4.20), we have that

$$\mathcal{R} = \bigcup_{\tau\in\mathcal{R}[0,t)} \mathcal{R}_\tau[t,1]$$

Then,

$$\begin{aligned}
\mathbb{E}_{\xi\in\mathcal{R}}\left[G(t, x_\xi(t), \xi(t))\right] &= \frac{1}{|\mathcal{R}_\tau[t,1]|\cdot|\mathcal{R}[0,t)|}\sum_{\tau\in\mathcal{R}[0,t)}\sum_{\xi\in\mathcal{R}_\tau[t,1]} G(t, x_\xi(t), \xi(t)) \\
&= \frac{1}{|\mathcal{R}[0,t)|}\sum_{\tau\in\mathcal{R}[0,t]}\frac{1}{|\mathcal{R}_\tau[t,1]|}\sum_{\xi\in\mathcal{R}_\tau[t,1]} G(t, x_\xi(t), \xi(t)) \\
&= \mathbb{E}_{\tau\in\mathcal{R}[0,t)}\mathbb{E}_{\xi\in\mathcal{R}_\tau[t,1]}\left[G(t, x_\xi(t), \xi(t))\right].
\end{aligned}$$

□

LEMMA 4.12. *Let $F : [0,1]_{\mathbb{H}} \times \mathbb{R}^* \to \mathbb{R}^*$ be an internal function such that $|F(t,x)| \le M,\ M\in\mathbb{R}$. Then, for every $t\in[0,1]$*

$$\begin{aligned}
\mathbb{E}_{\xi\in\mathcal{R}}\left[F(t, x_\xi(t))\cdot\xi(t)\right] &\sim 0. \\
\mathbb{E}_{\xi\in\mathcal{R}}\left[F(t, x_\xi(t))\cdot\xi(t)^2\right] &\sim \alpha\cdot\mathbb{E}_{\xi\in\mathcal{R}}\left[F(t, x_\xi(t))\right]
\end{aligned}$$

PROOF. By lemma 4.11, we have that

$$\mathbb{E}_{\xi\in\mathcal{R}}\left[F(t, x_\xi(t))\cdot\xi(t)\right] = \mathbb{E}_{\tau\in\mathcal{R}[0,t)}\mathbb{E}_{\xi\in\mathcal{R}_\tau[t,1]}\left[F(t, x_\xi(t))\cdot\xi(t)\right]$$

Since $x_\xi(t)$ does not depend on $\xi(s)$ for $s > t$, we have that

$$\begin{aligned}\mathbb{E}_{\xi\in\mathcal{R}_\tau[t,1]}\left[F(t,x_\xi(t))\cdot\xi(t)\right] &= \frac{1}{|\mathcal{R}_\tau[t,1]|}\sum_{\tau\in\mathcal{R}[0,t]}\left(F(t,x_\xi(t))\cdot\xi(t)\right)\\ &= F(t,x_\xi(t))\cdot\frac{1}{|\mathcal{R}_\tau[t,1]|}\sum_{\tau\in\mathcal{R}[0,t]}\xi(t)\\ &= F(t,x_\xi(t))\cdot\mathbb{E}_{\xi\in\mathcal{R}_\tau[t,1]}\left[\xi(t)\right]\end{aligned}$$

Then since F is bounded, by (4.22), we get the conclusion:

$$\mathbb{E}_{\xi\in\mathcal{R}}\left[F(t,x_\xi(t))\xi(t)\right] = \mathbb{E}_{\tau\in\mathcal{R}[0,t)}\left[F(t,x_\xi(t))\cdot\mathbb{E}_{\xi\in\mathcal{R}_\tau[t,1]}\left[\xi(t)\right]\right]\sim 0$$

Analogously, we have that

$$\mathbb{E}_{\xi\in\mathcal{R}}\left[F(t,x_\xi(t))\xi(t)^2\right] = \mathbb{E}_{\tau\in\mathcal{R}[0,t)}\left[F(t,x_\xi(t))\mathbb{E}_{\xi\in\mathcal{R}_\tau[t,1]}\left[\xi(t)^2\right]\right]$$

and by (4.23), we get that

$$F(t,x_\xi(t))\mathbb{E}_{\xi\in\mathcal{R}_\tau[t,1]}\left[\xi(t)^2\right] = F(t,x_\xi(t))\left(\alpha+\varepsilon_\tau\right)$$

where $\varepsilon_\tau \sim 0$. Then

$$\begin{aligned}\mathbb{E}_{\xi\in\mathcal{R}}\left[F(t,x_\xi(t))\xi(t)^2\right] &= \alpha\mathbb{E}_{\tau\in\mathcal{R}[0,t)}\left[F(t,x_\xi(t))\right] + \mathbb{E}_{\tau\in\mathcal{R}[0,t)}\left[F(t,x_\xi(t))\varepsilon_\tau\right]\\ &\sim \alpha\mathbb{E}_{\tau\in\mathcal{R}[0,t)}\left[F(t,x_\xi(t))\right]\end{aligned}$$

This concludes the proof. □

Now we see a basic property of the density function.

LEMMA 4.13. *Let $\varphi \in \mathcal{D}([0,1]\times\mathbb{R})$ and let $x_\xi(t)$, $\xi \in \mathcal{R}$, be the family of solutions of a grid stochastic ODE. Then*

$$\mathbb{E}_{\xi\in\mathcal{R}}\left[\varphi(t,x_\xi(t))\right] \sim ⓔ \sum_{x\in\mathbb{H}}\varphi(t,x)\rho(t,x).$$

In particular,

$$\mathbb{E}_{\xi\in\mathcal{R}}\left[\varphi(t,x_\xi(t))\right] \sim \int \varphi(t,x)\rho(t,x)\Delta x.$$

PROOF. We have

$$\begin{aligned}\mathbb{E}_{\xi\in\mathcal{R}}\left[\varphi(t,x_\xi(t))\right] &= \frac{1}{|\mathcal{R}|}\sum_{\xi\in\mathcal{R}}\varphi(t,x_\xi(t)) = \frac{1}{|\mathcal{R}|}\sum_{x\in\mathbb{H}}\left[\sum_{x\le x_\xi(t)<x+ⓔ}\varphi(t,x_\xi(t))\right]\\ &= \frac{1}{|\mathcal{R}|}\sum_{\xi\in\mathcal{R}}\varphi(t,x_\xi(t)) = \frac{1}{|\mathcal{R}|}\sum_{x\in\mathbb{H}}\left[\sum_{x\le x_\xi(t)<x+ⓔ}\left(\varphi(t,x)+\eta_\xi(t)\right)\right]\end{aligned}$$

where $\eta_\xi(t) := \varphi(t,x_\xi(t)) - \varphi(t,x)$ is infinitesimal. Hence

$$\frac{1}{|\mathcal{R}|}\sum_{\xi\in\mathcal{R}}\eta_\xi(t)\sim 0$$

and so

$$\begin{aligned}
\mathbb{E}_{\xi\in\mathcal{R}}\left[\varphi(t,x_\xi(t))\right] &= \frac{1}{|\mathcal{R}|}\sum_{x\in\mathbb{H}}\left[\sum_{x\le x_\xi(t)<x+\text{Ⓔ}}\varphi(t,x)\right] \\
&= \frac{1}{|\mathcal{R}|}\sum_{x\in\mathbb{H}}\left(\varphi(t,x)\cdot\left|\{x_\xi\in\mathcal{S}: x\le x_\xi(t)<x+\text{Ⓔ}\}\right|\right) \\
&= \frac{1}{|\mathcal{R}|}\sum_{x\in\mathbb{H}}\left[\varphi(t,x)\cdot\text{Ⓔ}\,|\mathcal{R}|\,\rho(t,x)\right] = \text{Ⓔ}\sum_{x\in\mathbb{H}}\left[\varphi(t,x)\rho(t,x)\right]
\end{aligned}$$

□

Now we can prove Theorem 4.5

PROOF. Given an arbitrary $\varphi\in\mathcal{C}([0,1]\times\mathbb{R})$ bounded in the second variable, we have that

$$\varphi(1,x_\xi(1))-\varphi(0,x_0)=\text{Ⓔ}\sum_{t\in[0,1-\text{Ⓔ}]_{\mathbb{H}}}\frac{\Delta\varphi}{\Delta t}(t,x_\xi(t)).$$

Now we assume that $\varphi\in\mathcal{D}([0,1)\times\mathbb{R}))$. Since x_ξ solves eq. (4.14), we can apply Theorem 3.4, and we obtain

$$\begin{aligned}
-\varphi(0,x_0) &\sim \text{Ⓔ}\sum_{t\in[0,1)_{\mathbb{H}}}\left[\varphi_t+\varphi_x\cdot\frac{\Delta x}{\Delta t}+\frac{\text{Ⓔ}}{2}\varphi_{xx}\cdot\left(\frac{\Delta x}{\Delta t}\right)^2\right]= \\
&\sim \text{Ⓔ}\sum_{t\in[0,1)_{\mathbb{H}}}\left[\varphi_t+\varphi_x\cdot(f+h\xi)+\frac{\text{Ⓔ}}{2}\varphi_{xx}\cdot(f+h\xi)^2\right]= \qquad (4.24)\\
&= \text{Ⓔ}\sum_{t\in[0,1)_{\mathbb{H}}}(\varphi_t+f\varphi_x)+(\varphi_x h+\text{Ⓔ}\,\varphi_{xx}f)\,\xi+\frac{\text{Ⓔ}}{2}\varphi_{xx}f+\frac{\text{Ⓔ}}{2}\varphi_{xx}h^2\xi^2
\end{aligned}$$

Now we want to compute $\mathbb{E}_{\xi\in\mathcal{R}}$ of each piece of the right hand side of the above equation: by lemma 4.13,

$$\mathbb{E}_{\xi\in\mathcal{R}}\left[\varphi_t+f\varphi_x\right]=\text{Ⓔ}\sum_{x\in\mathbb{H}}\left[\varphi_t+f\varphi_x\right]\rho \qquad (4.25)$$

for every $t\in[0,1]$.

Let us consider the second piece: $(\varphi_x h+\text{Ⓔ}\,\varphi_{xx}f)\,\xi$. If we set

$$F(t,x_\xi(t))=\varphi_x(t,x_\xi(t))h(t,x_\xi(t))+\text{Ⓔ}\,\varphi_{xx}(t,x_\xi(t))f(t,x_\xi(t))$$

it turns out that $F(t,x_\xi(t))$ is bounded. In fact if $x_\xi(t)$ is bounded, $\varphi_x(t,x_\xi(t))$, $h(t,x_\xi(t))$, $\varphi_{xx}(t,x_\xi(t))$ and $f(t,x_\xi(t))$ are bounded since they are standard functions; if $x_\xi(t)$ is unbounded, $\varphi_x(t,x_\xi(t))=0$. Then we can apply lemma 4.12 and we get that

$$\mathbb{E}_\xi\left[(\varphi_x h+\text{Ⓔ}\,\varphi_{xx}f)\,\xi\right]\sim 0. \qquad (4.26)$$

Moreover

$$\mathbb{E}_\xi\left[\frac{\text{Ⓔ}}{2}\varphi_{xx}f\right]\sim 0 \qquad (4.27)$$

since $\frac{\text{Ⓔ}}{2}\varphi_{xx}f\sim 0$.

Finally, let us see the last term: $\frac{\textcircled{\varepsilon}}{2}\varphi_{xx}h^2\xi^2$. We can see that $\frac{\textcircled{\varepsilon}}{2}\varphi_{xx}h^2$ is bounded (actually, it is infinitesimal); then we can apply lemma 4.12 with $F(t, x_\xi(t)) = \frac{\textcircled{\varepsilon}}{2}\varphi_{xx}(t, x_\xi(t))h(t, x_\xi(t))^2$:

$$\begin{aligned}
\mathbb{E}_\xi\left[\frac{\textcircled{\varepsilon}}{2}\varphi_{xx}h^2\xi^2\right] &\sim \alpha\mathbb{E}_\xi\left[\frac{\textcircled{\varepsilon}}{2}\varphi_{xx}h^2\right] \\
&= \frac{1}{2}\mathbb{E}_\xi\left[\varphi_{xx}h^2\right] \\
&= \frac{1}{2|\mathcal{R}|}\sum_{\xi\in\mathcal{R}}\varphi_{xx}(t, x(t))h(t, x(t))^2
\end{aligned} \tag{4.28}$$

Then, by lemma 4.13

$$\mathbb{E}_\xi\left[\frac{\textcircled{\varepsilon}}{2}\varphi_{xx}h^2\xi^2\right] \sim \frac{\textcircled{\varepsilon}}{2}\sum_{x\in\mathbb{H}}\varphi_{xx}(t, x)h(t, x)^2\rho(t, x)$$

Then, by (4.24),...,(4.28)

$$\begin{aligned}
-\varphi(0, x_0) &= \mathbb{E}_{t,\xi}\left[-\varphi(0, x_0)\right] \\
&\sim \textcircled{\varepsilon}\sum_{t\in[0,1)_\mathbb{H}}\Big(\mathbb{E}_\xi\left[\varphi_t + f\varphi_x\right] + \mathbb{E}_\xi\left[\left(\varphi_x h + \textcircled{\varepsilon}\varphi_{xx}f\right)\xi\right]\Big) + \\
&\quad + \textcircled{\varepsilon}\sum_{t\in[0,1)_\mathbb{H}}\left(\mathbb{E}_\xi\left[\frac{\textcircled{\varepsilon}}{2}\varphi_{xx}f\right] + \mathbb{E}_\xi\left[\frac{\textcircled{\varepsilon}}{2}\varphi_{xx}h^2\xi^2\right]\right) \\
&\sim \textcircled{\varepsilon}^2\sum_{t\in[0,1)_\mathbb{H}}\left(\sum_{x\in\mathbb{H}}\left(\varphi_t + f\varphi_x\right)\rho + \varphi_{xx}h^2\rho\right) \\
&\sim \int\!\!\int\left(\varphi_t + f\varphi_x + \varphi_{xx}h^2\right)\rho\, dxdt
\end{aligned}$$

Since the first and the last term of this equation are standard, we get eq. (4.18). □

5. Conclusion

In this section we will make some comments suggested by the results of this paper.

The first comment is that,

- *in many applications of Nonstandard Analysis, only elementary facts and techniques of nonstandard calculus seem to be necessary.*

α-theory provides an entry to nonstandard methods which is much simpler than the usual ones, but it seems absolutely adequate to treat the kind of problems considered here.

However, if you compare this paper with [**2**], [**8**] and [**11**], the reason why this paper is much simpler lies in the fact that we have not made a nonstandard theory of the *stochastic differential equations*, but rather, we have replaced them with the *stochastic grid equations* which are much simpler mathematical objects; in this context the Ito integral is replaced by the α-integral and the proof of the key point of this theory, the Ito formula, reduces to an exercise.

The reason why a much simpler object can represent the processes of diffusion at the microscopic level is that we have taken the infinitesimals seriously and we have used them to model an aspect of the "physical reality". We think that in general,

- *the advantages of a theory which includes infinitesimals rely more on the possibility of making new models rather than in the demonstration techniques.*

In the case considered in this paper, the way the model has been constructed appears quite natural: the *stochastic grid equations*, which need the notion of infinitesimals, describe the diffusion processes at the microscopic level; the Fokker-Plank equation describes the diffusion processes at the macroscopic level and uses standard differential equations (and the theory of distributions when the data are not regular).

The connection of these two levels is given by Eq. (3.3) which relates grid functions (microscopic level) with distributions (macroscopic level).

A final remark concerns the *theory of probability.* We have not used the notion of probability to show that everything can be kept to a very elementary level; our variant of the Ito formula makes sense also in a context where probability does not appear.

However probability can be introduced in a very elementary way. We may think of the stochastic class $\mathcal{R}$ as a sample space. The events are the hyperfinite sets $E \subset \mathcal{P}(\mathcal{R})^*$ and the probability P of an event is given by

$$P(E) = \frac{|E|}{|\mathcal{R}|}. \tag{5.1}$$

For example E might be the event that at a time t_0, the moving particle lies in the interval $[a, b]$, namely

$$E = \{x_\xi \in \mathcal{S} : x_\xi(t_0) \in [a, b]_{\mathbb{H}}\}$$

In this case we have that

$$P(E) \sim \int_a^b \rho(t, x)\, \Delta x$$

Obviously, this is the most natural extension to infinite sample spaces of the classical definition of probability. There is no doubt that definition (5.1), is the most simple and intuitive definition of probability. The price which you pay is that P takes its values in $\mathbb{R}^*$ and not in $\mathbb{R}$; thus, a probability space $(\mathcal{R}, P)$ is an internal object and, if it is infinite, it is not standard. The problems arise if you want to connect the nonstandard world with the standard one. This operation can be done in a very elegant way via the Loeb integral (see [**9**] or also [**8**]). A different way to make the connection between this two worlds easy has been proposed by Nelson [**11**]. However, we think that, in most cases, it is not necessary to make this connection. Only at the very end do you need to consider the shadows of the numbers which you have obtained. At least this can be done in the theory which we have exposed in this paper. So, if we accept a mathematical description in which the infinitesimal exist,

- *the probability of an event is given by a hyperrational number,* and many theorems become simpler. This scheme also avoids some facts in probability theory which are in contrast with common sense: for example the fact that the union of impossible events gives a possible one (the probability of a non-denumerable set might be non-null even if the probability of each singleton is null).

Concluding the final remark of this paper could be the following one:

- *the infinitesimals should be taken seriously.*

References

[1] S. Albeverio, J. E. Fenstad, R. Hoegh-Krohn, and T. Lindstrøm, *Non-standard Methods in Stochastic Analysis and Mathematical Physics*, Academic Press, New York, 1986.

[2] Anderson, Robert M. A nonstandard representation for Brownian motion and Itô integration. Bull. Amer. Math. Soc. 82 (1976), no. 1, 99–101.

[3] Benci, V. A Construction of a Nonstandard Universe, in *Advances in Dynamical System and Quantum Physics*, S. Albeverio, R. Figari, E. Orlandi, A. Teta ed.,(Capri, 1993), 11–21, World Scientific, (1995).

[4] Benci, V., An algebraic approach to nonstandard analysis, in *Calculus of Variations and Partial Differential Equations*, G. Buttazzo, A. Marino and M.K.V. Murthy, eds., Springer-Verlag, 1999, 285-326.

[5] Benci, V., Di Nasso, M. Alpha-theory: an elementary axiomatics for nonstandard analysis. Expo. Math. 21 (2003), no. 4, 355–386.

[6] Benci, Vieri; Di Nasso, Mauro A ring homomorphism is enough to get nonstandard analysis. Bull. Belg. Math. Soc. Simon Stevin 10 (2003), no. 4, 481–490.

[7] Benci, Vieri; Forti, Marco; Di Nasso, Mauro The eightfold path to nonstandard analysis. Nonstandard methods and applications in mathematics, 3–44, Lect. Notes Log., 25, Assoc. Symbol. Logic, La Jolla, CA, 2006.

[8] Keisler, H. J. An infinitesimal approach to stochastic analysis. Mem. Amer. Math. Soc. 48 (1984), no. 297, x+184 pp.

[9] Loeb, Peter A. Conversion from nonstandard to standard measure spaces and applications in probability theory. Trans. Amer. Math. Soc. 211 (1975), 113–122.

[10] Rago, E. Una deduzione dell'equazione di Fokker-Planck con metodi Nonstandard, Thesis, University of Pisa, (2001).

[11] Nelson, E. *Radically elementary probability theory*, Princeton University Press, Princeton, New Jersey, (1987).

Dipartimento di Matematica Applicata "U.Dini", Università di Pisa, Via F. Buonarroti 1/c, 56127 Pisa

E-mail address: benci@dma.unipi.it

Dipartimento di Matematica Applicata "U.Dini", Università di Pisa, Via F. Buonarroti 1/c, 56127 Pisa

E-mail address: galatolo@dm.unipi.it

Dipartimento di Matematica e Applicazioni, Università di Milano Bicocca, Via Cozzi, 53 - 20125 Milano

E-mail address: marco.ghimenti@unimib.it

Contemporary Mathematics
Volume **530**, 2010

Ultrafilters, IP sets, Dynamics, and Combinatorial Number Theory

Vitaly Bergelson

ABSTRACT. We survey the connection between ultrafilters, ergodic theory, and combinatorics.

1. Introduction

The main theme of this survey is the multifaceted and mutually perpetuating connection between ultrafilters, ergodic theory, and combinatorics. In this short introductory section we collect some general definitions and facts about ultrafilters. The reader will find more detail and discussion in [**Ber2**], [**Ber4**], and [**Ber6**]. For a comprehensive treatment of topological algebra in the Stone-Čech compactification the reader is referred to [**HS**].

Given a set S, a filter on S is a nonempty family $\mathcal{S} \subset \mathcal{P}(S)$ (where $\mathcal{P}(S)$ denotes the power set of S) with the properties

(i) $\emptyset \notin \mathcal{S}$,
(ii) $A \in \mathcal{S}$ and $A \subseteq B$ implies $B \in \mathcal{S}$,
(iii) $A \in \mathcal{S}$ and $B \in \mathcal{S}$ implies $A \cap B \in \mathcal{S}$.

A filter $\mathcal{S}$ is an *ultrafilter* if, in addition, it satisfies

(iv) For any $A \subseteq S$ either $A \in \mathcal{S}$ or $A^c \in \mathcal{S}$ (but not both, in view of (i) and (iii)).

It is not hard to see that property (iv) is equivalent to

(v) If, for some integer r, $S = A_1 \cup \ldots \cup A_r$, then for some i, $1 \leq i \leq r$, one has $A_i \in \mathcal{S}$.

Indeed, (v) immediately implies (iv) by considering $S = A \cup A^c$. On the other hand, (iv) implies (v) since if, say, $A_1 \notin \mathcal{S}$ then $A_1^c \in \mathcal{S}$ which implies, by (ii), that $S_1 = A_2 \cup \ldots \cup A_r \in \mathcal{S}$, and the rest of the argument is clear.

It is easy to come up with numerous examples of filters (just take your favorite family of sets with the finite intersection property). On the other hand, when S is infinite, the only explicit example of an ultrafilter is of a somewhat degenerate nature and can be described as follows. Fix an element $a \in S$ and let

---

2010 *Mathematics Subject Classification.* Primary 05D10; Secondary 37A45.

The author acknowledges support received from the National Science Foundation via Grants DMS-0600042 and DMS-0901106.

$$A = \{R \subseteq S : a \in R\}.$$

Ultrafilters of this form are in 1-1 correspondence with elements of S and are called *principal.* All the other ultrafilters on S are called nonprincipal. When S is infinite, the proof of existence of nonprincipal ultrafilters requires Zorn's lemma - see [**CoN**], pp. 161-162. So, we have a somewhat strange situation: on one hand, one can show that the cardinality of the set of ultrafilters on S is $2^{2^{|S|}}$ (where $|S|$ denotes the cardinality of S, see [**Pos**]) but, on the other hand, it is not clear how to envision them.

Still, recognizing some members of a given ultrafilter (or, at least, knowing something about properties of members of an ultrafilter) often provides useful insights and leads to interesting applications.

A typical statement in *partition* Ramsey theory has the following (admittedly, vague) form:

(P) For any finite partition of an infinite "well organized" set S, one of the cells of the partition is also "well organized".[1]

For example, put $S \subseteq \mathbb{N} = \{1, 2, \ldots\}$ and interpret "well organized" as :
(i) containing arbitrarily long arithmetic progressions;
(ii) containing a *finite sums set*, that is a set of the form
$FS((x_i)_{i\in\mathbb{N}}) = \{x_{i_1} + x_{i_2} + \ldots + x_{i_k} : i_1 < i_2 < \ldots < i_k; k \in \mathbb{N}\}$[2],
where $(x_i)_{i\in\mathbb{N}}$ is (for convenience) an increasing sequence in $\mathbb{N}$.

Then the statement (P) holds for each of the two interpretations of "well organized" and corresponds to (i) van der Waerden's theorem ([**vdW**]), and (ii) Hindman's finite sums theorem ([**H1**]).

The following observation due to Hindman shows that there is a natural connection between (correct) statements as above and ultrafilters.

THEOREM ([**H2**], Theorem 6.7). *Let $\mathcal{A}$ be a family of subsets of a set S. The following statements are equivalent.*

(i) For any finite partition $S = \bigcup_{i=1}^{r} C_i$, one of the C_i belongs to $\mathcal{A}$

(ii) There exists an ultrafilter $\mathcal{S}$ on S such that for every $A \in \mathcal{S}$ there exists $G \in \mathcal{A}$ with $G \subseteq A$.

Assume now that $(S, \cdot)$ is a countably infinite discrete semigroup. The semigroup operation on S allows one to naturally transform the set of ultrafilters on S into a left topological semigroup. This left topological semigroup is nothing else but βS, the Stone-Čech compactification of S. See [**HS**] for more information and discussion. To define said operation it will be convenient to modify our point of view on ultrafilters.

Namely, rather than viewing an ultrafilter as a *maximal filter* (that is, a family of sets satisfying the properties (i),(ii), (iii), and (iv) or (v) above), we will find it useful to interpret the ultrafilters as $\{0, 1\}$-valued finitely additive probability measures on $\mathcal{P}(S)$. Indeed, given an ultrafilter $\mathcal{S}$ on S, assign to a set $A \subseteq S$ measure 1 if and only if $A \in \mathcal{S}$. It is easy to see that this indeed defines a finitely additive probability measure on $\mathcal{P}(S)$ (the formula (v) being responsible for its

[1] Compare with statement (D) in Section 7.

[2] Such sets are also called *IP sets*, a term introduced in [**FW**].

finite additivity). In the other direction, it is easy to see that given any $\{0,1\}$-valued finitely additive probability measure μ on $\mathcal{P}(S)$, the set of μ-large sets (that is, sets having measure 1) forms an ultrafilter.

At this point we, in accordance with a well established tradition, will switch notation and denote ultrafilter measures by lower case letters p, q, r, s, t, etc.

The perspicacious reader will observe that the following definition of product of ultrafilters is nothing but convolution of measures.

DEFINITION. Let p, q be ultrafilters on a semigroup $(S,\cdot)$. The ultrafilter $p \cdot q$ is defined by

$$A \in p \cdot q \Leftrightarrow \{x \in S : Ax^{-1} \in p\} \in q.$$

(The set Ax^{-1} is defined by the rule $y \in Ax^{-1}$ iff $yx \in A$).

It is a routine exercise to verify that this operation is well defined and associative (see, for example, [**H2**], Lemma 8.4). We also would like to remark that, when restricted to principal ultrafilters, the above operation coincides with the operation on $(S,\cdot)$ (we identify the principal ultrafilters with the elements of S).

Finally, we describe the topology on βA. Given $A \subseteq S$, let $\bar{A} = \{p \in \beta S : A \in p\}$. The collection $\{\bar{A} : A \subseteq S\}$ forms a basis for the open sets (and for closed sets as well) and makes βS into a (non-metrizable) compact Hausdorff space. One can also verify that for any fixed $p \in \beta S$, the function $\lambda_p(q) = p\cdot q$ is a continuous self-map of βS (see [**H2**] and [**HS**] for more information).

With the operation introduced above, $(\beta S,\cdot)$ becomes a compact *left topological* semigroup. By Ellis' Theorem, which we will presently formulate, $(\beta S,\cdot)$ has idempotents, that is, elements satisfying $x \cdot x = x$.

This result will be utilized numerous times during the course of this survey. When $S = \mathbb{N}$, one actually has two semigroup structures $(\mathbb{N},+)$ and $(\mathbb{N},\cdot)$ and, consequently, one has additive idempotents and multiplicative idempotents in $\beta\mathbb{N}$. The additive idempotents are elements of $(\beta\mathbb{N},+)$ and satisfy the equation $p + p = p$, where $+$ denotes the extension of the operation $+$ to $\beta\mathbb{N}$. Similarly, the multiplicative idempotents are elements of $(\beta\mathbb{N},\cdot)$ and satisfy the equation $q\cdot q = q$. It is the interplay between the two operations on $\beta\mathbb{N}$ which is behind many Ramsey-theoretical results that will be discussed below.

THEOREM (Ellis, [**E**]). *If $(G,\star)$ is a compact left topological semigroup (i.e. for any $x \in G$ the function $\lambda_x(y) = x \star y$ is continuous), then G has an idempotent.*

Acknowledgement. The author thanks Neil Hindman for enlightening discussions and plenty of useful input. Thanks also go to Cory Christopherson for invaluable technical help. Finally, the author would like to thank the anonymous referee for numerous helpful remarks.

2. Hindman's Finite Sums Theorem

We start our discussion with Hindman's (by now classical) finite sums theorem.

THEOREM 2.1 ([**H1**]). *For any finite partition $\bigcup_{i=1}^{r} C_i$, there exist an infinite sequence $(n_i)_{i\in\mathbb{N}} \subset \mathbb{N}$ and $j \in \{1,2,\ldots,r\}$ such that C_j contains the set*

$$FS((n_i)_{i\in\mathbb{N}}) = \{n_{i_1} + n_{i_2} + \ldots + n_{i_k} : i_1 < i_2 < \ldots < i_k;\ k \in \mathbb{N}\}.$$

The original proof of Theorem 2.1 in [**H1**] was elementary but very complicated. The short proof via ultrafilters which we will present below was found somewhat later and is due to F. Galvin and S. Glazer (see the account of the history of the ultrafilter proof in [**H5**] and [**HS**], pp. 102-103). To stress the dynamical nature of the ultrafilter proof of Theorem 2.1, we will make first a short digression into some basic ergodic theory stemming from the classical work of Poincaré on celestial mechanics ([**Poi1**], [**Poi2**]). This digression will also allow us to motivate the introduction of certain notions of largeness which will be utilized in subsequent sections.

Let $(X, \mathcal{B}, \mu)$ be a probability space and $T : X \to X$ a measure preserving transformation[3], meaning that for any set $A \in \mathcal{B}$ one has $\mu(T^{-1}A) = \mu(A)$.

The classical Poincaré recurrence theorem[4] asserts that for any $A \in \mathcal{B}$ with $\mu(A) > 0$ almost every point $x \in A$ returns to A under some nonzero power of T. More formally, Poincaré's recurrence theorem states that $\forall A \in \mathcal{B}$ with $\mu(A) > 0$, one has

$$\mu(\{x \in A : \exists n \in \mathbb{N} \text{ such that } T^n x \in A\}) = \mu(A). \tag{2.1}$$

Poincaré's recurrence theorem immediately follows from the following statement.

PROPOSITION 2.2. [5] *For any measure preserving system* $(X, \mathcal{B}, \mu, T)$*, and any* $A \in \mathcal{B}$ *with* $\mu(A) > 0$*, there exists* $n \in \mathbb{N}$ *such that*

$$\mu(A \cap T^{-n}A) > 0. \tag{2.2}$$

PROOF. Let $(m_i)_{i \in \mathbb{N}} \subseteq \mathbb{N}$ be an arbitrary increasing sequence, and consider the sets $T^{-m_i}A$, $i \in \mathbb{N}$. Since T is measure preserving, one has $\mu(T^{-m_i}A) = \mu(A)\ \forall i \in \mathbb{N}$. If $k > \frac{1}{\mu(A)}$ then, due to the fact that μ is an additive function on $\mathcal{B}$ and $\mu(X) = 1$, there exist $1 \le i < j \le k$ such that $\mu(T^{-m_i}A \cap T^{-m_j}A) = \mu(A \cap T^{-(m_j - m_i)}A) > 0$, and so $n = m_j - m_i$ satisfies (2.2). □

REMARK 2.3. The above proof works for any finitely additive probability measure. This rather trivial observation will be utilized below in the ultrafilter proof of Hindman's finite sums theorem.

REMARK 2.4. Given r integers $n_1 < n_2 < \ldots < n_r$, the set of differences $\{n_j - n_i : 1 \le i < j \le r\}$ is called a Δ_r set. A set $E \subseteq \mathbb{N}$ is called Δ_r^* if it has nontrivial intersection with any Δ_r set.[6] What was actually shown in the course of

---

[3]The quadruple $(X, \mathcal{B}, \mu, T)$, where $(X, \mathcal{B}, \mu)$ is a probability space and $T : X \to X$ is measure-preserving, is called a measure-preserving system.

[4]cf. [**Poi1**], § 8 and [**Poi2**], §§ 291-296.

[5]To derive (2.1) from Proposition 2.2, one argues as follows. Let A_0 be the (measurable!) set $\{x \in A : (\forall n \in \mathbb{N})\ T^n x \notin A\}$. If $\mu(A_0) > 0$ then for some $n \in \mathbb{N}$ one will have $\mu(A_0 \cap T^{-n}A_0) > 0$. But then for any $x \in A_0 \cap T^{-n}A_0$ one will have $T^n x \in A_0$, which gives a contradiction.

[6]We have here an instance of a natural way of introducing a notion of largeness. More generally, given a family $\mathcal{A}$ of subsets of a set A, one defines a dual family $\mathcal{A}^* = \{S \subseteq A : \forall B \in \mathcal{A}, S \cap B \neq \emptyset\}$. We will encounter many examples of important dual families, the most important of which is the family of *IP** sets, that is, the family of sets having nontrivial intersection with any IP set.

the proof of Proposition 2.2 is that the set $R_A = \{n : \mu(A \cap T^{-n}A) > 0\}$ is a Δ_r^* set for any $r > 1/\mu(A)$. This, in turn, implies that R_A is a *syndetic* set, that is, a set which has a nontrivial intersection with any long enough interval. Indeed, if this was not the case, R_A^c would contain arbitrarily long intervals, which leads to a contradiction since, as it is not hard to see, for any fixed r, any sufficiently long interval contains a Δ_r set.[7]

Let $(X, \mathcal{B}, \mu, T)$ be a measure preserving system. Let $A \in \mathcal{B}$ with $\mu(A) > 0$. By Poincaré's recurrence theorem, we can find $n_1 \in \mathbb{N}$ such that $\mu(A \cap T^{-n_1}A) > 0$. Applying Poincaré's recurrence theorem again to the set $A_1 = A \cap T^{-n_1}A$, we can find $n_2 > n_1$ such that

$$\begin{aligned} \mu(A_1 \cap T^{-n_2}A_1) &= \mu((A \cap T^{-n_1}A) \cap T^{-n_2}(A \cap T^{-n_1}A)) \\ &= \mu(A \cap T^{-n_1}A \cap T^{-n_2}A \cap T^{-(n_1+n_2)}A) > 0. \end{aligned}$$

Continuing in this manner we will obtain an infinite sequence $(n_i)_{i\in\mathbb{N}} \subseteq \mathbb{N}$ such that for each element $m \in FS((n_i)_{i\in\mathbb{N}})$ one has $\mu(A \cap T^{-m}A) > 0$. We see that the finite sums sets naturally appear in the process of repeated applications of Poincaré's recurrence theorem. Since ultrafilters on $(\mathbb{N}, +)$ are finitely additive probability measures, one could use this "iterative" approach to get a proof of Hindman's finite sums theorem if, in addition, the operation $n \mapsto n+1$ could be interpreted as "measure-preserving". We will momentarily see that, for the idempotent ultrafilters in $(\beta\mathbb{N}, +)$, something like this is the case.

Let $p \in (\beta\mathbb{N}, +)$ satisfy $p + p = p$. By the definition of the operation $+$ in $\beta\mathbb{N}$ (see Introduction), we have

$$A \in p \Leftrightarrow A \in p + p \Leftrightarrow \{n \in \mathbb{N} : (A - n) \in p\} \in p. \tag{2.3}$$

Formula (2.3) implies that if A is p-large, then, for p-many $n \in \mathbb{N}$, the set $A - n$ is also p-large. This is the translation-invariance we were looking for.

Proof of Theorem 2.1. Let a partition $\bigcup_{i=1}^{r} C_i$ be given. Let $p = p + p$ be an idempotent ultrafilter. Then one of the cells of the partition, call it C, is p-large. By (2.3), the set $\{n : (C - n) \in p\}$ is also p-large and hence one can find $n_1 \in C \cap \{n : (C - n) \in p\}$ such that $C_1 = C \cap (C - n_1) \in p$. Repeating this procedure, let $n_2 \in C_1 \cap \{n : (C_1 - n) \in p\}$ be such that $n_2 > n_1$ and $C_2 = C_1 \cap (C_1 - n_2) = C \cap (C - n_1) \cap (C - n_2) \cap (C - (n_1 + n_2)) \in p$. Note that $n_1, n_2, n_1 + n_2 \in C$. Choosing $n_3 \in C_2 \cap \{n : (C_2 - n) \in p\}$ will give us $FS((n_i)_{i=1}^3) \subseteq C$. Continuing in this way, we will obtain an increasing sequence $(n_i)_{i\in\mathbb{N}}$ such that, for any $k \in \mathbb{N}$, $FS((n_i)_{i=1}^k) \subseteq C$. We are done. □

The (proof of) Theorem 2.1 tells us that if $p \in (\beta\mathbb{N}, +)$ is an idempotent, then any p-large set A contains an IP set $FS((n_i)_{i\in\mathbb{N}})$. The proof, however, does not guarantee that the set $FS((n_i)_{i\in\mathbb{N}})$ obtained in the course of the proof is itself p-large. Moreover, it is easy to see that there are IP sets in A which cannot be p-large. The following proposition (attributed in [**HS**] to F. Galwin) shows that nevertheless, for any IP set E there exists an idempotent $q \in (\beta\mathbb{N}, +)$ such that $E \in q$.

[7]The sets in $(\mathbb{N}, +)$ (or in $(\mathbb{Z}, +)$) which contain arbitrarily long intervals are called *thick*. If $\mathcal{T}$ denotes the family of thick sets and $\mathcal{S}$ denotes the family of syndetic sets, then, clearly, $\mathcal{T}^* = \mathcal{S}$ and $\mathcal{S}^* = \mathcal{T}$. We will see below that these notions can be meaningfully defined in any semigroup.

THEOREM 2.5. *Given any sequence $(n_i)_{i\in\mathbb{N}} \subseteq \mathbb{N}$, there is an idempotent $p \in (\beta\mathbb{N}, +)$ such that, for any $m \in \mathbb{N}$, $FS((n_i)_{i=m}^{\infty}) \in p$.*

PROOF. Let $\overline{FS((n_i)_{i=m}^{\infty})}$ denote the closure in $\beta\mathbb{N}$, and let

$$S = \bigcap_{m=1}^{\infty} \overline{FS((n_i)_{i=m}^{\infty})}.$$

S is an intersection of a decreasing sequence of compact sets and hence is compact and nonempty. We will show now that S is a semigroup. Let $p, q \in S$. To show that $p + q \in S$ one needs to verify that, for any $m \in \mathbb{N}$, $A = FS((n_i)_{i=m}^{\infty}) \in p + q$, which is equivalent to showing that $\{x \in \mathbb{N} : (A - x) \in p\} \in q$. Let $a \in A$. Then $a = n_{i_1} + n_{i_2} + \ldots + n_{i_l}$, where $m \leq n_{i_1} < n_{i_2} < \ldots < n_{i_l}$. Let $k = l + 1$. Then $FS((n_i)_{i=k}^{\infty} \subseteq A - a$. But $FS((n_i)_{i=k}^{\infty} \in p$ which implies that $A - a \in p$. So

$$A \subseteq \{x \in \mathbb{N} : (A - x) \in p\} \in q,$$

and we are done. □

Theorem 2.5 gives an easy answer to another important question: which ultrafilters (besides the idempotent ones) have the property that their members contain IP sets?

Let Γ be the closure in $\beta\mathbb{N}$ of the (nonempty!) set of idempotents:

$$\Gamma = \text{cl}\{p \in (\beta\mathbb{N}, +) : p + p = p\}.$$

THEOREM 2.6. *An ultrafilter p belongs to Γ if and only if every p-large set contains an IP set.*

PROOF. $\Rightarrow$: Let $p \in \Gamma$ and let $A \in p$. Then $\bar{A}$ is a neighborhood of p in $\beta\mathbb{N}$ so there is $q \in \beta\mathbb{N}$ such that $q = q + q$ and $q \in \bar{A}$, or, which is the same, $A \in q$. Then, by Theorem 2.1, A has to contain an IP set.

$\Leftarrow$: Let p be given and assume that every $A \in p$ contains an IP set. We have to show that $p \in \Gamma$. Fix $A \in p$ and let $E \subseteq A$ be an IP set. Then, by Theorem 2.5 there is an idempotent $q = q + q$ such that $E \in q$. This implies that $q \in \bar{E}$ and hence $q \in \bar{A}$. So we see that, for any $A \in p$, $\bar{A} \cap \{q \in (\beta\mathbb{N}, +) : q + q = q\} \neq \emptyset$. This implies that $p \in \Gamma$. □

3. Many Equivalent Forms of Hindman's Finite Sums Theorem

We start this section with the observation that from Theorem 2.1 one can easily derive its multiplicative analog.

THEOREM 3.1. *For any finite partition $\mathbb{N} = \bigcup_{i=1}^{r} B_i$, one of the B_i contains a finite products set, namely a set of the form*

$$FP((n_i)_{i\in\mathbb{N}}) = \{n_{i_1} \cdot n_{i_2} \cdot \ldots \cdot n_{i_k} : i_1 < i_2 < \ldots < i_k; k \in \mathbb{N}\}.^8$$

PROOF. Let $C_i = \{n \in \mathbb{N} : 2^n \in B_i\}$ and apply Theorem 2.1. □

REMARK 3.2. Another approach to Theorem 3.1 is to invoke the existence of idempotents in $(\beta\mathbb{N}, \cdot)$ and to mimic the proof of Theorem 2.1. This approach leads to a stronger result since it shows that for any multiplicative idempotent $p \in (\beta\mathbb{N}, \cdot)$, any $A \in p$ contains a multiplicative IP set.

[8] We will also call such a set a *multiplicative IP set* .

The above remark can be applied to any semigroup $(S,\cdot)$, since, by Ellis' theorem, $(\beta S,\cdot)$ always has an idempotent. So we have the following general result.

THEOREM 3.3. *Let $(S,\cdot)$ be a semigroup. For any finite partition $S=\bigcup_{i=1}^{r}C_i$, one of the C_i contains a set of the form*

$$FP((x_i)_{i\in\mathbb{N}})=\{x_{i_k}\cdot x_{i_{k-1}}\cdot\ldots\cdot x_{i_1}: i_1<i_2<\ldots<i_k; k\in\mathbb{N}\}.^{9}$$

We will introduce now one more, set-theoretical, version of Theorem 2.1, which is often utilized in various applications.

Let $\mathcal{F}$ denote the family of all finite nonempty subsets of $\mathbb{N}$.[10] $\mathcal{F}$ forms a natural semigroup with respect to the operation of taking unions. Applying Theorem 3.3 to $(\mathcal{F},\cup)$, one obtains the fact that for any finite partition $\mathcal{F}=\bigcup_{i=1}^{r}C_i$, one of the C_i has to contain a *finite unions set* of the form

$$FU((\alpha_i)_{i\in\mathbb{N}})=\{\alpha_{i_1}\cup\alpha_{i_2}\cup\ldots\cup\alpha_{i_k}: i_1<i_2<\ldots<i_k; k\in\mathbb{N}\}.$$

Unfortunately, this formulation, due to the idempotent nature of the operation $\cup$, is not strong enough to be useful. The following enhanced version is free of this flaw.

THEOREM 3.4 (cf. [**Ba**]). *For any partition $\mathcal{F}=\bigcup_{i=1}^{r}C_i$ there exist $j\in\{1,2,\ldots,r\}$ and a sequence $(\alpha_i)_{i\in\mathbb{N}}\subseteq\mathcal{F}$ such that*

(i) $\min\alpha_{k+1}>\max\alpha_k$ *for each* $k\in\mathbb{N}$.

(ii) $\forall\beta\in\mathcal{F}$, $\bigcup_{t\in\beta}\alpha_t\in C_j$.

An ostensibly stronger version of the finite sums theorem states that given an IP set $A=FS((n_i)_{i\in\mathbb{N}})\subseteq\mathbb{N}$ and a finite coloring $A=\bigcup_{i=1}^{r}C_i$, one of the C_i contains an IP set.

This fact, however, is just one more equivalent form of the finite sums theorem.

THEOREM 3.5 ([**BerHi3**], Lemma 2.1). *The following statements are equivalent.*

(i) Let $(S,\cdot)$ be a semigroup, let $r\in\mathbb{N}$ and let $(x_n)_{n\in\mathbb{N}}\subseteq S$. If $FP((x_n)_{n\in\mathbb{N}})=\bigcup_{i=1}^{r}C_i$, then there is $i\in\{1,2,\ldots,r\}$ and a sequence $(y_n)_{n\in\mathbb{N}}\subseteq S$, such that $FP((y_n)_{n\in\mathbb{N}})\subseteq C_i$.

(ii) Let $r\in\mathbb{N}$ and let $\mathbb{N}=\bigcup_{i=1}^{r}C_i$. There exist $i\in\{1,2,\ldots,r\}$ and a sequence $(x_n)_{n\in\mathbb{N}}\subseteq\mathbb{N}$ such that $FS((x_n)_{n\in\mathbb{N}})\in C_i$.

(iii) Let $r\in\mathbb{N}$ and let $\mathcal{F}=\bigcup_{i=1}^{r}C_i$. There exist $i\in\{1,2,\ldots,r\}$ and a sequence $(\alpha_n)_{n\in\mathbb{N}}\subseteq\mathcal{F}$ such that $\min\alpha_{n+1}>\max\alpha_n$ for each $n\in\mathbb{N}$ and $\cup_{n\in\beta}\alpha_n\in C_i$ whenever $\beta\in\mathcal{F}$.

PROOF. (i) $\Longrightarrow$ (ii): This implication immediately follows from the fact that $(\mathbb{N},+)$ is $\mathbb{N}=FS((2^{n-1})_{n\in\mathbb{N}})$.

---

[9]Note that the finite products set $FP((x_i)_{i\in\mathbb{N}})$ is made of products $x_{i_k}\cdot x_{i_{k-1}}\cdot\ldots\cdot x_{i_1}$ in decreasing order of indices. By switching the operation in $(S,\cdot)$ from $x\cdot y$ to $y\cdot x$ (which affects the operation in βS as well), one can guarantee the products in the increasing order as well. Of course, when S is commutative, one does not have to care about such things.

[10]Note that the elements of $FS((n_i)_{i\in\mathbb{N}})$ are naturally indexed by the elements of $\mathcal{F}$: for any $\alpha=\{i_1,i_2,\ldots,i_k\}\in\mathcal{F}$, let $n_\alpha=\sum_{i\in\alpha}n_i=n_{i_1}+\ldots+n_{i_k}$.

This observation applies of course to multiplicative IP sets as well.

(ii) $\implies$ (iii): Let $\mathcal{F} = \bigcup_{i=1}^{r} C_i$. Let, for $i \in \{1, 2, \ldots, r\}$, $A_i = \{\sum_{n\in\alpha} 2^n : \alpha \in C_i\}$, and let A_0 be the set of odd natural numbers. Pick $i \in \{1, 2, \ldots, r\}$ and a sequence $(x_n)_{n\in\mathbb{N}} \subseteq \mathbb{N}$ with $FS((x_n)_{n\in\mathbb{N}}) \subseteq A_i$.

Let $\gamma_1 = \{1\}$ and $y_1 = \gamma_1$. Inductively, given $\gamma_n \in \mathcal{F}$ and $y_n = \sum_{i\in\gamma_n} x_i$, pick $\alpha_n \in \mathcal{F}$ such that $y_n = \sum_{i\in\alpha_n} 2^i$, let $l = \max \alpha_n$ and $m = \max \gamma_n$. Let γ_{n+1} consist of 2^{l+1} members of $\{m+1, m+2, \ldots\}$ such that for any $t, s \in \gamma_{n+1}$ one has $x_t \equiv x_s \mod 2^{l+1}$. Then, letting $y_{n+1} = \sum_{i\in\gamma_{n+1}} x_i$, one has that 2^{n+1} divides y_{n+1}, so if $\alpha_{n+1} \in \mathcal{F}$ is chosen so that $y_{n+1} = \sum_{i\in\alpha_{n+1}} 2^i$, one will have $\min \alpha_{n+1} > \max \alpha_n$. Let now $\beta \in \mathcal{F}$ and let $\gamma = \bigcup_{n\in\beta} \gamma_n$, $\alpha = \bigcup_{n\in\beta} \alpha_n$. Then $\sum_{n\in\beta} y_n = \sum_{n\in\gamma} x_n \in A_i$ and $\sum_{n\in\beta} y_n = \sum_{i\in\alpha} 2^i$, so $\alpha \in C_i$.

(iii) $\implies$ (i): Assume that $FP((x_n)_{n\in\mathbb{N}}) = \bigcup_{i=1}^{r} C_i$ and, for each $i \in \{1, 2, \ldots, r\}$ let $C_i = \{\alpha \in \mathcal{F} : \prod_{n\in\alpha} x_n \in C_i\}$ (note that $\prod_{n\in\alpha} x_n$ denotes the product taken in decreasing order of indices). Choose $i \in \{1, 2, \ldots, r\}$ and a sequence $(\alpha_n)_{n\in\mathbb{N}} \subseteq \mathcal{F}$ as guaranteed by (iii). For each $n \in \mathbb{N}$, let $y_n = \prod_{i\in\alpha_n} x_i$. Then, given $\beta \in \mathcal{F}$, $\prod_{n\in\beta} y_n = \prod_{i\in\alpha} x_i$, where $\alpha = \bigcup_{n\in\beta} \alpha_n$. Since $\alpha \in C_i$, we have $\prod_{n\in\beta} y_n \in C_i$. □

4. Additive and Multiplicative IP Sets in One Cell of a Partition

As we have seen in the previous section, for any finite coloring $\mathbb{N} = \bigcup_{i=1}^{r} C_i$ there must exist $i, j \in \{1, 2, \ldots, r\}$ such that C_i contains an additive IP set $FS((n_i)_{i\in\mathbb{N}})$ and C_j contains a multiplicative IP set $FP((m_i)_{i\in\mathbb{N}})$. This leads to the natural question whether one can have $i = j$. We will see in this section that the answer to this question is YES.[11] Actually, we will present two proofs of this interesting fact. The first proof, due to Hindman [**H3**], utilizes the topological algebra in $\beta\mathbb{N}$. The second one, obtained in [**BerHi3**], utilizes the combinatorial richness of IP* sets.

Let

$$\Gamma = \{p \in (\beta\mathbb{N}, +) : \text{any } A \in p \text{ contains an IP set}\}.$$

We have seen already (see Theorem 2.6) that

$$\Gamma = \text{cl}\{p \in (\beta\mathbb{N}, +) : p + p = p\}.$$

The following lemma shows that Γ is a right ideal of $(\beta\mathbb{N}, \cdot)$ meaning that, for any $p \in \Gamma$, $p \cdot \beta\mathbb{N} \subseteq \Gamma$.

Lemma 4.1. *Γ is a right ideal in $(\beta\mathbb{N}, \cdot)$.*

Proof. Γ is certainly nonempty. Let $p \in \Gamma$ and $q \in \beta\mathbb{N}$, and let us show that $p \cdot q \in \Gamma$. Let $A \in p \cdot q$. By the definition of the operation in $(\beta\mathbb{N}, \cdot)$, we have $\{n \in \mathbb{N} : A/n \in p\} \in q$. Take any $m \in \mathbb{N}$ with $A/m \in p$ and let $FS((n_i)_{i\in\mathbb{N}})$ be an (additive) IP set contained in A/m. (The existence of such a set follows from the fact that $A/m \in p \in \Gamma$). This implies that A contains an IP set, and we are done. □

Theorem 4.2. *For an arbitrary finite partition $\mathbb{N} = \bigcup_{i=1}^{r} C_i$ there exist $j \in \{1, 2, \ldots, r\}$ and two increasing sequences $(n_i)_{i\in\mathbb{N}}$, $(m_i)_{i\in\mathbb{N}}$ such that $FS((n_i)_{i\in\mathbb{N}}) \subseteq C_j$ and $FP((m_i)_{i\in\mathbb{N}} \subseteq C_j$.*

[11] Encouraged by this answer, one may ask if it is also always possible to have $n_i = m_i$, $i \in \mathbb{N}$. This time the answer is NO. See [**H4**].

PROOF. We know that $\Gamma = \mathrm{cl}\{p \in (\beta\mathbb{N}, +) : p+p = p\}$ is a closed right ideal in $(\beta\mathbb{N}, \cdot)$ and hence, by Ellis' theorem, contains a multiplicative idempotent $q = q \cdot q$. Let $j \in \{1, 2, \ldots, r\}$ be such that $C_j \in q$. Then, since $q \in \Gamma$, C_j contains an additive IP set $FS((n_i)_{i\in\mathbb{N}})$. On the other hand, since $q = q \cdot q$, C_j has to contain a multiplicative IP set $FP((m_i)_{i\in\mathbb{N}})$ as well. □

We will now present an elementary proof of Theorem 4.2. Before doing so we will introduce and briefly discuss some important notions of largeness.

DEFINITION 4.3. Let $(S, \cdot)$ be a semigroup. A set $A \subseteq S$ is called an IP* set if it has nontrivial intersection with any IP set contained in S.

We collect some useful facts about IP* sets in the following lemma.

LEMMA 4.4. *(i) Let $(S, \cdot)$ be a semigroup and assume that $A \subseteq S$ is an IP* set. Then for any IP set $E \subseteq S$, $A \cap E$ contains an IP set.*

(ii) Let $(S, \cdot)$ be a semigroup. A set $A \subseteq S$ is an IP set if and only if $A \in p$ for every idempotent $p = p \cdot p$ in $(\beta S, \cdot)$.*

(iii) Let $(S, \cdot)$ be any semigroup, $k \in \mathbb{N}$, and let $A_1, A_2, \ldots, A_k$ be IP sets in S. Then $\bigcap_{i=1}^{k} A_i$ is also an IP* set.*

PROOF. To prove (i), consider the partition $E = (A \cap E) \cup (A^c \cap E)$. If $A \cap E$ does not contain an IP set, then $A^c \cap E$ does, but this contradicts the fact that A is an IP* set.

To prove (ii), assume first that A is an IP* set in S. If, for some idempotent p, $A \notin p$, then $A^c \in p$ and hence there is an IP set $E \subseteq A^c$, which contradicts (i). In the other direction, let us assume that $A \in p$ for any $p = p \cdot p$. If A is not an IP* set, then there exists an IP set E such that $A \cap E = \emptyset$. But then A^c contains the IP set E, and by theorem 2.5 there exists an idempotent p such that $E \in p$. Hence $A^c \in p$ and $A \notin p$. Contradiction.

As for (iii), it immediately follows from (ii). □

DEFINITION 4.5. Let $(S, \cdot)$ be a semigroup. A set $A \subseteq S$ is *thick* if it contains a translate of every finite set $F \subseteq S$. Formally, A is thick if, for every finite $F \subseteq S$, there exists $t \in S$ such that $tF \subseteq S$.[12]

It is easy to see that a set $A \subseteq (\mathbb{N}, +)$ is thick if and only if it contains arbitrarily long intervals. Equivalently, $A \subseteq (\mathbb{N}, +)$ is thick if and only if, for any $n \in \mathbb{N}$, one has $A \cap (A-1) \cap (A-2) \cap \ldots \cap (A-n) \neq \emptyset$. Similarly, $B \subseteq (\mathbb{N}, \cdot)$ is thick if and only if, for any $n \in \mathbb{N}$, $B \cap B/2 \cap B/3 \cap \ldots \cap B/n \neq \emptyset$.

LEMMA 4.6. *Let $A \subseteq (\mathbb{N}, +)$ be an IP* set. Then A is multiplicatively thick.*

PROOF. It is enough to verify that, for any $k \in \mathbb{N}$, the set A/k is also IP*. (The result in question will then follow from the fact that, $\forall n \in \mathbb{N}$,
$A \cap A/2 \cap A/3 \cap \ldots \cap A/n$ is, by Lemma 4.4, an IP* set and hence is nonempty).

Let $E \subseteq (\mathbb{N}, +)$ be an IP set. Then kE is also an IP set and, by Lemma 4.4, there exists an IP set $E_0 \subseteq E$ such that $kE_0 \subseteq kE \cap A$. Then $E_0 \subseteq A/k$ and we are done. □

LEMMA 4.7. *If $A \subseteq \mathbb{N}$ is multiplicatively thick, then A contains a multiplicative IP set $FP((n_i)_{i\in\mathbb{N}})$.*

---

[12]To be more precise such a set ought to be called left thick (the right thick sets being the sets which contain a right translate of any finite set).

PROOF. The proof goes along the same lines as the proof of Theorem 2.1. Let $n_1 \in A$. Then $A \cap A/n_1 \neq \emptyset$. Let $n_2 \in A \cap A/n_1$. Clearly

$$A \cap A/n_1 \cap A/n_2 \cap A/n_1n_2 \neq \emptyset.$$

And so on. □

Second Proof of Theorem 4.2 (Cf [**BerHi3**], Thm. 2.4.) Let $\mathbb{N} = \bigcup_{i=1}^{r} C_i$ and let $I = \{i \in \{1, 2, \ldots, r\} : C_i$ contains an additive IP set$\}$. Let $A = \bigcup_{i \in I} C_i$. Clearly, A is an IP* set. By Lemma 4.6, A is multiplicatively thick and by Lemma 4.7 contains a multiplicative IP set E. Now, $E \subseteq \bigcup_{i \in I} C_i$, so by Theorem 3.5, one of the C_i, $i \in I$, has to contain a multiplicative IP set. Since for every $i \in I$, C_i contains an additive IP set, we are done. □

5. Additively and Multiplicatively Central Sets

In this section we will introduce the notion of centrality, one more useful notion of largeness. It was originally introduced by Furstenberg via the notions of proximality and uniform recurrence (see [**F2**], Def 8.3, p.161) and only somewhat later was shown to have an equivalent form in terms of ultrafilters (see [**BerHi1**]).

A topological dynamical system (with "time" $\mathbb{N}_0 = \mathbb{N} \cup \{0\}$) is a pair (X, T) where X is a compact (not necessarily metrizable) space and $T : X \to X$ a continuous map. The system (X, T) is *minimal* if for any $x \in X$ one has $\overline{\{T^n x : n \in \mathbb{N}_0\}} = X$. One can show by a simple application of Zorn's lemma that any topological dynamical system (X, T) has a minimal subsystem (Y, T), where Y denotes a T-invariant nonempty closed subset of X (and, by slight abuse of notation, the restriction of T to Y is denoted by the same symbol). Let $\sigma : \mathbb{N} \to \mathbb{N}$ denote the shift operation: $\sigma(x) = x + 1$, $x \in \mathbb{N}$. In Section 1.1 we have observed that Hindman's finite sums theorem can be viewed as an instance of application of Poincaré's recurrence theorem to the "measure-preserving system" $(\mathbb{N}, \mathcal{P}(\mathbb{N}), p, \sigma)$, where p is an arbitrary idempotent in $(\beta\mathbb{N}, +)$. As we will momentarily see, a certain subclass of idempotent ultrafilters allows one to make a useful connection between minimal right ideals in $(\beta\mathbb{N}, +)$ and minimal topological systems.

Extend the shift operation σ from $\mathbb{N}$ to $\beta\mathbb{N}$ by the rule $\sigma(q) = q + 1$, $q \in \beta\mathbb{N}$ (here 1 is identified with the principal ultrafilter of sets containing the integer 1). This makes the pair $(\beta\mathbb{N}, \sigma)$ a topological dynamical system.

THEOREM 5.1. *The family of minimal closed σ-invariant subsets of $\beta\mathbb{N}$ coincides with the family of minimal right ideals of $(\beta\mathbb{N}, +)$.*

PROOF. We first observe that closed σ-invariant sets in $\beta\mathbb{N}$ coincide with right ideals. Indeed, if I is a right ideal, i.e. satisfies $I + \beta\mathbb{N} \subseteq I$, then for any $p \in I$ one has $p+1 \in I + \beta\mathbb{N} \subseteq I$, so that I is σ-invariant. On the other hand, if S is a closed σ-invariant set in $\beta\mathbb{N}$ and $p \in S$, then $p + \beta\mathbb{N} = p + \bar{\mathbb{N}} = \overline{p + \mathbb{N}} \subseteq \overline{S} = S$, which implies $S + \beta\mathbb{N} \subseteq S$.

Now the theorem follows from the simple general fact that any minimal right ideal in a compact left-topological semigroup $(G, \cdot)$ is closed. Indeed, if R is a right ideal in $(G, \cdot)$ and $x \in R$, then xG is compact as the continuous image of G and is an ideal. Hence the minimal ideal containing x is compact as well. (The fact that R contains a minimal ideal follows by a routine application of Zorn's lemma to the non-empty family $\{I : I$ is a closed right ideal of G and $I \subseteq R\}$). □

Observe now that any minimal right ideal in $(\beta\mathbb{N}, +)$, being a compact left-topological semigroup, contains, by Ellis' theorem, an idempotent.

DEFINITION 5.2. An idempotent $p \in (\beta\mathbb{N}, +)$ is called *minimal* if p belongs to a minimal right ideal.

It is not hard to show that any minimal right ideal R of $(\beta\mathbb{N}, +)$ is of the form $q + \beta\mathbb{N}$ for some $q \in R$. Indeed, for any $q \in R$, $q + \beta\mathbb{N} \subseteq R + \beta\mathbb{N} = R$. Since R is minimal, we get $q + \beta\mathbb{N} = R$. Note that since $q + \beta\mathbb{N}$ is the continuous image of $\beta\mathbb{N}$ under the function $\lambda_q(p) = q + p$, minimal right ideals in $(\beta\mathbb{N}, +)$ are compact. It follows that one can choose q to be an idempotent. This gives the following result.

THEOREM 5.3. *Any minimal subsystem of $(\beta\mathbb{N}, \sigma)$ is of the form $(p + \beta\mathbb{N}, \sigma)$, where p is a minimal idempotent in $(\beta\mathbb{N}, +)$.*

We are going to show that, if p is a minimal idempotent in $(\beta\mathbb{N}, +)$ and $A \in p$, then A is a *piecewise syndetic* set, namely, a set of the form $S \cap T$, where T is a thick set and S is syndetic (i.e. has bounded gaps). A useful equivalent definition of piecewise syndeticity is given by the following lemma, the proof of which is left to the reader.

LEMMA 5.4. *A set $A \subseteq (\mathbb{N}, +)$ is piecewise syndetic if and only if there exists a finite set $F \subseteq \mathbb{N}$ such that the family*

$$\Big\{\bigcup_{t\in F}(A - t) - n : n \in \mathbb{N}\Big\}$$

has the finite intersection property.

THEOREM 5.5. *Let p be a minimal idempotent in $(\beta\mathbb{N}, +)$.*
(i) For any $A \in p$, the set $B = \{n \in \mathbb{N} : (A - n) \in p\}$ is syndetic.
(ii) Any $A \in p$ is piecewise syndetic.

PROOF. Statement (i) follows immediately from the fact that $(p + \beta\mathbb{N}, \sigma)$ is a minimal system. Indeed, note that the assumption $A \in p$ just means that $p \in \bar{A}$, i.e. $\bar{A}$ is a (clopen) neighborhood of p. Now, by minimality, every point in $(p + \beta\mathbb{N}, \sigma)$ is *uniformly recurrent*, i.e. visits any of its neighborhoods V along a syndetic set. This implies that the set $\{n : p + n \in \bar{A}\} = \{n : A \in p + n\} = \{n : A - n \in p\}$ is syndetic.

(ii) Since the set $B = \{n : A - n \in p\}$ is syndetic, the union of finitely many shifts of B covers $\mathbb{N}$, i.e. for some finite set $F \subseteq \mathbb{N}$ one has $\bigcup_{t\in F}(B - t) = \mathbb{N}$. So, for any $n \in \mathbb{N}$ there exists $t \in F$ such that $n \in B - t$, or $n + t \in B$. By the definition of B this implies that $(A - (n + t)) \in p$. It follows that for any n the set $(\bigcup_{t\in F}(A - t)) - n$ belongs to p, and consequently the family $\{(\bigcup_{t\in F}(A - t)) - n : n \in \mathbb{N}\}$ has the finite intersection property. By Lemma 5.4, this is equivalent to piecewise syndeticity of A, and we are done. □

At this point we want to make a simple but important observation. Namely, all the definitions, results and proofs in this section which pertain to $(\mathbb{N}, +)$ can be transferred (usually verbatim) to the more general situation where the semigroup $(\mathbb{N}, +)$ is replaced by a (discrete) semigroup $(S, \cdot)$. In particular, this remark applies to the semigroup $(\mathbb{N}, \cdot)$.

We collect for the reader's convenience some definitions and results related to $(\beta\mathbb{N}, \cdot)$.

Definition 5.6. (i) An idempotent $p \in (\beta\mathbb{N}, \cdot)$ is *minimal* if it belongs to a minimal right ideal of $(\beta\mathbb{N}, \cdot)$.

(ii) A set $A \subseteq (\beta\mathbb{N}, \cdot)$ is *syndetic* if there exists a finite set $F \subseteq \mathbb{N}$ such that $\bigcup_{n\in F} A/n = \mathbb{N}$.

(iii) A set $A \subseteq (\beta\mathbb{N}, \cdot)$ is *piecewise syndetic* if A is of the form $A = S \cap T$, where A is multiplicatively syndetic and T is multiplicatively thick.

Theorem 5.7. *Let p be a minimal idempotent in $(\beta\mathbb{N}, \cdot)$.*

(i) For any $A \in p$, the set $B = \{n \in \mathbb{N} : (A/n \in p\}$ is (multiplicatively) syndetic.

(ii) Any $A \in p$ is (multiplicatively) piecewise syndetic.

Definition 5.8. (i) A set $A \subseteq (\mathbb{N}, +)$ is *additively central* if it is a member of a minimal idempotent $p \in (\beta\mathbb{N}, +)$.

(ii) A set $A \subseteq (\mathbb{N}, \cdot)$ is *multiplicatively central* if it is a member of a minimal idempotent $p \in (\beta\mathbb{N}, \cdot)$.

(iii) A set $A \subseteq (\mathbb{N}, +)$ is *additively central** (or AC*) if for any central set $S \subseteq (\mathbb{N}, +)$, $A \cap S \neq \emptyset$.

(iv) A set $A \subseteq (\mathbb{N}, \cdot)$ is *multiplicatively central** (or MC*) if for any central set $S \subseteq (\mathbb{N}, \cdot)$, $A \cap S \neq \emptyset$.

Remark 5.9. (i) One can show (see for example the proof of Theorem 5.4 in [**BerHi1**]) that if p is a minimal idempotent in $(\beta\mathbb{N}, +)$, then so is np for any $n \in \mathbb{N}$. This implies that if A is a central set in $(\mathbb{N}, +)$, then, for any $n \in \mathbb{N}$, A/n is also central.

(ii) It is easy to see that a set $A \subseteq \mathbb{N}$ is additively (multiplicatively) central* if and only if A is a member of any minimal additive (multiplicative) idempotent.

The usefulness of minimal idempotents in Ramsey theory stems from the fact that their members, central sets, are both large (in particular, are piecewise syndetic) and combinatorially rich. For example, one can show that any central set in $(\mathbb{N}, +)$ not only contains an IP set, but also contains arbitrarily long arithmetic progressions, and, more generally, contains a solution of any partition regular system of linear equations. (See [**F2**], Ch. 8). Similarly, any central set in $(\mathbb{N}, \cdot)$ contains a multiplicative IP set, as well as, for any $k \in \mathbb{N}$, *geoarithmetic* configurations of the form $\{b(a + id)^j : 0 \leq i, j \leq k\}$, where $a, b, d \in \mathbb{N}$. (See [**Ber5**], [**BeiBerHS**], [**Bei1**], [**M**]).

Lemma 5.10. *Any additively thick set in $\mathbb{N}$ is additively central and any multiplicatively thick set is multiplicatively central.*

Proof. We will deal with the multiplicative case, the other being practically identical. Let $A \subseteq \mathbb{N}$ be a multiplicatively thick set. Since this is equivalent to the fact that for any $n \in \mathbb{N}$, $A \cap A/2 \cap \ldots \cap A/n \neq \emptyset$, which, in turn, implies that there is $x_n \in A$ such that $\{x_n, 2x_n, \ldots, nx_n\} \subseteq A$, we will assume that for some infinite sequence $(x_n)_{n\in\mathbb{N}} \subseteq \mathbb{N}$, $A = \bigcup_{n\in\mathbb{N}}\{x_n, 2x_n, \ldots, nx_n\}$. Now, any infinite subset of $\mathbb{N}$ is a member of *some* nonprincipal ultrafilter, so let $p \in \beta\mathbb{N} \setminus \mathbb{N}$ be such that $\{x_n : \mathbb{N} \in \mathbb{N}\} \in p$.

We claim that $p \cdot \beta\mathbb{N} \subseteq \bar{A}$. Indeed, since, for any $n \in \mathbb{N}$, $\{x_m : m \geq n\} \subseteq A/n$, we have $A/n \in p$, and so $\{n : A/n \in p\} = \mathbb{N} \in q$ for any idempotent $q \in \beta\mathbb{N}$. Note that $p \cdot \beta\mathbb{N}$ is a right ideal of $(\beta\mathbb{N}, \cdot)$ so by Zorn's lemma it contains a minimal ideal R which, as was already remarked above, is necessarily closed. So, by Ellis' theorem, R contains an idempotent q. Then $A \in q$ which implies that A is multiplicatively central. $\square$

LEMMA 5.11. *Any AC* set in $\mathbb{N}$ is multiplicatively thick.*

PROOF. The proof is similar to that of Lemma 4.6. Let $A \subseteq \mathbb{N}$ be an AC* set. It is enough to check that, for any $k \in \mathbb{N}$, A/k is also an AC* set. (Indeed, it will follow that $A \cap A/2 \cap \ldots \cap A/n$ is AC* and hence nonempty). Now, to see that A/k is an AC* set, one argues as follows. By Remark 5.9, if p is a minimal idempotent, then so is kp for any $k \in \mathbb{N}$ and since A is an AC* set, it is a member of kp which implies $A/k \in p$. So $A \cap A/2 \cap \ldots \cap A/n \in p$ for any minimal idempotent $p \in (\beta\mathbb{N}, +)$. We are done. $\square$

COROLLARY 5.12. *Any AC* set in $\mathbb{N}$ is multiplicatively central.*

PROOF. Follows immediately from Lemma 5.11. $\square$

In view of Theorem 2.5, Theorem 4.2 says that for any finite partition $\mathbb{N} = \bigcup_{i=1}^{r} C_i$, there exist an additive idempotent in $(\beta\mathbb{N}, +)$, a multiplicative idempotent $q \in (\beta\mathbb{N}, \cdot)$ and $j \in \{1, 2, \ldots, r\}$ such that $C_j \in p$ and $C_j \in q$. The following theorem is a strengthening of this fact.

THEOREM 5.13 (cf [**BerHi1**], Corollary 5.5). *For any finite partition $\mathbb{N} = \bigcup_{i=1}^{r} C_i$, one of the C_i is both additively and multiplicatively central.*

PROOF. Similar to the proof of Theorem 4.2, there are (at least) two possible approaches. The first one utilizes the fact that the set

$$M = \mathrm{cl}\{p \in \beta\mathbb{N} : p \text{ is a minimal idempotent in } (\beta\mathbb{N}, +)\}$$

is a right ideal in $(\beta\mathbb{N}, \cdot)$. (See [**BerHi1**], Theorem 5.4). By Zorn's lemma M contains a minimal right ideal R which contains a minimal idempotent $q = q \cdot q$. So, for some $j \in \{1, 2, \ldots, r\}$, $C_j \in q$. But then, by the definition of M, C_j is also a member of some additive minimal idempotent. So, C_j is both additively and multiplicatively central.

The other approach follows the lines of our second proof of Theorem 4.2. Namely, let $I = \{i \in \{1, 2, \ldots, r\} : C_i \text{ is additively central}\}$. Then $A = \bigcup_{i \in I} C_i$ is an AC* set. Now, by Lemma 5.11, A is multiplicatively thick and hence, by Lemma 5.10, is multiplicatively central. But then one of the C_i, $i \in I$ is multiplicatively central and since every C_i, $i \in I$ is additively central, we are done. $\square$

Now that we know that, for any finite partition of $\mathbb{N}$, one of the cells of the partition is both additively and multiplicatively central, it is natural to ask whether all additively central sets must contain rich multiplicative structure and similarly whether all multiplicatively central sets must contain rich additive structure. The following two results show that the answers turn out to be NO (Proposition 5.14) and YES (Theorem 5.15) respectively.

PROPOSITION 5.14 ([**BerHi2**], Theorem 3.4). *There is an additively central set $A \subseteq \mathbb{N}$ such that for no $x, y \in \mathbb{N}$ is $\{x, y, x \cdot y\} \subseteq A$.*

PROOF. One can actually construct an additively thick set $A = \bigcup_{n=1}^{\infty} \{x_n, x_n + 1, \ldots, x_n + y_n\}$ which satisfies the requirements. To make it work one has just to choose $x_1 \geq 2$, to make sure that x_n grows fast enough so that for no $i, j < n$ will one have $x_i x_j \in \{x_n, x_n + 1, \ldots, x_n + y_n\}$ and to pick (increasing) $y_n < x_2 x_n$. $\square$

THEOREM 5.15 ([**BerHi2**], Theorem 3.5). *Let $A \subseteq \mathbb{N}$ be a multiplicatively central set. Then for each m there exists an m-element sequence $(y_n)_{n=1}^m$ such that $FS((y_n)_{n=1}^m) \subseteq A$.*

PROOF. Let $T = \{p \in \beta\mathbb{N}:$ for all $B \in p$ and all $m \in \mathbb{N}$ there exists $(y_n)_{n=1}^m$ with $FS((y_n)_{n=1}^m) \subseteq B\}$. Now all additive idempotents are in T so $T \neq \emptyset$. We claim that T is a two sided ideal of $(\beta\mathbb{N}, \cdot)$. To this end let $p \in T$ and let $q \in \beta\mathbb{N}$. To see that $p \cdot q \in T$, let $B \in p \cdot q$ and $m \in \mathbb{N}$ be given. Then $\{n \in \mathbb{N} : B/n \in p\} \in q$ so pick $n \in \mathbb{N}$ with $B/n \in p$. Pick $(y_t)_{t=1}^m$ with $FS((y_t)_{t=1}^m) \subseteq B/n$. Then $FS((n \cdot y_t)_{t=1}^m) \subseteq B$.

To see that $q \cdot p \in T$, let $B \in q \cdot p$ and $m \in \mathbb{N}$ be given. Then $\{n \in \mathbb{N} : B/n \in q\} \in p$, so pick $(y_t)_{t=1}^m$ with

$$FS((y_t)_{t=1}^m) \subseteq \{n \in \mathbb{N} : B/n \in q\}.$$

Since $FS((y_t)_{t=1}^m)$ is finite we have $\bigcap\{B/n : n \in FS((y_t)_{t=1}^m)\} \in q$ so pick $a \in \bigcap\{B/n : n \in FS((y_t)_{t=1}^m)\}$. Then $FS((a \cdot y_t)_{t=1}^m) \subseteq B$.

Now A is multiplicatively central so pick a minimal idempotent $p \in (\beta\mathbb{N}, \cdot)$ with $A \in p$. Pick a minimal right ideal R of $(\beta\mathbb{N}, \cdot)$ with $p \in R$. Since T is a two sided ideal, $R \subseteq T$. (Since T is a left ideal $T \cap R \neq \emptyset$ and hence $T \cap R$ is a right ideal so $T \cap R = R$.) Then $p \in T$. Since $A \in p$, we are done. □

REMARK 5.16. It is natural to ask whether any multiplicatively central set in $\mathbb{N}$ contains an infinite additive IP set. The answer is NO: one can construct a multiplicatively central $A \subseteq \mathbb{N}$ such that for no $(y_n)_{n\in\mathbb{N}} \subseteq \mathbb{N}$ is $FS((y_n)_{n\in\mathbb{N}}) \subseteq A$. See [**BerHi2**], Theorem 3.6.

The following result will be needed in the next section.

THEOREM 5.17 (cf. [**BerHi1**], Theorem 5.6). *There is a minimal idempotent $q \in (\beta\mathbb{N}, \cdot)$ such that every member of q is additively central.*

PROOF. Let $M = \mathrm{cl}\{p \in \beta\mathbb{N} : p$ is a minimal idempotent in $(\beta\mathbb{N}, +)\}$. As was already mentioned in the proof of Theorem 5.13, M is a right ideal in $(\beta\mathbb{N}, \cdot)$ and hence contains a minimal right ideal R. Let $q \in R$ be a minimal multiplicative idempotent. Let $A \in q$. Then $q \in \bar{A} \cap M$, which implies that A is additively central. □

6. An Application: Partition Regularity of the Equation a+b = cd.

In this short section we will utilize Theorem 5.17 to show the partition regularity of the equation $a + b = cd$, thereby providing an affirmative answer to a question posed in [**CsGSa**]. For another solution to this question see [**H6**].

THEOREM 6.1. *For any finite coloring $\mathbb{N} = \bigcup_{i=1}^r C_i$, one of the C_i contains arbitrarily large and distinct a, b, c, d such that $a + b = cd$.*

PROOF. Let $p \in \beta\mathbb{N}$ be a minimal multiplicative idempotent with the property that any member of p is additively central (see Theorem 5.17). Let a partition $\mathbb{N} = \bigcup_{i=1}^r C_i$ be given and let $i \in \{1, 2, \ldots, r\}$ be such that $C_i \in p$. For convenience of notation we will denote this C_i by C. Since $C \in p = p \cdot p$, we have $\{n : C/n \in p\} \in p$. So there exists $d \in C$ such that $C/d \in p$ and hence $C \cap C/d \in p$. (Note that there are "many" such d's).

Now since any member of p is additively central, there exists an additive idempotent q such that $C \cap C/d \in q$. This implies that

$$\{n : (C \cap C/d) - n \in q\} \in q.$$

Let $b' \in C \cap C/d$ be such that $(C \cap C/d) - b' \in q$. Then

$$(C \cap C/d) \cap ((C \cap C/d) - b') \in q$$

and hence is nonempty. Note now that it follows from the choice of b' that $b = b'd \in C$. Now, since $(C \cap C/d) \bigcap((C \cap C/d) - b') \neq \emptyset$, we obtain $E = (dC \cap C) \cap ((dC \cap C) - b) \neq \emptyset$. Choose $a \in E$. Then in particular $a \in C$, $b \in C$, and $a + b \in dC$, so that for some $c \in C$ we get $a + b = cd$.

It is clear from the proof that a, b, c, d can be chosen arbitrarily large and distinct. We are done. □

7. Ultrafilters and Diophantine Approximation

Let X be a topological space, and let $p \in (\beta\mathbb{N}, +)$. Given a sequence $(x_n)_{n\in\mathbb{N}}$ in X, we shall write $p\text{-}\lim_{n\in\mathbb{N}} x_n = y$ if, for every neighborhood U of y one has

$$\{n \in \mathbb{N} : x_n \in U\} \in p.$$

It is easy to see that $p\text{-}\lim_{n\in\mathbb{N}} x_n$ exists and is unique in any compact Hausdorff space.

THEOREM 7.1. *Let X be a compact Hausdorff space, let $p, q \in \beta\mathbb{N}$ and let $(x_n)_{n\in\mathbb{N}}$ be a sequence in X. Then*

$$(q + p)\text{-}\lim_{r\in\mathbb{N}} x_r = p\text{-}\lim_{t\in\mathbb{N}} q\text{-}\lim_{s\in\mathbb{N}} x_{s+t}.$$

In particular, if p is an idempotent and $p = q$ one has

$$p\text{-}\lim_{r\in\mathbb{N}} x_r = p\text{-}\lim_{s\in\mathbb{N}} p\text{-}\lim_{t\in\mathbb{N}} x_{s+t}.$$

PROOF. Recall that

$$q + p = \{A \subseteq \mathbb{N} : \{n \in \mathbb{N} : (A - n) \in q\} \in p\}.$$

Let $x = (q + p)\text{-}\lim_{r\in\mathbb{N}} x_r$. It will suffice for us to show that for any neighborhood U of x, we have that for p-many t, $q\text{-}\lim_{s\in\mathbb{N}} x_{s+t} \in U$. Fix such a U. We have $\{r : x_r \in U\} \in q + p$, so that

$$\{t : \{x : x_s \in U\} - t \in q\} = \{t : \{x : x_{s+t} \in U\} \in q\} \in p.$$

This implies, in particular, that for p-many t, $q\text{-}\lim_{s\in\mathbb{N}} x_{s+t} \in U$. □

As an immediate application of Theorem 7.1, let X be the one dimensional torus $\mathbb{T} = \mathbb{R}/\mathbb{Z}$ and let, for some $a \in \mathbb{T}$, $x_n = na$. (It is convenient to interpret $\mathbb{T}$ as the unit interval $[0, 1]$ with the ends glued up and $x_n = na$ as corresponding to the real sequence $na \mod 1 \in [0, 1)$).

We claim that for any idempotent $p \in (\beta\mathbb{N}, +)$ one has $p\text{-}\lim_{n\in\mathbb{N}} na = 0$. To see this, let $c = p\text{-}\lim_{n\in\mathbb{N}} na$. Then

$$\begin{aligned} c = (p+p)\text{-}\lim_{n\in\mathbb{N}} na &= p\text{-}\lim_{n\in\mathbb{N}} p\text{-}\lim_{m\in\mathbb{N}} (n+m)a \\ &= p\text{-}\lim_{n\in\mathbb{N}} (c+na) \\ &= 2c, \end{aligned}$$

and it follows that $c = 0$.

It is now easy to inductively extend this observation to polynomial sequences of the form $x_n = a_1 n + \ldots + a_k n^k$, where $a_i \in \mathbb{T}$, $i = 1, 2, \ldots, k$. For example, if $x_n = n^2 a$ and $p = p + p$, one has

$$\begin{aligned} p\text{-}\lim_{n\in\mathbb{N}} n^2 a &= p\text{-}\lim_{n\in\mathbb{N}} p\text{-}\lim_{m\in\mathbb{N}} (n^2 + 2nm + m^2)a \\ &= p\text{-}\lim_{n\in\mathbb{N}} (n^2 a + 2n(p\text{-}\lim_{m\in\mathbb{N}} ma) + p\text{-}\lim_{m\in\mathbb{N}} m^2 a) \\ &= p\text{-}\lim_{n\in\mathbb{N}} n^2 a + p\text{-}\lim_{m\in\mathbb{N}} m^2 a \\ &= 2(p\text{-}\lim_{n\in\mathbb{N}} n^2 a). \end{aligned}$$

which implies $p\text{-}\lim_{n\in\mathbb{N}} n^2 a = 0$. (Note that we used the "linear" fact that for any fixed m and $a \in \mathbb{T}$, $p\text{-}\lim_{n\in\mathbb{N}} 2nma = 0$). So, modulo the completely trivial details of a routine inductive proof, we have established the following result.

THEOREM 7.2. *For any $k \in \mathbb{N}$, $a_i \in \mathbb{T}$, $i = 1, \ldots, k$, and $p \in \beta\mathbb{N}$ with $p = p+p$, one has $p\text{-}\lim_{n\in\mathbb{N}}(a_1 n + \ldots + a_k n^k) = 0$.*

Let $\|\cdot\|$ denote the distance to a closest integer in $\mathbb{R}$. Utilizing the characterization of IP* given by Lemma 4.4 (ii), we have the following immediate corollary.

COROLLARY 7.3 (cf. [**F2**], Theorem 2.19). *For any $\epsilon > 0$, $l \in \mathbb{N}$, and any real polynomials g_i satisfying $g_i(0) = 0$, $i = 1, \ldots, l$, the set*

$$R_\epsilon = \{n \in \mathbb{N} : \|g_i(n)\| < \epsilon, \ i = 1, \ldots, l\}$$

is IP*.

We will discuss now a strengthening of Corollary 7.3 which involves multiplicatively central sets. First, we need a definition.

DEFINITION 7.4. Given any r integers $n_1, \ldots, n_r \in \mathbb{N}$, call the finite sums set $FS((n_i)_{i=1}^r)$ an IP$_r$ set. A set $A \subseteq \mathbb{N}$ is IP$_r^*$ set if for any IP$_r$ set E one has $A \cap E \neq \emptyset$.

The following theorem is an immediate consequence of Theorem 5.15.

THEOREM 7.5. *If $r \in \mathbb{N}$ and $A \subseteq \mathbb{N}$ is an IP$_r^*$ set then A is an MC* set.*

One can show that the set R_ϵ appearing in the formulation of Corollary 7.3 is an IP$_r^*$ set for some r (which depends only on ϵ, on k, and on the maximal degree of the polynomials g_i, $i = 1, 2, \ldots, k$). To give a flavour of the reasoning leading to this statement, let us show for example that for any real number x and any $\epsilon > 0$ there exists r such that the set $\{n \in \mathbb{N} : \|n^2 x\| \leq \epsilon\}$ is an IP$_r^*$ set.

We will use the following special case of the Hales-Jewett theorem (see the discussion of various equivalent forms of the Hales-Jewett thoerem in [**BerL2**] and [**Ber2**], Section 4). Given a finite set F, let $\mathcal{P}(F)$ denote the set of all subsets of F.

THEOREM 7.6. *For any $t \in \mathbb{N}$ there exists $r = r(t)$ such that for any t-coloring*

$$\mathcal{P}(\{1, 2, \ldots, r\}) \times \mathcal{P}(\{1, 2, \ldots, r\}) = \bigcup_{i=1}^{t} C_i,$$

one of the C_i contains a configuration of the form

$$\{(\alpha_1, \alpha_2), (\alpha_1 \cup \gamma, \alpha_2), (\alpha_1, \alpha_2 \cup \gamma), (\alpha_1 \cup \gamma, \alpha_2 \cup \gamma)\},$$

where $\gamma \subseteq \{1, 2, \ldots, r\}$ is nonempty and disjoint from α_1 and α_2.

To show that the set $\{n \in \mathbb{N} : \|n^2 x\| < \epsilon\}$ is an IP_r^* set for some r, one argues as follows. First, assume for convenience and without loss of generality that $\epsilon = 1/t$ for some $t \in \mathbb{N}$ and that $x \notin \mathbb{Q}$. Let now $r = r(t)$, as in Theorem 7.6, and let an r-element subset $\{n_1, \ldots, n_r\} \subseteq \mathbb{N}$ be given. For any nonempty $\alpha \subseteq \{1, 2, \ldots, r\}$ we will write $n_\alpha = \sum_{i \in \alpha} n_i$. Also set $n_\emptyset = 0$. Let us take the partition of $[0, 1)$ into t semiopen intervals $I_1, I_2, \ldots, I_t$ (of length $1/t$ each) and correspond to each pair

$$(\alpha, \beta) \subseteq \{1, 2, \ldots, r\} \times \{1, 2, \ldots, r\}$$

the unique subinterval I_j for which $n_\alpha n_\beta x \mod 1 \in I_j$. This induces a t-coloring of

$$\mathcal{P}(\{1, 2, \ldots, r\}) \times \mathcal{P}(\{1, 2, \ldots, r\})$$

and by Theorem 7.6 we have that for some $\alpha_1, \alpha_2, \gamma \subseteq \{1, 2, \ldots, r\}$ where γ is nonempty and disjoint from α_1 and α_2, and some $j \in \{1, 2, \ldots, t\}$, the four numbers $n_{\alpha_1} n_{\alpha_2} x \mod 1$, $(n_{\alpha_1} + n_\gamma) n_{\alpha_2} x \mod 1$, $n_{\alpha_1}(n_{\alpha_2} + n_\gamma) x \mod 1$, and $(n_{\alpha_1} + n_\gamma)(n_{\alpha_2} + n_\gamma) x \mod 1$ are all in I_j. Applying the identity

$$n_{\alpha_1} n_{\alpha_2} - (n_{\alpha_1} + n_\gamma) n_{\alpha_2} - n_{\alpha_1}(n_{\alpha_2} + n_\gamma) + (n_{\alpha_1} + n_\gamma)(n_{\alpha_2} + n_\gamma) = n_\gamma^2$$

and taking into account that the length of I_j is $1/t$ and that $x \notin \mathbb{Q}$, we get $\|n_\gamma^2 x\| < 1/t$. Since $\gamma \subseteq \{1, 2, \ldots, r\}$, we established the fact that $\{n \in \mathbb{N} : \|n^2 x\| < \epsilon\}$ is an IP_r^* set.

A similar argument shows that for any $k \in \mathbb{N}$, any $x \in \mathbb{R}$, and any $\epsilon > 0$, the set $\{n \in \mathbb{N} : \|n^k x\| < \epsilon\}$ is an IP_r^* set for some r. Now, one can show that for any $\mathrm{IP}_{r_i}^*$ sets A_i, $i = 1, 2, \ldots, k$, there exists $r \in \mathbb{N}$ such that the set $A_1 \cap A_2 \cap \ldots \cap A_k$ is an IP_r. This implies that for any $\epsilon > 0$, $k \in \mathbb{N}$, and $x_1, x_2, \ldots, x_k \in \mathbb{R}$, the set $\{n \in \mathbb{N} : \|n x_1 + n^2 x_2 + \ldots + n^k x_k\| < \epsilon\}$ is an IP_r set for some r. Moreover, and for the same reason, this is also true for any finite set of polynomials. We summarize this in the following theorem. (Note that the last claim of this theorem follows from Theorem 7.5).

THEOREM 7.7. *For any $\epsilon > 0$, $l \in \mathbb{N}$, and any real polynomials g_i satisfying $g_i(0) = 0$, $i = 1, 2, \ldots, l$, the set*

$$R_\epsilon = \{n \in \mathbb{N} : \|g_i(n)\| < \epsilon, \ i = 1, 2, \ldots, l\}$$

is an IP_r^* set for some r. Moreover, R_ϵ is a multiplicatively central set.*

8. Ultrafilters and Measure Preserving Systems

As we have seen in previous sections, the usefulness of ultrafilters in *partition* Ramsey theory stems from the fact that for any ultrafilter $p \in (\beta S, \cdot)$ and any finite partition $S = \bigcup_{i=1}^{r} C_i$ one (and only one) of the C_i is a member of p. If it is known that members of p always posses a certain property, then one cell of the partition will have this property as well.

For example, one can show that if p is a minimal idempotent in $(\beta\mathbb{N}, +)$ then every $A \in p$ is *AP-rich*, that is, contains arbitrarily long arithmetic progressions (see [**BerHi1**] Section 3, [**BerFHiK**], and [**Ber4**], Theorem 2.10). This immediately implies the classical van der Waerden theorem stating that for any finite partition $\mathbb{N} = \bigcup_{i=1}^{r} C_i$, one of the C_i is AP-rich.

This leads to the question whether there is any way to tell which cell of a given partition has the property of being AP-rich. Questions of this kind are dealt with by *density* Ramsey theory, which "upgrades" the results of the form (P) (see Introduction) to the following:

(D) Any "large" subset of an infinite "well organized" set S is "well organized".

For example, if $S = \mathbb{N}$, "well organized" means being AP-rich, and "large" is interpreted as the property of a set $E \subseteq \mathbb{N}$ to have positive upper Banach density, $d^*(E) = \limsup_{N-M\to\infty} \frac{|E\cap\{M+1,\ldots,N\}|}{N-M}$, then (D) is the celebrated Szemerédi theorem on arithmetic progressions ([**Sz**]).[13]

In [**F**] Furstenberg proved Szemerédi's theorem by deriving it as a corollary of the following beautiful and far reaching extension of Poincaré's recurrence theorem.

THEOREM 8.1. *For any probability measure preserving system $(X, \mathcal{B}, \mu, T)$, any $A \in \mathcal{B}$ with $\mu(A) > 0$, and any $k \in \mathbb{N}$, there exists $n > 0$ such that*

$$\mu(A \cap T^{-n}A \cap T^{-2n}A \cap \ldots \cap T^{-kn}A) > 0. \tag{8.1}$$

Note that if for $E \subseteq \mathbb{N}$ one has $d^*(E) > 0$ then the fact that E contains a $(k+1)$-element arithmetic progression just means that for some $n > 0$

$$E \cap (E-n) \cap (E-2n) \cap \ldots \cap (E-kn) \neq \emptyset. \tag{8.2}$$

It is not hard to see that, under the hypotheses of Theorem 8.1, the validity of (8.2) for some $n > 0$ implies the ostensibly stronger conclusion

$$d^*(E \cap (E-n) \cap (E-2n) \cap \ldots \cap (E-kn)) > 0. \tag{8.3}$$

Note that for any $n \in \mathbb{N}$ one has $d^*(E-n) = d^*(E)$. So we see that both Szemerédi's theorem and Furstenberg's ergodic Szemerédi theorem are about iterations of a "size" preserving transformation. To derive Szemerédi's theorem from Theorem 8.1 one can use the following form of Furstenberg's correspondence principle (see for example [**Ber1**] or [**Ber3**]).

THEOREM 8.2 (Furstenberg's correspondence principle). *Given a set $E \subseteq \mathbb{Z}$ with $d^*(E) > 0$ there is a probability measure preserving system $(X, \mathcal{B}\mu, T)$ and a*

---

[13]For more discussion and examples see [**Ber1**], Section 1, and [**Ber6**].

set $A \in \mathcal{B}$ with $\mu(A) = d^*(E)$ such that for any $k \in \mathbb{N}$ and any $n_1, n_2, \dots, n_k \in \mathbb{Z}$ one has:

$$d^*(E \cap (E-n_1) \cap (E-n_2) \cap \dots \cap (E-n_k)) \geq \mu(A \cap T^{-n_1}A \cap T^{-n_2}A \cap \dots \cap T^{-n_k}A).$$

Fix a probability measure preserving system $(X, \mathcal{B}, \mu, T)$, a set $A \in \mathcal{B}$ with $\mu(A) > 0$, an integer $k \in \mathbb{N}$, and consider the following set:

$$R_{k,A} = \{n \in \mathbb{N} : \mu(A \cap T^{-n}A \cap T^{-2n}A \cap \dots \cap T^{-kn}A) > 0\}.$$

When $k = 1$ this is just the set R_A which we encountered in Section 1. As we have seen, R_A is large in quite a strong sense, namely, R_A is a Δ_r^* set for any $r > \frac{1}{\mu(A)}$, and, in particular, is syndetic.[14]

It is natural to inquire to what extent the largeness properties of R_A generalize to $R_{k,A}$. The fact that $R_{k,A}$ is syndetic was already contained in Furstenberg's original paper [**F**]. Indeed, Furstenberg actually proved in [**F**] that

$$\liminf_{N-M\to\infty} \frac{1}{N-M} \sum_{n=M}^{N-1} \mu(A \cap T^{-n}A \cap T^{-2n}A \cap \dots \cap T^{-kn}A) > 0,$$

which implies that for any $A \in \mathcal{B}$ with $\mu(A) > 0$ there is a constant $a > 0$ such that

$$\{n \in \mathbb{N} : \mu(A \cap T^{-n}A \cap T^{-2n}A \cap \dots \cap T^{-kn}A) > a\} \text{ is syndetic.} \tag{8.4}$$

We remark in passing that neither the original combinatorial proof in [**Sz**], nor the more recent harmonic analysis proof by Gowers [**G**], leads to the syndeticity of the set $R_{k,A}$.

A much stronger result in this direction was obtained by H. Furstenberg and Y. Katznelson in [**FK1**] where they showed that $R_{k,A}$ is an IP* set. As a matter of fact they established an even stronger fact. Recall that a set $E \subseteq \mathbb{N}$ is called IP$_r^*$ if for any r-element set $\{n_1, n_2, \dots, n_r\} \subseteq \mathbb{N}$, E has nontrivial intersection with the set $FS((n_i)_{i=1}^r)$.

THEOREM 8.3 ([**FK1**], [**FK2**]). *Let $k \in \mathbb{N}$ and let $T_1, T_2, \dots, T_k$ be commuting measure preserving transformations of a probability space $(X, \mathcal{B}, \mu)$. Then for any $A \in \mathcal{B}$ with $\mu(A) > 0$, there exist $c > 0$ and $r \in \mathbb{N}$ such that*

$$\{n \in \mathbb{N} : \mu(A \cap T_1^{-n}A \cap T_2^{-n}A \cap \dots \cap T_k^{-n}A) > c\}$$

is an IP$_r^*$ set.

In view of Theorem 7.5, we have the following corollary.

COROLLARY 8.4. *Under the conditions and notation of Theorem 8.3, the set*

$$\{n \in \mathbb{N} : \mu(A \cap T_1^{-n}A \cap T_2^{-n}A \cap \dots \cap T_k^{-n}A) > c\}$$

is a multiplicatively central* set.

We will formulate now (a special case of) the IP polynomial Szemerédi theorem obtained in [**BerM2**]. It is an open question whether the set R appearing in the formulation is an IP$_r^*$ set.

[14] Note that in view of Furstenberg's correspondence principle this fact implies that for any set $E \subseteq \mathbb{N}$ with $d^*(E) > 0$, the set of differences $E - E = \{x - y : x, y \in E\}$ is syndetic.

THEOREM 8.5. *For any $k \in \mathbb{N}$, let $T_1, T_2, \dots, T_k$ be commuting measure preserving transformations of a probability space $(X, \mathcal{B}, \mu)$ and let g_i be polynomials satisfying $g_i(\mathbb{Z}) \subseteq \mathbb{Z}$ and $g_i(0) = 0$, $i = 1, 2, \dots, k$. Then, for any $A \in \mathcal{B}$ with $\mu(A) > 0$, there exists $c > 0$ such that the set*

$$R = \left\{ n \in \mathbb{N} : \mu(A \cap T_1^{g_1(n)} A \cap T_2^{g_2(n)} A \cap \ldots \cap T_k^{g_k(n)} A) > c \right\}$$

is an IP set.*

The fact that the set R appearing above is IP* is useful (via Furstenberg's correspondence principle) in various combinatorial applications. To formulate one such application, we need to define the notion of *multiplicatively large* sets.

DEFINITION 8.6. A set $A \subseteq \mathbb{N}$ is *multiplicatively large* if for some sequence of positive integers $(a_n)_{n \in \mathbb{N}}$ one has

$$\limsup_{n \to \infty} \frac{|A \cap a_n F_n|}{|F_n|} > 0,$$

where $F_n = \{p_1^{i_1} \cdot p_2^{i_2} \cdot \ldots \cdot p_n^{i_n} : 0 \le i_j \le n, 1 \le j \le n\}$ and where the sequence $(p_i)_{i \in \mathbb{N}}$ consists of the primes in some arbitrary order.

REMARK 8.7. (i) The notion of mulitplicatively large is a natural analog of the corresponding additive property of having positive upper Banach density.

(ii) It is not hard to see that multiplicatively syndetic, and more generally, multiplicatively piecewise syndetic, sets are multiplicatively large.

(iii) The notions of largeness based on additive and multiplicative densities do not overlap. For example, the set $2\mathbb{N} - 1$ of odd natural numbers has density $1/2$ along any sequence of intervals $[a_n, b_n]$ with $b_n - a_n \to \infty$. On the other hand, it is not hard to see that this set has density zero with respect to any averaging scheme in $(\mathbb{N}, \cdot)$. In the other direction, consider the set $S = \bigcup_{n=1}^{\infty} a_n F_n$, where F_n are defined above and the integers a_n satisfy $a_n > |F_n|$, $n = 1, 2, \dots$. Then it is not had to check that S has zero upper Banach density. At the same time, S has multiplicative density one with respect to the sequence $(a_n F_n)_{n \in \mathbb{N}}$.

It turns out that multiplicatively large sets are much richer than the sets having positive density in $(\mathbb{N}, +)$. In particular, any multiplicatively large set contains not only arbitrarily long geometric progressions (as could be expected by mere analogy), but also arbitrarily long arithmetic progressions and more general configurations of mixed type such as
$\{q^i(a + jd) : 0 \le i, j \le k\}$ (see [**Ber5**], Theorem 1.5).

The following theorem (in the proof of which Theorem 8.5 plays a decisive role) is yet another manifestation of the combinatorial richness of multiplicatively large sets.

THEOREM 8.8 ([**Ber5**], Theorem 3.15). *Let $E \subseteq \mathbb{N}$ be a multiplicatively large set. For any $k \in \mathbb{N}$, there exist $a, b, d \in \mathbb{N}$ such that*

$$\left\{ b(a + id)^j : 0 \le i, j \le k \right\} \subseteq E.$$

As was already mentioned in Section 6, for any sequence $(x_n)_{n \in \mathbb{N}} \subseteq X$ in a compact Hausdorff space, $p\text{-}\lim_{n \in \mathbb{N}} x_n$ exists for any $p \in \beta\mathbb{N}$. Since the unit ball in a separable Hilbert space is compact in the weak topology, and since the unit ball is preserved under the action of unitary operators, this opens interesting possibilities of applications of p-limits to measure preserving dynamics (and hence, via Furstenberg's correspondence principle, to combinatorics). We will describe now some examples of such applications.

THEOREM 8.9 ([**Ber2**], Theorem 3.12. See also [**BerFM**] and [**BerHåM**]). *Let $q(t) \in \mathbb{Q}[t]$ with $q(\mathbb{Z}) \subseteq \mathbb{Z}$ and $q(0) = 0$. Let U be a unitary operator on a Hilbert space $\mathcal{H}$ and let $p \in (\beta\mathbb{N}, +)$ be an idempotent. Then, letting $p\text{-}\lim_{n\in\mathbb{N}} U^{q(n)} f = P_p(f)$, where the limit is in the weak topology, P_p is an orthogonal projection onto a subspace of $\mathcal{H}$.*

COROLLARY 8.10. *Let $E \subseteq \mathbb{N}$ satisfy $d^*(E) > 0$. Then, for any $\epsilon > 0$, for any polynomial $q(t) \in \mathbb{Q}[t]$ with $q(\mathbb{Z}) \subseteq \mathbb{Z}$ and $q(0) = 0$, the set*

$$\left\{n \in \mathbb{N} : d^*(E \cap (E - q(n))) > (d^*(E))^2 - \epsilon\right\}$$

is an IP set.*

PROOF. We will show that it follows from Theorem 8.10 that for any invertible measure preserving system $(X, \mathcal{B}, \mu, T)$ and any $A \in \mathcal{B}$ with $\mu(A) > 0$,

$$p\text{-}\lim \mu(A \cap T^{q(n)} A) \geq \mu(A)^2.$$

The result in question will follow then from Furstenberg's correspondence principle.

Take $\mathcal{H} = L^2(X, \mathcal{B}, \mu)$, and take U to be the unitary operator induced by T, that is, $Ug(x) = g(Tx)$, and let $f = \mathbf{1}_A$. We have:

$$\begin{aligned} p\text{-}\lim_{n\in\mathbb{N}} \mu(A \cap T^{q(n)} A) &= p\text{-}\lim_{n\in\mathbb{N}} \langle U^{q(n)} f, f\rangle = \langle P_p f, f\rangle \\ &= \langle P_p f, P_p f\rangle \langle 1, 1\rangle \geq (\langle P_p f, 1\rangle)^2 = (\langle f, 1\rangle)^2 \\ &= (\langle \mathbf{1}_A, 1\rangle)^2 = (\mu(A))^2. \end{aligned}$$

□

In recent years, the class of the so-called *essential* (see [**BerD**]) idempotent ultrafilters in $(\beta\mathbb{N}, +)$, which is broader than that of minimal ones, has started to gain importance. The defining property of essential idempotents is that all their members have positive upper Banach density. For example, one can show that members of essential idempotents, called $\mathcal{D}$ sets, share much in the way of combinatorial richness with central sets (see [**BeiBerDF**]).

In [**BerM4**], convergence along essential idempotents was employed to obtain an extension of the polynomial Szemerédi theorem (see [**BerL1**] and [**BerM1**]). Before formulating it, we have to introduce the notion of generalized polynomials. A function $f : \mathbb{Z} \to \mathbb{Z}$ is a *generalized polynomial* if it can be obtained regular polynomials with the help of the greatest integer function $[\cdot]$ and the usual arithmetic operators. Thus the functions given by expressions like $[n^2\alpha][n^5\beta] - [n^3\delta]$ are generalized polynomials.

Note that, unlike conventional polynomials, generalized polynomials need not be eventually monotone (consider $[[n\alpha]n\beta] - [n^2\alpha\beta]$), may take only finitely many values (for example, $[(n+1)\alpha] - [n\alpha] - [\alpha]$ takes only the values zero and one), and may vanish on sets of positive density while growing without bound on other such sets (multiply the previous example by n).

Despite such oddities, new evidence has begun to emerge that generalized polynomials do possess certain strong regularities. In particular, it was shown in [**BL3**] that any bounded generalized polynomial g can be expressed as $g(n) = f(T^n x)$, where T is a translation on a nilmanifold X (that is, $X = N/\Gamma$, where N is a nilpotent group and Γ is a cocompact lattice) and f is a Riemann integrable function on X.[15]

Here now is the promised formulation of the extension of the polynomial Szemerédi theorem.

THEOREM 8.11 ([**BerM4**]). *Let $k \in \mathbb{N}$, let $q_i(x)$ be generalized polynomials, $1 \le i \le k$, and let $p \in (\beta\mathbb{N}, +)$ be an essential idempotent. Then there exist constants c_i, $1 \le i \le k$, such that if $E \subseteq \mathbb{N}$ satisfies $d^*(E) > 0$, then the set*

$$\Big\{n \in \mathbb{N} : \exists a \in \mathbb{N} : \{a, a + q_1(n) - c_1, \ldots, a + q_k(n) - c_k\} \subseteq E\Big\}$$

belongs to p.

The reader will find additional interesting applications of ultrafilters in [**BerM3**].

9. Beiglböck's Proof of Jin's Sumsets Theorem

Let $A, B \subseteq \mathbb{R}$ satisfy $\lambda(A) > 0$, $\lambda(B) > 0$, where λ denotes Lebesgue measure, and consider the sumset $A + B = \{x + y : x \in A, y \in B\}$. The classical lemma of Steinhaus states that $A + B$ has to contain an open interval. This result is an instance of the sumset phenomenon, which manifests itself in results where the sum of two "large" sets is "very large".

Another example of the sumset phenomenon is provided by the following very interesting theorem of R. Jin proved in [**J1**]. (See also [**J2**], [**BerFW**], and [**BeiBerF**]). For a set $C \subseteq \mathbb{Z}$, the upper Banach density, $d^*(C)$, is defined by

$$d^*(C) = \limsup_{N-M\to\infty} \frac{|C \cap \{M+1, \ldots, N\}|}{N-M}.$$

THEOREM 9.1 ([**J1**]). *For any $A, B \subseteq \mathbb{Z}$ satisfying $d^*(A) > 0$, $d^*(B) > 0$, the sumset $A + B$ is piecewise syndetic.*

The original proof of Theorem 9.1 in [**J1**] utilized nonstandard analysis. This proof was converted to a standard one in [**J2**]. Later on, additional approaches were found, which allow one to strengthen Jin's result and to extend it to general amenable groups (see [**JK**], [**BerFW**], [**BeiBerF**]). We will present now a most recent proof of Jin's theorem due to Beiglböck [**Bei2**]. This proof is short and sweet and makes a nice use of ultrafilters.

For a set $A \subseteq \mathbb{Z}$ and an ultrafilter $p \in \beta\mathbb{Z}$, let

$$A - p = \{k \in \mathbb{Z} : A - k \in p\}.$$

(Note that when p is a principal ultrafilter, this reduces to the usual definition of a shifted set).

LEMMA 9.2. *For any $A, B \subseteq \mathbb{Z}$ there exists $p \in \beta\mathbb{Z}$ such that $d^*(A \cap (B - p)) \ge d^*(A)d^*(B)$.*

To see that Theorem 9.1 follows from Lemma 9.2, assume that $d^*(A)$, $d^*(B) > 0$. Then by the Lemma, there exists $p \in \beta\mathbb{Z}$ such that the set $C = (-A) \cap (B - p)$ has $d^*(C) > 0$. By Footnote 14, $C - C$ is syndetic and so is $S := A + (B - p) \supseteq C - C$.

---

[15] One can show that if T is a translation on a nilmanifold X then for any idempotent $p \in (\beta\mathbb{N}, +)$ and any $x \in X$, one has $p\text{-}\lim_{n\in\mathbb{N}} T^n x = x$. This leads to interesting Diophantine applications. See, for example, Theorem D in [**BL3**].

Note that if $s \in A+(B-p)$ then $A+B-s \in p$. (Indeed, if $s \in A+\{k \in \mathbb{Z} : B-k \in p\}$ then, for some $a \in A$, $B-(s+a) \in p$, which implies $A+B-s \in p$).

So, for every finite set $\{s_1, s_2, \ldots, s_n\} \subseteq A+(B-p)$ we have $\bigcap_{i=1}^n (A+B-s_i) \in p$, and hence this intersection is nonempty. This, in turn, implies that for some $t \in \mathbb{Z}$, $t+\{s_1, s_2, \ldots, s_n\} \subseteq A+B$. So we see that $A+B$ contains shifts of all finite subsets of the syndetic set $A+(B-p)$ which implies that $A+B$ is piecewise syndetic.

It remains to prove Lemma 9.2. Before doing so, we summarize some facts which will be used in the proof.

First, we note that, given a set $A \subseteq \mathbb{Z}$, one can always find an *invariant mean*, i.e. a shift-invariant finitely additive probability measure m on $\mathbb{Z}, \mathcal{P}(\mathbb{Z}))$, such that $m(A) = d^*(A)$. To see this, let finite intervals $I_n \subseteq \mathbb{Z}$ be such that $d^*(A) = \lim_{n\to\infty} m_n(A)$, where for $B \in \mathcal{P}(\mathbb{Z})$, $m_n(B) := \frac{|B\cap I_n|}{|I_n|}$. Now take m to be a cluster point of the set $\{m_n : n \in \mathbb{N}\}$ in the (compact) set $[0,1]^{\mathcal{P}(\mathbb{Z})}$.

Since $B(\mathbb{Z})$, the space of bounded functions on $\mathbb{Z}$, is isomorphic to $C(\beta\mathbb{Z})$, it follows from the Riesz representation theorem that there exists a regular Borel probability measure μ on $\beta\mathbb{Z}$ such that $m(A) = \mu(\bar{A})$ for all $A \subseteq \mathbb{Z}$ (here $\bar{A} = \{p \in \beta\mathbb{Z} : A \in p\}$).

To prove the lemma, pick a sequence of intervals $I_n \subseteq \mathbb{Z}$, $n \in \mathbb{N}$, with $|I_n| \to \infty$ and $d^*(B) = \lim_{n\to\infty} \frac{|B\cap I_n|}{|I_n|}$, and pick an invariant mean m such that $m(A) = d^*(A)$. Finally, define $f_n : \beta\mathbb{Z} \to [0,1]$ by

$$f_n(p) := \frac{|I_n \cap B \cap \{k \in \mathbb{Z} : (A-k) \in p\}|}{|I_n|} = \frac{1}{|I_n|} \sum_{k \in I_n \cap B} \mathbf{1}_{\overline{A-k}}(p),$$

and let $f(p) = \limsup_{n\to\infty} f_n(p) \leq d^*\big(B \cap \{k \in \mathbb{Z} : (A-k) \in p\}\big)$. By Fatou's lemma,

$$\begin{aligned} \int f d\mu \geq \limsup_{n\to\infty} \int \frac{1}{|I_n|} \sum_{k\in I_n\cap B} \mathbf{1}_{\overline{A-k}} d\mu &= \limsup_{n\to\infty} \frac{1}{|I_n|} \sum_{k\in I_n \cap B} m(A-k) \\ &= d^*(A) \cdot d^*(B) \end{aligned}$$

This implies that for some $p \in \beta\mathbb{Z}$, $d^*(A) \cdot d^*(B) \leq f(p)$, and we are done.

References

Ba. Baumgartner, J. *A short Proof of Hindman's Theorem*, J. Combinatorial Theory Ser. A 17 (1974), 384–386.

Bei1. Beiglböck, M. *A variant of the Hales-Jewett theorem*, Bull. Lond. Math. Soc. 40 (2008), no. 2, 210–216.

Bei2. Beiglböck, M. *An ultrafilter approach to Jin's Theorem*, Israel Journal of Mathematics, to appear.

BeiBerDF. Beiglböck, M.; Bergelson, V.; Downarowicz, T.; Fish, A. *Solvability of Rado Systems in D-sets*, Topology Appl. 156 (2009), no. 16, 2565–2571

BeiBerF. Beiglböck, M.; Bergelson, V.; Fish, A. *Sumset phenomenon in countable amenable groups*, Adv. Math. 223 (2010), no. 2, 416–432

BeiBerHS. Beiglböck, M.; Bergelson, V.; Hindman, N.; Strauss, D. *Some new results in multiplicative and additive Ramsey theory*, Trans. Amer. Math. Soc. 360 (2008), no. 2, 819–847

Ber1. Bergelson, V. *Ergodic Ramsey theory*, Logic and combinatorics (Arcata, Calif., 1985), 63–87, Contemp. Math., 65, Amer. Math. Soc., Providence, RI, 1987.

Ber2. Bergelson, V. *Ergodic Ramsey theory – an update*, Ergodic theory of Z^d actions (Warwick, 1993–1994), 1–61, London Math. Soc. Lecture Note Ser., 228, Cambridge Univ. Press, Cambridge, 1996.

Ber3. Bergelson, V. *Ergodic theory and Diophantine problems*, Topics in symbolic dynamics and applications (Temuco, 1997), 167–205, London Math. Soc. Lecture Note Ser., 279, Cambridge Univ. Press, Cambridge, 2000.

Ber4. Bergelson, V. *Minimal idempotents and ergodic Ramsey theory*, Topics in dynamics and ergodic theory, 8–39, London Math. Soc. Lecture Note Ser., 310, Cambridge Univ. Press, Cambridge, 2003.

Ber5. Bergelson, V. *Multiplicatively large sets and ergodic Ramsey theory*, Probability in mathematics. Israel J. Math. 148 (2005), 23–40.

Ber6. Bergelson, V. *Combinatorial and Diophantine applications of ergodic theory*, (Appendix A by Leibman, A. and Appendix B by Quas, A. and Wierdl, M.) Handbook of dynamical systems. Vol. 1B, 745–869, Elsevier B. V., Amsterdam, 2006.

BerD. Bergelson, V,; Downarowicz, T. *Large sets of integers and hierarchy of mixing properties of measure preserving systems*, Colloq. Math. 110 (2008), no. 1, 117–150.

BerFHiK. Bergelson, V.; Furstenberg, H.; Hindman, N.; Katznelson, Y. *An algebraic proof of van der Waerden's theorem*, Enseign. Math. (2) 35 (1989), no. 3-4, 209–215.

BerFM. Bergelson, V.; Furstenberg, H.; McCutcheon, R. *IP-sets and polynomial recurrence*, Ergodic Theory Dynam. Systems 16 (1996), no. 5, 963–974.

BerFW. Bergelson, V.; Furstenberg, H.; Weiss, B. *Piecewise-Bohr sets of integers and combinatorial number theory*, Topics in discrete mathematics, 13–37, Algorithms Combin., 26, Springer, Berlin, 2006.

BerHåM. Bergelson, V.; Håland Knutson, I. J.; McCutcheon, R. *IP-systems, generalized polynomials and recurrence*, Ergodic Theory Dynam. Systems 26 (2006), no. 4, 999–1019.

BerHi1. Bergelson, V.; Hindman, N. *Nonmetrizable topological dynamics and Ramsey theory* Trans. Amer. Math. Soc. 320 (1990), no. 1, 293–320.

BerHi2. Bergelson, V.; Hindman, N. *On* IP* *sets and central sets*, Combinatorica 14 (1994), no. 3, 269–277.

BerHi3. Bergelson, V.; Hindman, N. *Additive and multiplicative Ramsey theorems in N–some elementary results*, Combin. Probab. Comput. 2 (1993), no. 3, 221–241.

BerL1. Bergelson, V.; Leibman, A. *Polynomial extensions of van der Waerden's and Szemerédi's theorems*, J. Amer. Math. Soc. 9 (1996), no. 3,725–753.

BerL2. Bergelson, V.; Leibman, A. *Set-polynomials and polynomial extension of the Hales-Jewett theorem*, Ann. of Math. (2) 150 (1999), no. 1, 33–75.

BL3. Bergelson, V.; Leibman, A. *Distribution of values of bounded generalized polynomials*, Acta Math. 198 (2007), no. 2, 155–230.

BerM1. Bergelson, V.; McCutcheon, R. *Uniformity in the polynomial Szemerédi theorem*, Ergodic theory of Z^d actions (Warwick, 1993–1994), 273–296, London Math. Soc. Lecture Note Ser., 228, Cambridge Univ. Press, Cambridge, 1996.

BerM2. Bergelson, V.; McCutcheon, R. *An ergodic IP polynomial Szemerédi theorem*, Mem. Amer. Math. Soc. 146 (2000), no. 695, viii+106 pp.

BerM3. Bergelson, V.; McCutcheon, R. *Central sets and a non-commutative Roth theorem*, Amer. J. Math. 129 (2007), no. 5, 1251–1275.

BerM4. Bergelson, V.; McCutcheon, R. *Idempotent ultrafilters, multiple weak mixing and Szemerédi's theorem for generalized polynomials*, Journal D'Analyse, to appear.

CoN. Comfort, W.; Negrepontis, S. *The theory of ultrafilters*, Die Grundlehren der mathematischen Wissenschaften, Band 211. Springer-Verlag, New York-Heidelberg, 1974. x+482 pp.

CsGSa. Csikvári, P. Sárközy, A. Gyarmati, K. *Density and Ramsey type results on algebraic equations with restricted solution sets*, to appear.

E. Ellis, R. *Distal transformation groups*, Pacific J. Math. 8 1958 401–405.

F. Furstenberg, H. *Ergodic behavior of diagonal measures and a theorem of Szemerédi on arithmetic progressions*, J. Analyse Math. 31 (1977), 204–256.

F2. Furstenberg, H. *Recurrence in ergodic theory and combinatorial number theory*, M. B. Porter Lectures. Princeton University Press, Princeton, N.J., 1981. xi+203 pp.

FK1. Furstenberg, H.; Katznelson, Y. *An ergodic Szemerédi theorem for IP-systems and combinatorial theory*, J. Analyse Math. 45 (1985), 117–168.

FK2. Furstenberg, H.; Katznelson, Y. *IP_r-sets, Szemerédi's theorem, and Ramsey theory*, Bull. Amer. Math. Soc. (N.S.) 14 (1986), no. 2, 275–278.

FW. Furstenberg, H.; Weiss, B. *Topological dynamics and combinatorial number theory*, J. Analyse Math. 34 (1978), 61–85.

G. Gowers, W. *A new proof of Szemerédi's theorem*, Geom. Funct. Anal. 11 (2001), no. 3, 465–588.

H1. Hindman, N. *Finite sums from sequences within cells of a partition of N*, J. Combinatorial Theory Ser. A 17 (1974), 1–11.

H2. Hindman, N. *Ultrafilters and combinatorial number theory*, Number theory, Carbondale 1979 (Proc. Southern Illinois Conf., Southern Illinois Univ., Carbondale, Ill., 1979), pp. 119–184, Lecture Notes in Math., 751, Springer, Berlin, 1979.

H3. Hindman, N. *Partitions and sums and products of integers*, Trans. Amer. Math. Soc. 247 (1979), 227–245.

H4. Hindman, N. *Partitions and sums and products - two counterexamples*, J. Combin. Theory Ser. A 29 (1980), no. 1, 113 - 120.

H5. Hindman, N. *The semigroup $\beta\mathbb{N}$ and its applications to number theory*, The analytical and topological theory of semigroups, 347–360, de Gruyter Exp. Math., 1, de Gruyter, Berlin, 1990.

H6. Hindman, N. *Monochromatic sums equal to products in N*, (to appear in Integers).

HS. Hindman, N,; Strauss, D. *Algebra in the Stone-Čech compactification*, Theory and applications. de Gruyter Expositions in Mathematics, 27. Walter de Gruyter & Co., Berlin, 1998. xiv+485 pp.

J1. Jin, R. *The sumset phenomenon*, Proc. Amer. Math. Soc. 130 (2002), no. 3, 855–861.

J2. Jin, R. *Standardizing nonstandard methods for upper Banach density problems*, Unusual applications of number theory, 109–124, DIMACS Ser. Discrete Math. Theoret. Comput. Sci., 64, Amer. Math. Soc., Providence, RI, 2004.

JK. Jin, R.; Keisler, H. *Abelian groups with layered tiles and the sumset phenomenon* Trans. Amer. Math. Soc. 355 (2003), no. 1, 79–97.

M. McCutcheon, R. *A variant of the density Hales-Jewett theorem*, to appear. Available at http://www.msci.memphis.edu/ randall/preprints/denbagel.pdf

Poi1. Poincaré, H. *Sur le problème des trois corps et les équations de la dynamique*, Acta. Math. , 13 (1890) pp. 1270.

Poi2. Poincaré, H. *New methods of Celestial Mechanics* (Translation of *Les méthodes nouvelles de la mécanique céleste* I (1892), II (1893), and III (1894)), D. Goroff, editor, Amer. Inst. of Physics, New York, 1993.

Pos. Pospíšil, B. *Remark on bicompact spaces*, Ann. of Math. (2) 38 (1937), no. 4, 845–846.

Sc. Schur, I. *Über die Kongruenz $x^m + y^m = z^m (\mod p)$*, Jahresber. Deutsche Math.-Verein. 25, 114-116, 1916.

Sz. Szemerédi, E. *On sets of integers containing no k elements in arithmetic progression*, Acta Arith. 27, 199-245, 1975.

vdW. van der Waerden, B. *Beweis einer Baudetschen Vermutung*, Nieuw. Arch. Wisk. 15 (1927), 212216.

Department of Mathematics, Ohio State University, Columbus, OH 43210, USA
E-mail address: `vitaly@math.ohio-state.edu`

Contemporary Mathematics
Volume **530**, 2010

Ultrafilters and Set Theory

Andreas Blass

ABSTRACT. This is a survey of some of the many set-theoretic aspects of ultrafilters.

1. Introduction

This paper is an expanded version of a survey talk given at the UltraMath 2008 conference. The subject, ultrafilters and set theory, is far too large for an hour's lecture or even for a paper such as this. I took several steps to fit my talk into the available time. First, I omitted all proofs, giving only definitions, theorems, and occasional comments. Second, I omitted those set-theoretic topics that were treated by other speakers at the conference: large cardinals (Neeman), the connection with dynamics and combinatorics (Bergelson and Hindman), and measure theory (Fremlin). Finally, I omitted Shelah's theory of possible cofinalities (pcf theory).

What remains? Quite a lot, as I hope the following makes clear.

2. What Is an Ultrafilter?

Let us begin with the simplest part of set theory, the Boolean operations such as union, intersection, and complement, acting on the subsets of a set X. More formally, we consider 2^X, the set of functions from X to $2 = \{0, 1\}$, or equivalently the power set of X, as an algebraic structure, whose k-ary operations are induced by all of the k-ary functions $2^k \to 2$ on the set 2.

Given some algebraic structures, in this case the Boolean algebras 2^X, it is natural to ask what the homomorphisms between them are.

PROPOSITION 2.1. *A homomorphism $h : 2^X \to 2^Y$ is given by a Y-indexed family $\langle \mathcal{U}_y : y \in Y \rangle$ of ultrafilters on X. The correspondence is*

$$y \in h(A) \iff A \in \mathcal{U}_y$$

for $y \in Y$ and $A \subseteq X$.

2010 *Mathematics Subject Classification.* Primary 03E05; Secondary 03C20, 03E02, 03E17, 03E40, 03E50, 03E60, 03E75, 54D35.

Partially supported by NSF grant DMS-0653696.

In particular, an ultrafilter on X amounts to a homomorphism $2^X \to 2$. This characterization of ultrafilters is quite robust, in the following two senses.

We can add more information, in that an ultrafilter on X gives homomorphisms $n^X \to n$ for all finite n (where we identify n with $\{0, 1, \dots, n-1\}$ as is customary in set theory, and where n and n^X are considered as algebras with respect to all operations $n^k \to n$ on n). The homomorphism associated to an ultrafilter $\mathcal{U}$ sends any element $f \in n^X$ to the unique $v \in n$ whose pre-image $f^{-1}\{v\}$ is in $\mathcal{U}$.

We can also remove information. Instead of homomorphisms with respect to operations $2^k \to 2$ of arbitrary arity k, it suffices to consider operations of arity $k \leq 2$. This follows from the well-known fact that all Boolean operations of any arity can be expressed in terms of only the unary and binary operations, in fact only negation and conjunction.

In this connection, a small digression already leads to an open problem. Notice that, for a map $h : 2^X \to 2$, each of the following properties trivially implies the next:

- h preserves operations of $\leq k+1$ arguments.
- h preserves relations of $\leq k+1$ arguments.
- h preserves operations of $\leq k$ arguments.

(Here preservation of a relation (on 2) means that, if the relation holds componentwise between some elements of 2^X, then it holds between their h-images as well.) Combining the lists of these properties for all k, we obtain a tower of implications. But this infinite tower collapses because, as mentioned above, all operations can be expressed in terms of unary and binary ones. That is, all the properties in the tower from "preserves operations of ≤ 2 arguments" on up are equivalent. Immediately below these, however, is another interesting property, "preserves relations of arity ≤ 2." (The properties even further down seem uninteresting.)

It is easy to check that $h : 2^X \to 2$ preserves ≤ 2-ary relations if and only if the family $h^{-1}\{1\}$ of subsets of X sent to 1 is a maximal linked family. ("Linked" means that every two sets in the family have nonempty intersection.)

The existence of maximal linked families in all nondegenerate Boolean algebras is known to be a weak axiom of choice strictly weaker than the Boolean prime ideal theorem. That is, in ZF, the existence of maximal linked families in all nondegenerate Boolean algebras does not imply the existence of ultrafilters in all nondegenerate Boolean algebras, but it is not provable outright either.

Open Problem: Does the existence of maximal linked families in all nondegenerate Boolean algebras follow, in ZF, from the assumption that every set admits a linear ordering?

I expect the answer to be negative. The converse implication is known not to be provable [**15**, Prop. 33].

REMARK 2.2. An indication of the importance and ubiquity of ultrafilters: The Boolean prime ideal theorem has the largest number of equivalent forms in the book [**32**] of Howard and Rubin, among all of the weak axioms of choice treated there.

We close this section with a surprising observation of Lawvere [**46**]: If we think of ultrafilters in terms of the associated homomorphisms $3^X \to 3$, rather than the customary $2^X \to 2$, then we don't need the homomorphisms to preserve binary operations — unary ones suffice.

THEOREM 2.3 (Lawvere [**46**]). *Let $h : 3^X \to 3$ be a function such that, for every $m : 3 \to 3$ and every $f \in 3^X$, $h(m \circ f) = m(h(f))$. Then there is an ultrafilter $\mathcal{U}$ on X such that, for all $f \in 3^X$, $h(f)$ is the unique $v \in 3$ for which $f^{-1}\{v\} \in \mathcal{U}$.*

Because Lawvere stated this result without proof, we provide a proof in the appendix of this paper.

Before we continue, it will be useful to fix notations for certain functions and a "quantifier notation" for ultrafilters.

NOTATION 2.4. For any set X, its identity map will be denoted by id_X; the subscript will be omitted when X is clear from the context. For two sets X and Y, the projection maps from the cartesian product $X \times Y$ to X and Y will be denoted either by p_X and p_Y or by p_1 and p_2. Similar notation is used with the product of more than two factors. $\tau = \tau_{X,Y} : X \times Y \to Y \times X$ is the "twist" map, $\tau(x, y) = (y, x)$.

If $\mathcal{U}$ is an ultrafilter on X, the quantifier $(\mathcal{U}x)$ means "for almost all x with respect to $\mathcal{U}$." That is, if $\varphi(x)$ is any statement about x,

$$(\mathcal{U}x)\, \varphi(x) \iff \{x \in X : \varphi(x)\} \in \mathcal{U}.$$

The same definition could be given for an arbitrary family $\mathcal{U}$ of subsets of X; then $\mathcal{U}$ is an ultrafilter if and only if the quantifier $(\mathcal{U}x)$ distributes over all propositional connectives.

3. Images, Limits, Sums, and Products

In this section, we describe some of the standard ways of constructing new ultrafilters from old.

DEFINITION 3.1. If $\mathcal{U}$ is an ultrafilter on X and if $f : X \to Y$, then the *image* of $\mathcal{U}$ under f is the ultrafilter $f(\mathcal{U})$ defined by

$$f(\mathcal{U}) = \{A \subseteq Y : f^{-1}(A) \in \mathcal{U}\}$$

or equivalently in terms of quantifiers,

$$(f(\mathcal{U})y)\, \varphi(y) \iff (\mathcal{U}x)\, \varphi(f(x)).$$

REMARK 3.2. The same definition of $f(\mathcal{U})$ is used when $\mathcal{U}$ is not an ultrafilter but merely a filter, or indeed merely a family of subsets of X closed under supersets.

If one thinks of ultrafilters as finitely additive, 2-valued measures, sets in the ultrafilter having measure 1 and other sets having measure 0, then this definition of $f(\mathcal{U})$ is the usual definition of pushing measures forward along maps. If one thinks of ultrafilters on X as points in the Stone-Čech compactification βX of the discrete space X, then f applied to ultrafilters is the unique extension of $f : X \to Y$ to a continuous map $\beta X \to \beta Y$.

If f and g are two functions $X \to Y$ and if the ultrafilter $\mathcal{U}$ on X contains the set $\{x \in X : f(x) = g(x)\}$, then $f(\mathcal{U}) = g(\mathcal{U})$.

The action of functions on ultrafilters is functorial, in the sense that $(f \circ g)(\mathcal{U}) = f(g(\mathcal{U}))$ and $\mathrm{id}_X(\mathcal{U}) = \mathcal{U}$ (where $g : X \to Y$, $f : Y \to Z$, and $\mathcal{U}$ is an ultrafilter on X). This implies the reflexivity and transitivity of the Rudin-Keisler ordering, defined as follows.

DEFINITION 3.3. If $\mathcal{U}$ and $\mathcal{V}$ are ultrafilters on X and Y, respectively, then $\mathcal{V}$ is *Rudin-Keisler below* $\mathcal{U}$, written $\mathcal{V} \leq_{RK} \mathcal{U}$, if there is a function $f : X \to Y$ such that $f(\mathcal{U}) = \mathcal{V}$.

Like any reflexive, transitive relation, the Rudin-Keisler ordering induces an equivalence relation and a partial ordering of the equivalence classes. The second part of the following theorem characterizes the equivalence relation; the first part is included both for its own interest and because it serves as a step in the proof of the second part.

THEOREM 3.4. *Let $\mathcal{U}$ be an ultrafilter on X.*

(1) *If $f : X \to X$ and $f(\mathcal{U}) = \mathcal{U}$, then $(\mathcal{U}x)\, f(x) = x$.*
(2) *An ultrafilter $\mathcal{V}$ on a set Y is Rudin-Keisler equivalent to $\mathcal{U}$, i.e., $\mathcal{U} \leq_{RK} \mathcal{V} \leq_{RK} \mathcal{U}$, if and only if $\mathcal{V} = f(\mathcal{U})$ for some function $f : X \to Y$ that is one-to-one on some set in $\mathcal{U}$. If, furthermore, X and Y have the same cardinality, then $\mathcal{V}$ is Rudin-Keisler equivalent to $\mathcal{U}$ if and only if $\mathcal{V} = f(\mathcal{U})$ for some bijection $f : X \to Y$.*

Because of the last part of the theorem, Rudin-Keisler equivalent ultrafilters are usually called *isomorphic*, and one writes $\mathcal{U} \cong \mathcal{V}$. The terminology is justified, even when X and Y have different cardinalities, by considering a category of ultrafilters in which morphisms from X to Y are equivalences classes of functions, two functions being equivalent if they agree on a set in the domain ultrafilter. See [**8**] for details.

REMARK 3.5. The proof of part (1) of the theorem uses the following combinatorial fact that doesn't directly involve ultrafilters. For any function $f : X \to X$, there is a partition of $\{x \in X : f(x) \neq x\}$ into three pieces, each disjoint from its image under f. As far as I know, this fact was first stated in [**39**] and first applied to ultrafilters in [**37**]. The proof of this fact proceeds by a direct analysis of functions f, best viewed as directed graphs with an arrow from each x to $f(x)$. Were it not for cycles of odd length ≥ 3, partitions into two parts would suffice. The third part is used for exactly one point on each odd cycle; that decision essentially forces the rest of the construction of the partition.

Although, as noted above, if two functions $f, g : X \to Y$ are equal on some set in $\mathcal{U}$, then $f(\mathcal{U}) = g(\mathcal{U})$, and although part (1) of the theorem gives the converse when $g = \mathrm{id}_X$, and although the converse follows easily when g is one-to-one on a set in $\mathcal{U}$, the converse is not true in general. If $\mathcal{V}$ is any non-principal ultrafilter on Y, then there is an ultrafilter $\mathcal{U}$ on $X = \{(y, y') \in Y^2 : y \neq y'\}$ such that both of the projection maps $Y^2 \to Y$ (restricted to X) send $\mathcal{U}$ to $\mathcal{V}$ even though the projection maps differ at every point in X. The ultrafilters $\mathcal{U}$ for which the converse does hold have been studied under the names of "ultrafilters with property C" [**27**], "separating ultrafilters" [**36**], and "Hausdorff ultrafilters" [**28, 5**]. The existence of such ultrafilters is provable if the continuum hypothesis (CH) is assumed, but it is not known whether it is provable in ZFC.

We turn next to a more complicated construction of new ultrafilters from old, the limit construction.

DEFINITION 3.6. Let $\langle \mathcal{V}_i : i \in I \rangle$ be an indexed family of ultrafilters on X, and let $\mathcal{U}$ be an ultrafilter on the index set I. Then the *limit* of the (family of) $\mathcal{V}_i$'s along $\mathcal{U}$ is the ultrafilter

$$\mathcal{U}\text{-}\lim_i \mathcal{V}_i = \{A \subseteq X : (\mathcal{U}i)\, A \in \mathcal{V}_i\}.$$

This agrees, in βX, with the usual topological notion of limit along a filter.

This notion of limit subsumes the notion of image discussed above. If $f : X \to Y$ and if $\mathcal{U}$ is an ultrafilter on X, then, writing $\hat{y}$ for the principal ultrafilter at y, we have $f(\mathcal{U}) = \mathcal{U}\text{-}\lim_x \widehat{f(x)}$.

A particularly useful case of the limit construction occurs when the family $\langle \mathcal{V}_i : i \in I \rangle$ is strongly discrete[1] in the sense that there is a family $\langle A_i : i \in I \rangle$ of pairwise disjoint sets with each $A_i \in \mathcal{V}_i$. Up to isomorphism, the limit in such a case amounts to the sum of ultrafilters, defined as follows. (In the definition, the sets $\{x\} \times Y$ serve as the pairwise disjoint A's.)

DEFINITION 3.7. Let $\langle \mathcal{V}_x : x \in X \rangle$ be an X-indexed family of ultrafilters on Y, and let $\mathcal{U}$ be an ultrafilter on X. The *sum* of the (family of) $\mathcal{V}_i$'s over $\mathcal{U}$ is the ultrafilter

$$\begin{aligned} \mathcal{U}\text{-}\sum_x \mathcal{V}_x &= \{A \subseteq X \times Y : (\mathcal{U}x)(\mathcal{V}_x y)\,(x,y) \in A\} \\ &= \mathcal{U}\text{-}\lim_x i_x(\mathcal{V}_x) \end{aligned}$$

where $i_x : Y \to X \times Y : y \mapsto (x, y)$.

Note for future reference that the two projections $X \times Y \to X$ and $X \times Y \to Y$ send $\mathcal{U}\text{-}\sum_x \mathcal{V}_x$ to $\mathcal{U}$ and $\mathcal{U}\text{-}\lim_x \mathcal{V}_x$, respectively. Note also that, in the special case where $X = Y = \omega$ and all the $\mathcal{V}_x$ are non-principal, the sum $\mathcal{U}\text{-}\sum_x \mathcal{V}_x$ contains the "above diagonal" set $\{(x, y) : x < y\}$; in fact, it contains $\{(x, y) : f(x) < y\}$ for every function $f : \omega \to \omega$.

The (tensor) product of two ultrafilters is the special case of the sum where all summands are equal.

DEFINITION 3.8. The *tensor product* $\mathcal{U} \otimes \mathcal{V}$ of ultrafilters $\mathcal{U}$ on X and $\mathcal{V}$ on Y is $\mathcal{U}\text{-}\sum_x \mathcal{V}$, i.e., the sum of a constant family all of whose members are $\mathcal{V}$. It is an ultrafilter on $X \times Y$. Equivalently,

$$((\mathcal{U} \otimes \mathcal{V})(x, y))\,\varphi(x, y) \iff (\mathcal{U}x)(\mathcal{V}y)\,\varphi(x, y).$$

REMARK 3.9. The tensor product notation serves to avoid confusion with the cartesian product $\mathcal{U} \times \mathcal{V}$ of ultrafilters (or just filters) defined as the filter generated by the sets $A \times B$ where $A \in \mathcal{U}$ and $B \in \mathcal{V}$. The cartesian product of two ultrafilters is usually not an ultrafilter. In fact it is an ultrafilter if and only if, for every X-indexed family of sets $B_x \in \mathcal{V}$, there is an $A \in \mathcal{U}$ such that $\bigcap_{x \in A} B_x \in \mathcal{V}$. (Despite its appearance, this condition is symmetric between $\mathcal{U}$ and $\mathcal{V}$.) This happens, in particular, if there is a cardinal κ such that $\mathcal{U}$ contains a set of size κ while $\mathcal{V}$ is closed under intersections of κ sets. But it never happens when $\mathcal{U}$ and $\mathcal{V}$ are non-principal ultrafilters on ω; in this case, $\mathcal{U} \times \mathcal{V}$ contains neither the "above diagonal" set mentioned above nor its complement.

More generally, if $\mathcal{U}$ and $\mathcal{V}$ are ultrafilters on the same set X, then both $\mathcal{U} \otimes \mathcal{V}$ and $\tau(\mathcal{V} \otimes \mathcal{U})$ extend $\mathcal{U} \times \mathcal{V}$. (Recall that τ is the interchange map $(x, y) \mapsto (y, x)$.) For $\mathcal{U} \times \mathcal{V}$ to be an ultrafilter, it is therefore necessary that $\mathcal{U} \otimes \mathcal{V} = \tau(\mathcal{V} \otimes \mathcal{U})$. For non-principal ultrafilters on ω, though, all tensor products concentrate on the above-diagonal part and their τ-images on the below-diagonal part.

---

[1]In any topological space, a family of points is *discrete* if each of the points has a neighborhood containing none of the others. It is *strongly discrete* if there are pairwise disjoint neighborhoods of all the points. The two concepts are equivalent for countable families in regular spaces.

It is only a matter of bookkeeping to check that $\otimes$ is associative, provided we identify the underlying sets $X\times(Y\times Z)$ and $(X\times Y)\times Z$ in the obvious way. In fact, the same bookkeeping gives the following more general versions of associativity.

PROPOSITION 3.10. (1) *Let $\mathcal{U}$ be an ultrafilter on X, $\mathcal{V}_x$ an ultrafilter on Y for each $x\in X$, and $\mathcal{W}_y$ an ultrafilter on Z for each $y\in Y$. Then*

$$\mathcal{U}\text{-}\lim_x(\mathcal{V}_x\text{-}\lim_y \mathcal{W}_y) = (\mathcal{U}\text{-}\lim_x \mathcal{V}_x)\text{-}\lim_y \mathcal{W}_y.$$

(2) *Let $\mathcal{U}$ be an ultrafilter on X, $\mathcal{V}_x$ an ultrafilter on Y for each $x\in X$, and $\mathcal{W}_{xy}$ an ultrafilter on Z for each $(x,y)\in X\times Y$. Then*

$$\mathcal{U}\text{-}\sum_x(\mathcal{V}_x\text{-}\sum_y \mathcal{W}_{xy}) = (\mathcal{U}\text{-}\sum_x \mathcal{V}_x)\text{-}\sum_{(x,y)} \mathcal{W}_{xy}.$$

Part (2) of this proposition has a rather remarkable converse, due to M. E. Rudin; essentially the only possible equalities between sums of ultrafilters are those given by associativity.

THEOREM 3.11 (Rudin, [**60**]). *Suppose $\mathcal{U},\mathcal{U}',\mathcal{V}_i$ and $\mathcal{V}'_i$ are ultrafilters on ω such that the sequences $\langle\mathcal{V}_i : i\in\omega\rangle$ and $\langle\mathcal{V}'_i : i\in\omega\rangle$ are strongly discrete and one-to-one. If $\mathcal{U}\text{-}\lim_i \mathcal{V}_i = \mathcal{U}'\text{-}\lim_i \mathcal{V}'_i$ then one of the following three alternatives holds.*

- *There is a bijection $f:\omega\to\omega$ such that $f(\mathcal{U})=\mathcal{U}'$ and $(\mathcal{U}i)\,\mathcal{V}_i = \mathcal{V}'_{f(i)}$.*
- *There are ultrafilters $\mathcal{W}_j$ on ω such that $\mathcal{U}' = \mathcal{U}\text{-}\lim_j \mathcal{W}_j$ and $(\mathcal{U}j)\,\mathcal{V}_j = \mathcal{W}_j\text{-}\lim_i \mathcal{V}'_i$.*
- *There are ultrafilters $\mathcal{W}_j$ on ω such that $\mathcal{U} = \mathcal{U}'\text{-}\lim_j \mathcal{W}_j$ and $(\mathcal{U}'j)\,\mathcal{V}'_j = \mathcal{W}_j\text{-}\lim_i \mathcal{V}_i$.*

In the second and third cases, the sequence $\langle\mathcal{W}_j : j\in\omega\rangle$ can be chosen to be one-to-one and strongly discrete.

This theorem implies in particular that, if $\mathcal{U}\otimes\mathcal{V}\cong\mathcal{U}'\otimes\mathcal{V}'$, then either (1) $\mathcal{U}\cong\mathcal{U}'$ and $\mathcal{V}\cong\mathcal{V}'$ or (2) there is $\mathcal{W}$ such that $\mathcal{U}'\cong\mathcal{U}\otimes\mathcal{W}$ and $\mathcal{V}\cong\mathcal{W}\otimes\mathcal{V}'$ or (3) the same as (2) with primed and unprimed interchanged.

DEFINITION 3.12. An ultrafilter $\mathcal{U}$ is *Rudin-Frolík*-below another ultrafilter $\mathcal{V}$, written $\mathcal{U}\leq_{RF}\mathcal{V}$, if $\mathcal{V}$ is isomorphic to an ultrafilter of the form $\mathcal{U}\text{-}\sum \mathcal{W}_i$.

It is easy to check that this is a reflexive, transitive relation, that $\leq_{RF}$ implies $\leq_{RK}$, and that the associated equivalence relation is isomorphism, just as for $\leq_{RK}$. Rudin's Theorem 3.11 above implies that the Rudin-Frolík order (unlike the Rudin-Keisler order) of ultrafilters on ω is tree-like, i.e., the predecessors of each element are linearly ordered.

4. Special Ultrafilters

Non-principal ultrafilters on ω, being obtained by means of the axiom of choice, are difficult to visualize and, at first, all look quite similar. Of course, some of these ultrafilters can be distinguished from others by considering whether or not they contain specific sets, but these distinctions are obliterated if one applies permutations of ω to the ultrafilters, i.e, if one considers only "combinatorial" properties of ultrafilters, properties preserved by isomorphisms.

It is easy to see that not all non-principal ultrafilters on ω are isomorphic. Indeed, it is a classical result of Pospíšil [**57**] that there are $2^{2^{\aleph_0}}$ non-principal ultrafilters on ω (and, more generally, 2^{2^κ} ultrafilters on any infinite cardinal κ). Since there are only $2^{\aleph_0}$ permutations of ω, the group of permutations cannot act transitively on the non-principal ultrafilters; that is, some (indeed many) pairs of ultrafilters are not related by any permutation of ω.

Nevertheless, it is difficult to see what properties, invariant under permutations, might distinguish some ultrafilters from others. In this section, we describe some such properties and some of the combinatorial structures that exist in the world of ultrafilters. At the end of the section, we also comment briefly on connections with structures that are not combinatorial but algebraic. For more information about such algebraic structures, see the paper of Hindman and Strauss in this volume or their book [**31**].

CONVENTION 4.1. From now on, when we refer to ultrafilters on sets (as opposed to ultrafilters in general Boolean algebras), we mean non-principal ultrafilters unless the contrary is explicitly stated.

The ultrafilters with the best combinatorial properties are the minimal ones in the Rudin-Keisler ordering. The following equivalent characterizations are easy to check, given part (1) of Theorem 3.4.

PROPOSITION 4.2. *For any (non-principal) ultrafilter $\mathcal{U}$ on a set X, the following are equivalent:*

- *$\mathcal{U}$ is RK-minimal, i.e., every non-principal ultrafilter $\leq_{RK} \mathcal{U}$ is $\cong \mathcal{U}$.*
- *Every function with domain X becomes either one-to-one or constant when restricted to a suitable set in $\mathcal{U}$.*
- *Whenever X is partitioned into pieces that are not in $\mathcal{U}$, there is a selector in $\mathcal{U}$, i.e., a set that contains exactly one element from each piece.*

Because of part (3) of this proposition, RK-minimal ultrafilters are often called *selective*.

The following observation considerably restricts the places where one should look for selective ultrafilters.

PROPOSITION 4.3. *If κ is a cardinal number and $\mathcal{U}$ is an ultrafilter on X that is not closed under intersections of κ sets at a time, then there is an $f : X \to \kappa$ such that $f(\mathcal{U})$ is a non-principal ultrafilter on κ.*

PROOF. If $A_i \in \mathcal{U}$ for all $i < \kappa$ but $\bigcap_{i<\kappa} A_i \notin \mathcal{U}$, then we can define the required f by letting $f(x)$ be the smallest i with $x \notin A_i$ (and an arbitrary element of κ if $x \in \bigcap_{i<\kappa} A_i$). □

In particular, every countably incomplete ultrafilter is RK-above some non-principal ultrafilter on ω. As a consequence, a countably incomplete, selective ultrafilter $\mathcal{U}$ must contain a countable set; it is essentially an ultrafilter on ω, in the sense that, even if its underlying set X is uncountable, $\mathcal{U}$ has the form $g(\mathcal{V})$ for some ultrafilter $\mathcal{V}$ on ω and some embedding $g : \omega \to X$.

REMARK 4.4. Similarly, if a selective ultrafilter $\mathcal{U}$ is countably complete, then it is (in the same sense) essentially on some measurable cardinal κ, namely the smallest κ such that $\mathcal{U}$ is not closed under intersections of κ sets at a time. Furthermore, one can show that in this situation $\mathcal{U}$ is isomorphic to a normal ultrafilter on κ.

In keeping with the restricted scope of this paper, as described in the introduction, we end the discussion of measurable cardinals here. The selective ultrafilters we consider from now on will be on ω.

Much of the interest in selective ultrafilters comes from their strong partition properties. The first of these properties were discovered by Kunen and first published in [**22**] (the published version of a Ph.D. thesis written under Kunen's direction). To state them, we use the standard notation $[X]^\kappa$ for the collection of all subsets of X of cardinality κ.

THEOREM 4.5. *For any non-principal ultrafilter $\mathcal{U}$ on ω, the following properties are equivalent:*

- *$\mathcal{U}$ is selective.*
- *Whenever $[\omega]^2$ is partitioned into finitely many pieces, there is a set $A \in \mathcal{U}$ such that $[A]^2$ is included in one of the pieces.*
- *Whenever $[\omega]^k$, for some finite k, is partitioned into finitely many pieces, there is a set $A \in \mathcal{U}$ such that $[A]^k$ is included in one of the pieces.*

Because of this connection with Ramsey's theorem, selective ultrafilters are often called *Ramsey ultrafilters.* Yet another name for the same concept (in French) is *ultrafiltre absolu* [**26**].

There is an extension of this result to partitions of $[\omega]^\omega$, due to Mathias [**48**]. It needs, however, an assumption that the partition is somewhat well-behaved. The reason for this need is that one can (using the axiom of choice) partition $[\omega]^\omega$ into two pieces such that, whenever two infinite subsets of ω differ by a single element, they lie in different pieces of the partition. In particular, there is no infinite A with $[A]^\omega$ included in a single piece.

To formulate Mathias's theorem, we need the natural topology on $[\omega]^\omega$, as a subspace of the power set $\mathcal{P}(\omega) \cong 2^\omega$, which we topologize as a product space, using the discrete topology on 2. This makes $\mathcal{P}(\omega)$ a compact metric space; $[\omega]^\omega$ is a G_δ subset and therefore admits a complete metric. (An explicit complete metric is given by taking the distance between two infinite subsets X and Y of ω to be 2^{-n} where n is the first number such that the nth elements of X and of Y are distinct.) Recall that a set (in a complete, separable, metric space) is called analytic if it is the image of a Borel set under a continuous function.

THEOREM 4.6 (Mathias [**48**]). *If $\mathcal{U}$ is a selective ultrafilter on ω and if $[\omega]^\omega$ is partitioned into an analytic piece and its complement, then there is $A \in \mathcal{U}$ such that $[A]^\omega$ is included in one piece.*

Mathias also pointed out that, if one forces with the partial order[2] $([\omega]^\omega, \subseteq)$, the generic object is a selective ultrafilter on ω, and no new reals are added. As a result, Theorem 4.6 gave a new proof of Silver's theorem [**65**] that every partition of $[\omega]^\omega$ into an analytic piece and its complement has an infinite homogeneous set; this in turn extended the Galvin-Prikry theorem [**29**], which said the same for partitions into Borel pieces.

Mathias also showed that, in a model of set theory obtained by Lévy collapsing below a Mahlo cardinal, Theorem 4.6 continues to hold if the assumption "analytic"

[2]This order is not separative; to force with it is the same as to force with its separative quotient, the result of identifying any two sets whose symmetric difference is finite. This separative quotient is countably closed.

is weakened to "ordinal-definable from real numbers." This hypothesis seems to be the weakest form of good behavior that one could hope for in this context. It follows from this result that, in this "Lévy-Mahlo model," every selective ultrafilter is generic, with respect to the forcing $[\omega]^\omega$, over the submodel of sets hereditarily ordinal definable from reals. Since the forcing notion is homogeneous, no two selective ultrafilters in this model can be distinguished by isomorphism-invariant properties expressible with real and ordinal parameters. In this sense, selectivity can be regarded as complete combinatorial information about an ultrafilter.

All of the preceding properties of selective ultrafilters on ω do not answer the obvious question whether such ultrafilters exist (except when one explicitly forces one with $[\omega]^\omega$). There are existence results, but they necessarily use hypotheses that go beyond ZFC.

A straightforward inductive construction shows that the continuum hypothesis (CH) implies the existence of selective ultrafilters on ω. One lists, in a sequence of length $\aleph_1$, the requirements that a filter $\mathcal{F}$ should satisfy in order to be a selective ultrafilter. These requirements are:

- For each $A \subseteq \omega$, either A or $\omega - A$ is in $\mathcal{F}$.
- For each $f : \omega \to \omega$, there is a set $A \in \mathcal{F}$ such that $f \restriction A$ is constant or one-to-one.

Then one builds a filter, in $\aleph_1$ steps, starting with the filter of cofinite sets, and at each step adjoining one more generator so as to satisfy one more of the requirements. An appropriate set to adjoin is easily found once one knows that the filter under construction is, at each stage, generated by the cofinite filter plus one more set. Maintaining this information at limit stages is the key to the construction and is where CH is used to ensure that, at each limit stage, we have only countably many generators to replace with one. The point of countability is that, given countably many subsets of ω with the strong finite intersection property (i.e., the intersection of any finitely many of them is infinite), one can easily produce an infinite set that is almost (i.e., except for finitely many elements) included in each of the given sets.

Martin's axiom (MA) implies the same sort of reduction from many generators to one, even if the number of generators to be reduced is not necessarily countable but merely smaller than the cardinality $\mathfrak{c}$ of the continuum. Thus, MA also implies the existence of selective ultrafilters. In fact, since there will be $\mathfrak{c}$ stages during the inductive construction where one can make an arbitrary choice, MA implies the existence of $2^\mathfrak{c}$ selective ultrafilters. Note the immediate consequence that there are, under MA, $2^\mathfrak{c}$ non-isomorphic selective ultrafilters, because there are only $\mathfrak{c}$ permutations of ω that could serve as isomorphisms.

Notice that the constructions just described, under CH or MA, actually show somewhat more, namely that every filter on ω generated by (strictly) fewer than $\mathfrak{c}$ sets is included in a selective ultrafilter. The following theorem of Canjar [**25**] (also found slightly later but independently by Bartoszyński and Judah) isolates the exact hypothesis needed for such a construction.

THEOREM 4.7. *The following two statements are equivalent (provably in ZFC).*

- *Every filter on ω that is generated by $< \mathfrak{c}$ sets and contains no finite sets can be extended to a selective ultrafilter on ω.*
- *The real line cannot be covered by fewer than $\mathfrak{c}$ meager (i.e., first Baire category) sets.*

Not only do constructions of the sort described above fail in some models of ZFC (those in which fewer than $\mathfrak{c}$ meager sets suffice to cover $\mathbb{R}$), but some models simply have no selective ultrafilters. Kunen [**43**] showed that there are no selective ultrafilters in the random real model (obtained from a model of CH by adding at least $\aleph_2$ random reals).

There are extensions of Theorem 4.6 dealing with several selective ultrafilters at once. We give here only a sample of the results and refer to [**12**] for the general theorems.

THEOREM 4.8. *Suppose $\mathcal{U}$ and $\mathcal{V}$ are two non-isomorphic, selective ultrafilters on ω, and let $[\omega]^\omega$ be partitioned into an analytic piece and its complement. Then there are sets $A \in \mathcal{U}$ and $B \in \mathcal{V}$ such that one of the two pieces contains all sets of the form $\{a_0 < b_0 < a_1 < b_1 < \dots\}$, where the a's and b's alternate and where all $a_i \in A$ and all $b_i \in B$.*

Although general results of this sort depend on selectivity, this particular sample would actually work with somewhat weaker assumptions; see Theorem 5.4 below.

We turn now to a related but broader class of ultrafilters, the P-points. The definition is like one of the characterizations of selectivity but with "one-to-one" replaced by "finite-to-one." (A function f is finite-to-one if each point in its range is the image of only finitely many points in the domain; there need not be a single finite upper bound for the sizes of all these pre-image sets.)

DEFINITION 4.9. An ultrafilter $\mathcal{U}$ on a countable set X is a *P-point* if every function with domain X becomes finite-to-one or constant when restricted to a suitable set in $\mathcal{U}$.

CONVENTION 4.10. Whenever it is convenient, we shall pretend that P-points (and other ultrafilters on countable sets X) are on ω; this entails no loss of generality, since we can always apply a bijection between X and ω. On the other hand, whenever it is convenient (e.g., when considering sums of ultrafilters), we shall allow ultrafilters to live on other countable sets. We rely on the context to make clear which sort of convenience is in force at any particular moment.

Clearly, every selective ultrafilter is a P-point, so the existence theorems for selective ultrafilters immediately imply existence theorems for P-points. Furthermore, CH (or MA) allows inductive constructions of many P-points that are not selective. The extent of applicability of such inductive constructions is described by the following analog, due to Ketonen, of Theorem 4.7. It involves the cardinal characteristic $\mathfrak{d}$, the *dominating number*, defined as the minimum cardinality of any set $\mathcal{D}$ of functions $\omega \to \omega$ such that every function $\omega \to \omega$ is majorized by one in $\mathcal{D}$. (It doesn't matter, for $\mathfrak{d}$, whether "majorized" means at all inputs in ω or at all but finitely many.)

THEOREM 4.11 (Ketonen [**40**]). *The following two statements are provably equivalent (in ZFC).*

- *Every filter on ω that is generated by $< \mathfrak{c}$ sets and contains no finite sets can be extended to a P-point.*
- $\mathfrak{d} = \mathfrak{c}$.

Ketonen's proof also shows that any ultrafilter on ω that is generated by strictly fewer than $\mathfrak{d}$ sets is a P-point.

Since P-points need not (for example, under MA) be selective, they cannot be proved to have the Ramsey-like partition properties described in Theorems 4.5 and 4.6, but they come close. Here is the P-point analog of Mathias's theorem; a similar analog of Kunen's theorem is an immediate consequence.

THEOREM 4.12. *If $\mathcal{U}$ is a P-point on ω and if $[\omega]^\omega$ is partitioned into an analytic piece and its complement, then there is $A \in \mathcal{U}$ and there is $f : \omega \to \omega$ such that one piece of the partition contains all those infinite subsets $\{a_0 < a_1 < \dots\}$ of A in which $f(a_n) < a_{n+1}$ for all n.*

Unlike selectivity, the property of being a P-point is a topological property of an ultrafilter as a point in the Stone-Čech remainder $\beta\omega - \omega$. That is a consequence of the following, fairly straightforward reformulation of the definition.

PROPOSITION 4.13. *Let $\mathcal{U}$ be an ultrafilter on ω. The following are equivalent:*

- *$\mathcal{U}$ is a P-point.*
- *Given any countably many sets $A_n \in \mathcal{U}$, there is some $B \in \mathcal{U}$ almost included in each of them (i.e., each $B - A_n$ is finite).*
- *The intersection of any countably many neighborhoods of the point $\mathcal{U} \in \beta\omega - \omega$ is again a (not necessarily open) neighborhood of $\mathcal{U}$.*[3]

Note that sums $\mathcal{U}\text{-}\sum_i \mathcal{V}_i$ (of non-principal ultrafilters on ω) are never P-points, because the projection of such a sum to the first factor is neither finite-to-one nor constant on any set in the sum. Thus, any hypothesis (such as CH or MA) that implies the existence of P-points also implies that $\beta\omega - \omega$ is not homogeneous. This was the motivation for the original construction of P-points under CH, by W. Rudin [**61**], the first construction of ultrafilters on ω with special properties. Later, Frolík proved in ZFC that $\beta\omega - \omega$ is not homogeneous, using essentially the Rudin-Frolík ordering.

Shelah [**70, 62**] showed that the existence of P-points is unprovable in ZFC.

Numerous variants of selectivity and P-point-ness have been studied — too many to treat here in any detail, so we give just a few comments and pointers to the literature.

The Ramsey property of selective ultrafilters admits an infinite sequence of weakenings of the form "for each partition of $[\omega]^2$ into finitely many pieces, there is a set A in the ultrafilter such that $[A]^2$ meets only k of the pieces." It is perhaps surprising that any ultrafilter having this weak Ramsey property with $k \leq 3$ is necessarily a P-point, but this is not so when $k = 4$. In fact, the weak Ramsey property with $k = 4$ holds for (isomorphic copies on ω of) ultrafilters of the form $\mathcal{U}\text{-}\sum_i \mathcal{V}_i$ when $\mathcal{U}$ and all the $\mathcal{V}_i$ are non-isomorphic selective ultrafilters. There are analogous weakenings of the Ramsey property for exponents larger than 2. For more information about these weak Ramsey properties, see [**9, 27, 59, 19**] and the references there.

A different sort of weakening of the Ramsey property was studied in [**7**], where Baumgartner and Taylor defined k-arrow ultrafilters on ω to be those $\mathcal{U}$ such that, whenever $[\omega]^2$ is partitioned into two pieces, either the first piece contains $[A]^2$ for some $A \in \mathcal{U}$ or the second piece contains $[A]^2$ for some A of cardinality at least k.

[3]This last equivalent statement is the general definition of P-point in arbitrary topological spaces.

Further information about arrow ultrafilters and similar combinatorial properties is in [**9**], [**27**], and [**45**].

Baumgartner [**6**] introduced a family of weakenings of the P-point property, as instances of the following scheme. Let $\mathcal{I}$ be a family of subsets of some set T; the sets in $\mathcal{I}$ are to be considered "small" in some sense. An ultrafilter $\mathcal{U}$ on X is an $\mathcal{I}$-ultrafilter if, for every function $f : X \to T$, the image ultrafilter contains a set from $\mathcal{I}$. Among the families $\mathcal{I}$ considered are (with $T = \mathbb{R}$) the families of discrete sets, of scattered sets (i.e., every nonempty subset has an isolated point), and of sets whose closure has Lebesgue measure zero, and (with $T = \aleph_1$) the sets with order-type below some specified, countable ordinal α. P-points are characterized by an extremely strong property of this sort: Whenever ω is mapped to any linearly ordered set, the image of any P-point contains a set whose order-type is ω, ω^*, or 1. It follows that all P-points are $\mathcal{I}$-ultrafilters for all the $\mathcal{I}$'s listed above.

Looking at P-points from the topological point of view, namely as points whose neighborhood filter in $\beta\omega - \omega$ is closed under countable intersections, we see that a P-point is never a limit of a countable sequence of other (non-principal) ultrafilters. The latter property is the definition of *weak P-point*, and Kunen [**44**] showed that such points exist (provably in ZFC).

A different weakening of selectivity is the "complement" of the definition of P-point. An ultrafilter $\mathcal{U}$ is a *Q-point* or *rare ultrafilter* if every finite-to-one function becomes one-to-one when restricted to some set in $\mathcal{U}$. Thus, an ultrafilter is selective if and only if it is both a P-point and a Q-point. An ultrafilter isomorphic to a tensor product $\mathcal{U}\otimes\mathcal{V}$ is never a Q-point; the second projection is finite-to-one on the above-diagonal set, which is in $\mathcal{U}\otimes\mathcal{V}$, but it is not one-to-one on any set in $\mathcal{U}\otimes\mathcal{V}$. On the other hand, a sum of distinct ultrafilters can be a Q-point; for example, if all the $\mathcal{V}_i$ are selective and no two of them are isomorphic, then $\mathcal{U}\text{-}\sum_i \mathcal{V}_i$ is a Q-point. In particular, if (as is the case under CH or MA) there are infinitely many non-isomorphic selective ultrafilters, then there are non-selective Q-points also.

Miller [**52**] has shown that the existence of Q-points (or even semi-Q-points, defined below) is not provable in ZFC. It remains an open problem whether it is consistent with ZFC that neither P-points nor Q-points exist.

An ultrafilter $\mathcal{U}$ on ω is *rapid* or a *semi-Q-point* if every function $\omega \to \omega$ is majorized by the increasing enumeration of some set in $\mathcal{U}$. Because of its use of the ordering of ω, this definition does not appear to be invariant under isomorphism. It is, nevertheless, invariant, as follows from an equivalent characterization in [**52**]: $\mathcal{U}$ is rapid if and only if, for each sequence of finite subsets P_n of ω, there exists $A \in \mathcal{U}$ such that $|A \cap P_n| \le n$ for all n.

The last special ultrafilters to be described in this section are not given by purely combinatorial properties; they depend on an algebraic structure of the underlying set. Quite generally, any binary operation $*$ on a set X induces a binary operation on ultrafilters on X; we regard $*$ as a function $X \times X \to X$ and define

$$\mathcal{U} * \mathcal{V} := *(\mathcal{U} \otimes \mathcal{V}).$$

Equivalently, in terms of quantifiers,

$$((\mathcal{U} * \mathcal{V})z)\,\varphi(z) \iff (\mathcal{U}x)(\mathcal{V}y)\,\varphi(x * y).$$

Under the usual identification of elements of X with principal ultrafilters $\hat{x}$, the $*$ operation on ultrafilters extends the given $*$ operation on X, i.e., $\hat{x} * \hat{y} = \widehat{x * y}$. This extension of operations is easily seen to preserve associativity. But it does not preserve commutativity. Furthermore, $\mathcal{U} * \mathcal{V}$ is, with respect to the topology of βX, a continuous function of $\mathcal{U}$ for each fixed $\mathcal{V}$, but not in general a continuous function of $\mathcal{V}$ when $\mathcal{U}$ is fixed. This semi-continuity, together with the compactness of βX, is enough to show that, if the operation $*$ is associative, then there are ultrafilters $\mathcal{U}$ that are idempotent, i.e., $\mathcal{U} * \mathcal{U} = \mathcal{U}$. We refer to the paper of Hindman and Strauss in this volume and to their book [**31**] for more information about idempotent ultrafilters and their remarkable combinatorial properties. Here, we just point out a few connections with the combinatorial properties considered earlier, for the semigroup most often considered in connection with idempotent ultrafilters.

On the semigroup ω with the operation of addition, an idempotent ultrafilter cannot be a P-point (or even a weak P-point) or a Q-point. Nevertheless, the "nicest" idempotent ultrafilters, the strongly summable ones, are connected with P-points as follows. Identify each natural number n with a finite set $b(n)$ of natural numbers by means of the binary expansion, $n = \sum_{k \in b(n)} 2^k$. Then the functions $n \to \max(b(n))$ and $n \to \min(b(n))$ send any strongly summable ultrafilter on ω to P-points.

5. Cardinal Characteristics and Near Coherence

Cardinal characteristics of the continuum are cardinal numbers, usually between $\aleph_1$ and $\mathfrak{c} = 2^{\aleph_0}$ inclusive, that give information about properties of the real line $\mathbb{R}$ or similar spaces like $2^\omega \cong \mathcal{P}(\omega)$, $[\omega]^\omega$, and the Baire space ${}^\omega\omega$ of functions $\omega \to \omega$. The information in question may be topological, measure-theoretic, or combinatorial (though topological and measure-theoretic information usually admits a useful combinatorial reformulation). We have already encountered a few cardinal characteristics: The cardinal $\mathfrak{c}$ of the continuum and the dominating number $\mathfrak{d}$ have appeared explicitly. The covering number for Baire category, defined as the minimum number of meager sets needed to cover $\mathbb{R}$, appears implicitly in Theorem 4.7. In this section, we define and discuss some of the other cardinal characteristics that have important connections with ultrafilters.

The one most directly connected with ultrafilters, and therefore written $\mathfrak{u}$ and called the *ultrafilter number*, is the smallest cardinality of any base for an ultrafilter on ω. There are always ultrafilters on ω that have no base of size $< \mathfrak{c}$, but it is consistent with ZFC that some ultrafilters on ω admit smaller bases. In fact, there is considerable flexibility in building models of ZFC with various values of $\mathfrak{u}$; see for example [**21**]. But we always have $\aleph_1 \leq \mathfrak{u} \leq \mathfrak{c}$, and $\mathfrak{u}$ cannot have countable cofinality.

Closely related to $\mathfrak{u}$ is the *unsplitting* number $\mathfrak{r}$ (also called the *refining* or *reaping* number). A family $\mathcal{R} \subseteq [\omega]^\omega$ is called *unsplit* if, for each $X \subseteq \omega$, there exists $R \in \mathcal{R}$ such that one of $R \cap X$ and $R - X$ is finite, i.e., X fails to split R into two infinite parts. Then $\mathfrak{r}$ is defined as the minimum cardinality of an unsplit family. Since any ultrafilter base is unsplit, $\mathfrak{r} \leq \mathfrak{u}$. The converse inequality was an open problem for quite some time, until Goldstern and Shelah [**30**] produced a model with $\mathfrak{r} < \mathfrak{u}$. Later, Aubrey [**1**] showed that this strict inequality can happen only when $\mathfrak{d}$ is small; specifically, $\min\{\mathfrak{d}, \mathfrak{u}\} \leq \mathfrak{r}$.

Another connection between $\mathfrak{r}$ and ultrafilters was found by Balcar and Simon [**3**]. A π-base for an ultrafilter $\mathcal{U}$ on ω is a family $\mathcal{B} \subseteq [\omega]^\omega$ such that every set in

$\mathcal{U}$ has a subset in $\mathcal{B}$. (So if $\mathcal{B}$ were included in $\mathcal{U}$ it would be a base.) Then $\mathfrak{r}$ is the minimum cardinality of any ultrafilter π-base.

Another cardinal characteristic, closely related to $\mathfrak{d}$ and almost as closely related to ultrafilters, is the *bounding* number $\mathfrak{b}$ (more accurately called the unbounding number). A family $\mathcal{F} \subseteq {}^{\omega}\omega$ is *unbounded* if, for every $g \in {}^{\omega}\omega$, there is $f \in \mathcal{F}$ such that $g(n) < f(n)$ for infinitely many n; equivalently, no single g majorizes, at all but finitely many[4] n, all $f \in \mathcal{F}$. Then $\mathfrak{b}$ is defined as the smallest cardinality of any unbounded family. Since dominating families are obviously unbounded, we have $\mathfrak{b} \leq \mathfrak{d}$; in fact, it is easy to check that $\aleph_1 \leq \mathfrak{b} \leq \mathfrak{d} \leq \mathfrak{c}$, that $\mathfrak{b}$ is a regular cardinal and that $\mathfrak{d}$ has cofinality at least $\mathfrak{b}$.

The connection between $\mathfrak{b}$ and ultrafilters is given by Solomon's theorem [**66**] that $\mathfrak{b} \leq \mathfrak{u}$. In fact, his proof shows much more: Suppose $\mathcal{F}$ is a filter on ω extending the cofinite filter, and suppose $\mathcal{F}$ is generated by strictly fewer than $\mathfrak{b}$ sets. Then not only is $\mathcal{F}$ not an ultrafilter, but there is a finite-to-one function $f : \omega \to \omega$ such that $f(\mathcal{F})$, which would be an ultrafilter if $\mathcal{F}$ were, is in fact the cofinite filter.

REMARK 5.1. It is often convenient to use the word "nearly" to mean "after applying a suitable finite-to-one function." Thus, the result just quoted could be formulated as "a filter generated by $< \mathfrak{b}$ sets is nearly the cofinite filter." This gives the correct intuition; such a filter is about as far as possible from being an ultrafilter.

It is important not to confuse "nearly" with "almost"; the latter is often used to mean "after removing a suitable finite set from consideration." For example, a set A is almost included in another set B if $A - B$ is finite.

Ultrafilters have a close connection with another cardinal characteristic, the groupwise density number $\mathfrak{g}$. To introduce groupwise density, it is useful to simultaneously discuss (plain) density. Both concepts are concerned with families $\mathcal{D} \subseteq [\omega]^{\omega}$ that are closed under subsets. Such a family is called *dense* if every infinite $X \subseteq \omega$ has a subset in $\mathcal{D}$. (This is the usual notion of density used in forcing.) It is *groupwise dense* if it has the stronger property that, whenever ω is partitioned into a sequence of finite intervals, some infinite union of those intervals is in $\mathcal{D}$. (To verify "stronger" here, just pick a partition so that each of its intervals contains at least one point from X.) Then the *distributivity number* $\mathfrak{h}$ is defined as the smallest cardinal κ such that some κ dense families have empty intersection; the *groupwise density number* $\mathfrak{g}$ is defined the same way with "groupwise dense" in place of "dense." Since groupwise dense families are dense, we always have $\mathfrak{h} \leq \mathfrak{g}$; it is also provable that $\mathfrak{g} \leq \mathfrak{d}$ and $\mathfrak{h} \leq \mathfrak{b}$.

The distributivity number was introduced by Balcar, Pelant, and Simon [**2**] in connection with a question about the topology of $\beta\omega - \omega$; how many nowhere dense sets are needed to cover this space. But the name of this cardinal comes from its characterization as the smallest κ such that the forcing notion $[\omega]^{\omega}$ is not (κ, ∞) distributive, i.e., that this forcing adjoins a new κ-sequence of ordinals. In fact, combinatorial results in [**2**] imply that this forcing collapses $\mathfrak{c}$ to $\mathfrak{h}$.

The groupwise density number was introduced in [**17**] for a purely technical purpose. The inequality $\mathfrak{u} < \mathfrak{g}$ summarizes the essential combinatorial properties of a model constructed by Shelah [**20**]; such a summary became desirable when

[4] In contrast to the definition of $\mathfrak{d}$, we must not replace "at all but finitely many n" in the definition of $\mathfrak{b}$ with "at all n"; the resulting cardinal would be simply $\aleph_0$.

the same model turned out to be useful for new purposes. For some of these applications, see [**13**] and the references there. (Later, $\mathfrak{g}$ also turned out to be interesting apart from the applications of $\mathfrak{u} < \mathfrak{g}$. See, for example, [**14, 49, 50**].)

The hypothesis $\mathfrak{u} < \mathfrak{g}$ is a strong one, both in the sense that it has many interesting consequences and in the (related) sense that it has rather few known models. Many of its applications (including the one in [**20**] from which it was abstracted) involve the principle of near coherence of filters, defined as follows.

DEFINITION 5.2. Two filters $\mathcal{F}_1$ and $\mathcal{F}_2$ on ω, each extending the cofinite filter, are *coherent* if their union has the finite intersection property, i.e., if there is a filter that extends them both. They are *nearly coherent* if there is a finite-to-one $f : \omega \to \omega$ such that $f(\mathcal{F}_1)$ and $f(\mathcal{F}_2)$ are coherent. The principle of *near coherence of filters* (NCF) is the assertion that every two filters extending the cofinite filter are nearly coherent.

Note that, if some ultrafilters extending $\mathcal{F}_1$ and $\mathcal{F}_2$ are nearly coherent, then so are $\mathcal{F}_1$ and $\mathcal{F}_2$ themselves. Note also that two ultrafilters are coherent only if they are equal. So NCF is equivalent to the statement that every two ultrafilters have a common image under some finite-to-one function.

It is shown in [**17**] that each of the following three statements implies the next:

- $\mathfrak{u} < \mathfrak{g}$.
- *Filter dichotomy* (FD): For every filter $\mathcal{F}$ on ω extending the cofinite filter, there is a finite-to-one $f : \omega \to \omega$ such that $f(\mathcal{F})$ is either an ultrafilter or the cofinite filter.
- NCF.

The implication from FD to NCF is not reversible [**51**]; reversibility of the other implication, from $\mathfrak{u} < \mathfrak{g}$ to FD, remains an open problem. Even the weakest of these principles, NCF, has numerous applications, in surprisingly varied parts of mathematics; see [**10**] and especially [**11**]. For example, we have the following.

THEOREM 5.3. *The following are equivalent.*

(1) *NCF.*
(2) *Each non-principal ultrafilter on ω has a finite-to-one image that is generated by strictly fewer than $\mathfrak{d}$ sets.*
(3) *The ideal of compact operators on (separable, complex) Hilbert space is not the sum of two properly smaller ideals.*
(4) *The Stone-Čech remainder $[0,1)^*$ of the half-open interval $[0,1)$ has only one composant, i.e., each pair of points lies in a compact, connected, proper subset of $[0,1)^*$.*

The concept of near coherence also allows us to strengthen Theorem 4.8 by weakening the hypothesis as follows.

THEOREM 5.4 ([**16**]). *A pair of ultrafilters $\mathcal{U}$ and $\mathcal{V}$ satisfy the conclusion of Theorem 4.8 if and only if they are P-points and are not nearly coherent.*

We conclude this section with some cardinal numbers very directly related to ultrafilters on ω, the cofinalities of the ultraproducts $\mathcal{U}$-prod ω. If $\mathcal{U}$ is an ultrafilter on ω, then the set ${}^{\omega}\omega$ of functions from ω to itself is pre-ordered by the relation

$$f \leq_{\mathcal{U}} g \iff (\mathcal{U}x)\, f(x) \leq g(x),$$

and we write cof($\mathcal{U}$-prod ω) for the cofinality of this pre-order, i.e., the minimum size of a family $\mathcal{C} \subseteq {}^\omega\omega$ such that for each $f \in {}^\omega\omega$ there is some $g \in \mathcal{C}$ with $f \leq_\mathcal{U} g$. It is immediate from the definition that this cofinality is a regular cardinal in the range $\mathfrak{b} \leq$ cof($\mathcal{U}$-prod ω) $\leq \mathfrak{d}$. The following proposition summarizes additional information, from various sources, about cofinalities of ultraproducts.

PROPOSITION 5.5. (1) *There is an ultrafilter $\mathcal{U}$ with cof($\mathcal{U}$-prod ω) = cf($\mathfrak{d}$).* [Canjar [**24**]]

(2) *It is consistent that $\mathfrak{b} \ll \mathfrak{d}$ and every regular cardinal κ in the range $\mathfrak{b} \leq \kappa \leq \mathfrak{d}$ occurs as cof($\mathcal{U}$-prod ω) for some $\mathcal{U}$. This occurs, for example, in the model obtained by adding many Cohen reals to a model of CH.* [Canjar [**23**] and Roitman [**58**]]

(3) *If an ultrafilter $\mathcal{U}$ has a base (or even a π-base) of cardinality $< \mathfrak{d}$ then cof($\mathcal{U}$-prod ω) = $\mathfrak{d}$.* [Nyikos [**55**]]

(4) *cof($\mathcal{U}$-prod ω) $\geq \mathfrak{g}$ for all $\mathcal{U}$.* [Blass and Mildenberger [**18**]]

(5) *If $\mathcal{U}$ and $\mathcal{V}$ are ultrafilters such that $\mathcal{U}$ has a base of cardinality $<$ cof($\mathcal{V}$-prod ω), then $\mathcal{U}$ and $\mathcal{V}$ are nearly coherent.* [Banakh and Blass [**4**, Prop. 19]]

(6) *NCF is equivalent to the statement that cof($\mathcal{U}$-prod ω) $> \mathfrak{u}$ for all $\mathcal{U}$.* [Mildenberger [**49**, Props. 2.1(c) and 3.2(e)]] *More generally, if there are ultrafilters with cof($\mathcal{U}$-prod ω) $> \mathfrak{u}$, then all of them have cof($\mathcal{U}$-prod ω) = $\mathfrak{d}$ and all of them are nearly coherent.*

6. Ultrafilters, Descriptive Set Theory, and Determinacy

Ultrafilters provide examples of various pathologies in real analysis. View subsets of ω as points in the unit interval $[0, 1]$ by means of binary expansions; that is, identify each $A \subseteq \omega$ with $\sum_{n\in A} 2^{-n}$. (The fact that countably many real numbers come from two A's rather than one will make no essential difference and will be ignored in most of what follows.) Then an ultrafilter (or indeed any family of subsets of ω) becomes a subset of $[0, 1]$, and we can ask about its behavior with respect to notions like Baire category and Lebesgue measure. The answer is that the behavior is bad.

THEOREM 6.1 (Sierpiński [**64**]). *A non-principal ultrafilter on ω, regarded as a subset of $[0, 1]$ is not Lebesgue measurable.*

The proof uses the zero-one law for Lebesgue measure (see [**56**, Thm. 21.3]) to infer that, if a non-principal ultrafilter were measurable, its measure would be 0 or 1. But the measure-preserving reflection $x \mapsto 1 - x : [0, 1] \to [0, 1]$ maps the ultrafilter to its complement (except for countably many points), so the measure of the ultrafilter would have to be $1/2$.

A similar argument, using the Baire category analog of the zero-one law [**56**, Thm. 21.4], shows that a non-principal ultrafilter cannot have the Baire property. (A set has the *Baire property* if it differs from some open set by a meager set.)

It is well known that Lebesgue measurability and the Baire property can be deduced from determinacy hypotheses [**53, 54**]. Therefore, non-principal ultrafilters on ω lead to non-determined games. But in fact, there is a straightforward way[5]

[5] I do not know who first exhibited this construction. I found it in 1969 or 1970, but if I remember correctly it was already known to Eugene Kleinberg. I believe it was known considerably earlier to Fred Galvin.

to convert such ultrafilters into non-determined games, without going via Lebesgue measurability or the Baire property. It proceeds as follows.

Given the ultrafilter $\mathcal{U}$, define a perfect-information game between two players, I and II, as follows. The players move alternately, for an ω-sequence of moves, with I moving first, and the moves will ultimately partition ω into two pieces, which we regard as belonging to the two players. At each move, a player puts into his piece a finite, nonempty, initial segment of the set of numbers not previously put into either piece. After ω moves, the winner is the player whose piece is in $\mathcal{U}$.

PROPOSITION 6.2. *If $\mathcal{U}$ is a non-principal ultrafilter on ω, then neither player has a winning strategy in the game just described.*

PROOF. Let σ be a strategy for I; we shall find a play in which I uses σ yet loses. The proof for II is analogous.

We define, by simultaneous induction, two plays p and q of the game. In both plays, I uses the strategy σ. II's moves are chosen so that his pieces in the two plays cover all but finitely many of the natural numbers. After I's opening move, which is the same in both plays as prescribed by σ, II makes an arbitrary legal move in one of the plays, say taking the first yet unclaimed element in play p. Thereafter, as soon as I has made a move in p (resp. q), II switches to q (resp. p) and takes there (at least) all the numbers that I just took in p (resp. q).

The ultrafilter $\mathcal{U}$ will contain II's piece in one of the two plays, so I loses that play despite having used σ. □

This proposition shows that the existence of non-principal ultrafilters on ω contradicts the axiom of determinacy (AD). Perhaps surprisingly, AD implies the existence of non-principal ultrafilters on certain other sets.

Before describing some of these ultrafilters, we point out a consequence of Proposition 4.3 (whose proof does not use the axiom of choice, as long as κ is a well-ordered cardinal) in conjunction with the non-existence of non-principal ultrafilters on ω.

COROLLARY 6.3. *AD implies that every ultrafilter (on any set) is countably complete.*

Here are two elegant examples of filters which, under the assumption AD, turn out to be ultrafilters.

EXAMPLE 6.4. Let $\mathcal{D}$ be the set of Turing degrees, ordered by Turing reducibility. For any degree $\mathbf{d}$, define its *cone* $\mathcal{C}_{\mathbf{d}}$ to be the set of Turing degrees $\geq_T \mathbf{d}$. The cones constitute a basis for a countably complete filter on $\mathcal{D}$, the *cone filter.* (The proof of countable completeness uses the axiom of choice for countably many sets of reals; this weak form of choice is a consequence of AD.) Martin [**47**] showed that, under AD, this cone filter is an ultrafilter, i.e., every set of Turing degrees either includes a cone or is disjoint from one. The proof uses an arbitrary set A of Turing degrees to define a game in which player I wins if and only if the sequence of moves of both players together has Turing degree in A. If σ is a winning strategy for I (resp. II), then the Turing degree $\mathbf{d}$ of σ has its cone $\mathcal{C}_{\mathbf{d}}$ included in (resp. disjoint from) A.

EXAMPLE 6.5. The closed unbounded subsets (called *clubs* or *cubs* for short) of the ordinal $\aleph_1$ form a basis for a filter on $\aleph_1$, the *club filter.* Solovay [**67**] showed that, under AD, this club filter is an ultrafilter. (The countable completeness of the club filter follows, under AD, by Corollary 6.3. It also follows under AC from the easily proved fact that the intersection of countably many clubs is a club; AC is used to choose club subsets from countably many sets in the filter.)

Thus AD implies that $\aleph_1$ is a measurable cardinal. What about the well-known theorem that measurable cardinals must be very large (inaccessible and more)? The proof of that theorem uses the axiom of choice, so it does not apply in the world of AD. In the absence of AC, what survives of the proof shows that a measurable (well-ordered) cardinal κ is regular and cannot be mapped one-to-one into the power set of any $\lambda < \kappa$. In particular, under AD, we have that $\aleph_1$ cannot be mapped one-to-one into $\mathcal{P}(\omega)$; there do not exist $\aleph_1$ distinct real numbers. (This last conclusion can also be obtained by combining the result of [**54**] with a theorem of Shelah [**63**] that a one-to-one map of $\aleph_1$ into the reals gives rise to a set of reals that is not Lebesgue measurable.)

The largeness of measurable cardinals reasserts itself if we pass to an inner model satisfying the axiom of choice. In particular, if we assume AD and let $\mathcal{U}$ be the club ultrafilter on $\kappa = \aleph_1$, then, in the inner model $L[\mathcal{U}]$ of sets constructible (in the sense of Gödel) from $\mathcal{U}$, we have the axiom of choice holding while $\mathcal{U} \cap L[\mathcal{U}]$ is a countably complete non-principal ultrafilter on κ. Thus, κ, which was merely $\aleph_1$ in the original AD world, is a very large cardinal in the sense of $L[\mathcal{U}]$. This shows that the theory "ZFC + a measurable cardinal exists" is consistent relative to AD. In fact, AD yields inner models of far larger cardinals.

It is known that, under AD, the club ultrafilter on $\aleph_1$ is the image of the cone ultrafilter on the Turing degrees under the map that sends any Turing degree $\mathbf{d}$ to the countable ordinal $\omega_1^{\mathbf{d}}$, the first ordinal not obtainable as the order-type of an ordering of ω recursive in $\mathbf{d}$.

A similar construction produces a non-principal ultrafilter on $\aleph_2$. One maps each degree $\mathbf{d}$ to the next cardinal of $L[\mathbf{d}]$ after (true) $\aleph_1$. (Here $L[\mathbf{d}]$ means the universe constructible from any real of degree $\mathbf{d}$.) Under AD, the image of the cone ultrafilter under this map turns out to be a normal, $\aleph_1$-complete ultrafilter on $\aleph_2$.

The pattern that seems to be emerging here does not continue. Under AD, $\aleph_3$ is not measurable. In fact, all the cardinals of the form $\aleph_n$ for $3 \leq n < \omega$ are singular with cofinality $\aleph_2$. The next regular cardinal is $\aleph_{\omega+1}$, and it is measurable. For more about the complicated structure of well-ordered cardinals under AD, see [**38**] and [**33, 34**]. The work of Jackson in particular makes heavy use of ultrafilters and ultrapowers to analyze the projective ordinals, i.e., ordinals of the form "the smallest ordinal that is not the order-type of a Δ_n^1 pre-well-ordering of the reals" for $n \in \omega$. See also Kleinberg's [**41, 42**] for connections between ultrafilters and strong, infinitary partition properties that contradict AC but are fairly prevalent under AD.

7. Forcing

In the Boolean-valued approach to forcing, ultrafilters (in complete Boolean algebras $\mathcal{B}$) play two roles. One is in the formalization of what is true in a Boolean-valued model. The other is in converting Boolean-valued models to ordinary two-valued models. We discuss the two roles in turn.

The first role can be summarized by saying that a forcing extension consists of the sets from the ground model, a generic ultrafilter, and sets obtainable from these by absolute constructions. More precisely, suppose T is a theory (to be thought of as expressing the relevant properties of the ground model) that includes ZFC and proves "$\mathcal{B}$ is a complete Boolean algebra." Then the method of Boolean valued models proves the consistency, relative to T, of the following theory $T^{\mathcal{B}}$. The vocabulary consists of $\in$, G (a constant symbol), $\check{V}$ (a unary predicate symbol), and $\check{S}$ for all predicate and function symbols S of T (including constants regarded as 0-ary functions) except $\in$. The axioms of $T^{\mathcal{B}}$ are:

(1) the axioms of ZFC,
(2) $\check{V}$ is a transitive class containing all the ordinals,
(3) all axioms of T with their quantifiers restricted to $\check{V}$ and their predicate and function symbols S (except for $\in$) replaced with the corresponding $\check{S}$,
(4) G is a $\check{V}$-generic ultrafilter in $\check{\mathcal{B}}$, and
(5) every set is $\mathrm{Val}(x, G)$ for some $x \in \check{V}$,

where Val is a certain explicitly defined, absolute (i.e., Δ_1^{ZF}) operation. Here (3) formalizes the notion that the class $\check{V}$ in the forcing extension is just like the ground model, while (4) and (5) express that the whole forcing extension is obtained from this class by adjoining a generic ultrafilter in the prescribed Boolean algebra.

Although the relative consistency proof for $T^{\mathcal{B}}$ uses the Boolean-valued model $V^{\mathcal{B}}$, once the proof is done, one can simply work in the theory $T^{\mathcal{B}}$ just as in any other consistent theory, without having to pay attention to the Boolean-valuedness of the available model.

The second role of ultrafilters in forcing theory is to reduce a Boolean-valued model $V^{\mathcal{B}}$ to a two-valued one. Given an ultrafilter $\mathcal{U}$ in $\mathcal{B}$, one forms a quotient $V^{\mathcal{B}}/\mathcal{U}$ by identifying any two elements x and y of $V^{\mathcal{B}}$ for which the truth value $\|x = y\|$ is in $\mathcal{U}$. The membership relation E in $V^{\mathcal{B}}/\mathcal{U}$ is defined by

$$[x]E[y] \iff \|x \in y\| \in \mathcal{U},$$

where $[x]$ means the equivalence class in $V^{\mathcal{B}}/\mathcal{U}$ of the element x of $V^{\mathcal{B}}$.

Usually, this construction is carried out with a V-generic ultrafilter $\mathcal{U}$. In order to get such an ultrafilter, one often begins the construction with some countable transitive model of set theory, rather than the whole universe, as V. Genericity of $\mathcal{U}$ then implies that the membership relation E is well-founded, so the quotient $V^{\mathcal{B}}/\mathcal{U}$ is isomorphic to, and is usually identified with, a transitive model. Under this identification, the class $\check{V}$ in $V^{\mathcal{B}}$ becomes the original ground model V. In particular, the ordinals of $V^{\mathcal{B}}/\mathcal{U}$ are those of V.

Vopěnka has, however, shown (see [**68**] or [**69**]) that genericity of the ultrafilter $\mathcal{U}$ is not strictly needed for converting Boolean-valued models to two-valued models. One can form the quotient $V^{\mathcal{B}}/\mathcal{U}$ using any ultrafilter in $\mathcal{B}$, even one in the ground model (so, in particular, one can use the whole universe as V, not some countable model). The quotient will satisfy all the statements whose truth value in $V^{\mathcal{B}}$ is 1 (or is merely in $\mathcal{U}$). The only unpleasantness one encounters is that the quotient model need not be well-founded and may have non-standard ordinals, even non-standard natural numbers. For example, if the forcing adds a single Cohen real, one can arrange $\mathcal{U}$ so that the Cohen real in the quotient contains none of the standard natural numbers. This lack of well-foundedness may interfere somewhat with intuition, but it has no effect on consistency proofs.

Another theorem of Vopěnka (see [**35**, Thm. 65]) goes in the opposite direction, producing genericity where one might not have expected it. Recall that HOD is the class of hereditarily ordinal-definable sets.

THEOREM 7.1 (Vopěnka). *For each set x there is a HOD-complete Boolean algebra $\mathcal{B} \in HOD$ and there is a HOD-generic ultrafilter $\mathcal{U} \subseteq \mathcal{B}$ such that x is an element of the generic extension $HOD[\mathcal{U}]$.*

Thus (in ZFC) every set is obtainable from ordinals and ultrafilters (in suitable Boolean algebras). This result leads to the intuition that ultrafilters constitute a second fundamental building block, after ordinals, for the universe of sets.

Appendix: Proof of Lawvere's Result

In this appendix, we prove Theorem 2.3.

Let an h satisfying the hypotheses of the theorem be given. For any $A \subseteq X$, we use the same symbol A to denote its characteristic function (constantly 1 on A and constantly 0 on $X - A$), which we regard as a function $A : X \to 3$ (that happens to never take the value 2). Define

$$\mathcal{U} = \{A \subseteq X : h(A) = 1\}.$$

We shall show that $\mathcal{U}$ is an ultrafilter with the property required in the theorem.

Claim 1: $X \in \mathcal{U}$

Apply the hypothesis of the theorem with m being the constant 1 function. Since $m \circ X = X$, we get $h(X) = m(h(X)) = 1$, as claimed.

Claim 2: For each $A \subseteq X$, exactly one of A and $X - A$ is in $\mathcal{U}$.

We consider three cases, depending on the value of $h(A)$.

If this value is 1, then we have $A \in \mathcal{U}$ by definition. To see that we do not also have $X - A \in \mathcal{U}$, consider the permutation c of 3 that interchanges 0 with 1 while fixing 2. Thus, in terms of characteristic functions, we have $c \circ A = X - A$. From the theorem's assumption about h, we conclude that $h(X - A) = c(h(A)) = 0$, so $X - A \notin \mathcal{U}$. This completes the first case.

Suppose next that $h(A) = 0$. Then, with the same c as above, we get $h(X-A) = c(h(A)) = 1$. So $X - A \in \mathcal{U}$ and $A \notin \mathcal{U}$.

Finally, suppose $h(A) = 2$. Let m be the function that sends 0 and 1 to 1 and fixes 2. Since $m \circ A$ is the constant function 1, i.e., X, we know from Claim 1 that $h(m \circ A) = 1$. On the other hand, $m(h(A)) = m(2) = 2$. This contradicts the hypothesis of the theorem, so this third case cannot occur. Thus, Claim 2 is proved.

Claim 3: If $A \in \mathcal{U}$ and $A \subseteq B \subseteq X$, then $B \in \mathcal{U}$.

Consider the function $f : X \to 3$ that is constantly 1 on A, 2 on $B - A$, and 0 on $X - B$. We apply the hypothesis of the theorem twice, first with m being the function that sends 2 to 0 while fixing 0 and 1. Then $m \circ f = A$ and so $m(h(f)) = h(A) = 1$. In view of the definition of m, this means $h(f) = 1$. Now apply the hypothesis of the theorem again, using in place of m the function m' that sends 2 to 1 while fixing 0 and 1. Then $m' \circ f = B$ and so $h(B) = m'(h(f)) = m'(1) = 1$, as claimed.

Claim 4: If $A, B \in \mathcal{U}$ then $A \cap B \in \mathcal{U}$.

Consider the function f that is constantly 1 on $A \cap B$, 0 on $B - A$, and 2 on $X - B$. We shall again apply the hypothesis of the theorem with different m's. For the first application, let m send 2 to 1 while fixing 0 and 1. Then $m \circ f$ is (the characteristic function of) $A \cup (X - B)$, which, being a superset of A, is in $\mathcal{U}$ by Claim 3. So

$$m(h(f)) = h(m \circ f) = h(A \cup (X - B)) = 1.$$

By definition of m, we conclude that $h(f)$ is 1 or 2.

For the second application of the hypothesis, use the function m' sending 0 and 1 to 1 and sending 2 to 0. Then $m' \circ f = B$ and, since $B \in \mathcal{U}$, we have

$$m'(h(f)) = h(m' \circ f) = h(B) = 1.$$

By definition of m', we conclude that $h(f)$ is 0 or 1. Combining this with the conclusion of the preceding paragraph, we have $h(f) = 1$.

Finally, let m'' send 2 to 0 while fixing 0 and 1. Then $m'' \circ f = A \cap B$ and

$$h(A \cap B) = h(m'' \circ f) = m''(h(f)) = m''(1) = 1,$$

as claimed.

The claims proved so far establish that $\mathcal{U}$ is an ultrafilter on X. It remains to prove that h is determined from $\mathcal{U}$ as stated in the theorem. So consider any $f : X \to 3$; we must show that $h(f)$ is the v described in the theorem, i.e., that $f^{-1}\{h(f)\} \in \mathcal{U}$. Abbreviate $f^{-1}\{h(f)\}$ as A. Let $m : 3 \to 3$ be the map sending $h(f)$ to 1 and both of the other elements of 3 to 0. Then $m \circ f$ is (the characteristic function of) A, and so we have

$$h(A) = h(m \circ f) = m(h(f)) = 1,$$

as required.

References

[1] Jason Aubrey, "Combinatorics for the dominating and unsplitting numbers," *J. Symbolic Logic* 69 (2004) 482–498.

[2] Bohuslav Balcar, Jan Pelant, and Petr Simon, "The space of ultrafilters on **N** covered by nowhere dense sets," *Fund. Math.* 110 (1980) 11–24.

[3] Bohuslav Balcar and Petr Simon, "On minimal π-character of points in extremally disconnected compact spaces," *Topology Appl.* 41 (1991) 133–145.

[4] Taras Banakh and Andreas Blass, "The number of near-coherence classes of ultrafilters is either finite or $2^{\mathfrak{c}}$," in *Set Theory. Centre de Recerca Matemàtica, Barcelona, 2003–2004*, ed. by J. Bagaria and S. Todorčević, Birkhäuser, Trends in Mathematics (2006) 257–273.

[5] Tomek Bartoszyński and Saharon Shelah, "On the density of Hausdorff ultrafilters," *Logic Colloquium 2004*, Assoc. Symbol. Logic, Lect. Notes Log. 29 (2008) 18–32.

[6] James Baumgartner, "Ultrafilters on ω," *J. Symbolic Logic* 60 (1995) 624–639.

[7] James Baumgartner and Alan Taylor, "Partition theorems and ultrafilters," *Trans. Amer. Math. Soc.* 241 (1978) 283–309.

[8] Andreas Blass, "Two closed categories of filters," *Fund. Math.* 94 (1977) 129–143.

[9] ———, "A model-theoretic view of some special ultrafilters," in *Logic Colloquium '77*, ed. by L. Pacholski, A. Macintyre, and J. Paris, North-Holland (1978) 79–90.

[10] ———, "Near coherence of filters, I: Cofinal equivalence of models of arithmetic," *Notre Dame J. Formal Logic* 27 (1986) 579–591.

[11] ———, "Near coherence of filters, II: Applications to operator ideals, the Stone-Čech remainder of a half-line, order ideals of sequences, and slenderness of groups," *Trans. Amer. Math. Soc.* 300 (1987) 557–581.

[12] ———, "Selective ultrafilters and homogeneity," *Ann. Pure Appl. Logic* 38 (1988) 215–255.

[13] ———, "Applications of superperfect forcing and its relatives," in *Set Theory and Its Applications*, ed. by J. Steprāns and S. Watson, Springer-Verlag, Lecture Notes in Mathematics 1401 (1989) 18–40.

[14] ———, "Groupwise density and related cardinals," *Arch. Math. Logic* 30 (1990) 1–11.

[15] ———, "Voting rules for infinite sets and Boolean algebras," in *Advances in Logic (The North Texas Logic Conference)*, ed. by S. Gao, S. Jackson, and Y. Zhang, A. M. S., Contemp. Math. 425 (2007) 87–103.

[16] ———, "Homogeneous sets from several ultrafilters," *Topology and Its Appl.* 156 (2009) 2581–2594.

[17] Andreas Blass and Claude Laflamme, "Consistency results about filters and the number of inequivalent growth types," *J. Symbolic Logic* 54 (1989) 50–56.

[18] Andreas Blass and Heike Mildenberger, "On the cofinality of ultrapowers," *J. Symbolic Logic* 64 (1999) 727–736.

[19] Andreas Blass and Gugu Moche, "Finite preimages under the natural map from $\beta(\mathbb{N}\times\mathbb{N})$ to $\beta\mathbb{N}\times\beta\mathbb{N}$," *Topology Proceedings* 26 (2001–2002) 407–432.

[20] Andreas Blass and Saharon Shelah, "There may be simple $P_{\aleph_1}$ and $P_{\aleph_2}$ points and the Rudin-Keisler ordering may be downward directed," *Ann. Pure Appl. Logic* 33 (1987) 213–243.

[21] ———, "Ultrafilters with small generating sets," *Israel J. Math.* 65 (1989) 259–271.

[22] David Booth, "Ultrafilters on a countable set," *Ann. Math. Logic* 2 (1970) 1–24.

[23] R. Michael Canjar, *Model-Theoretic Properties of Countable Ultraproducts Without the Continuum Hypothesis,* Ph.D. thesis, University of Michigan (1982).

[24] ———, "Cofinalities of countable ultraproducts: the existence theorem," *Notre Dame J. Formal Logic* 30 (1989) 539–542.

[25] ———, "On the generic existence of special ultrafilters," *Proc. Amer. Math. Soc.* 110 (1990) 233–241.

[26] Gustave Choquet, "Deux classes remarquables d'ultrafiltres sur N," *Bull. Sci. Math. (2)* 92 (1968) 41–48.

[27] Maryvonne Daguenet, "Ultrafiltres à la façon de Ramsey," *Trans. Amer. Math. Soc.* 250 (1979) 91–120.

[28] Mauro DiNasso and Marco Forti, "Hausdorff ultrafilters," *Proc. Amer. Math. Soc.* 134 (2006) 1809–1818.

[29] Fred Galvin and Karel Prikry, "Borel sets and Ramsey's theorem," *J. Symbolic Logic* 38 (1973) 193–198.

[30] Martin Goldstern and Saharon Shelah, "Ramsey ultrafilters and the reaping number, Con($\mathfrak{r}<\mathfrak{u}$)," *Ann. Pure Appl. Logic* 49 (1990) 121–142.

[31] Neil Hindman and Dona Strauss, *Algebra in the Stone-Čech Compactification: Theory and Applications,* de Gruyter Expositions in Mathematics 27 (1998).

[32] Paul Howard and Jean Rubin, *Consequences of the Axiom of Choice,* American Mathematical Society, Mathematical Surveys and Monographs 59 (1998).

[33] Steve Jackson, "AD and the projective ordinals," *Cabal Seminar 81–85* ed. by A. S. Kechris, D. A. Martin, and J. R. Steel, Springer-Verlag, Lecture Notes in Math. 1333 (1988) 117–220.

[34] ———, "A computation of δ^1_5," *Mem. Amer. Math. Soc.* 140 (1999) no. 670.

[35] Thomas Jech, *Set Theory,* Academic Press (1978).

[36] Aki Kanamori and Alan Taylor, "Separating ultrafilters on uncountable cardinals," *Israel J. Math.* 47 (1984) 131–138.

[37] Miroslav Katětov, "A theorem on mappings," *Comm. Math. Univ. Carolinae* 8 (1967) 431–433.

[38] Alexander Kechris, "AD and projective ordinals," *Cabal Seminar 76–77* ed. by A. S. Kechris and Y. N. Moschovakis, Springer-Verlag, Lecture Notes in Math. 689 (1978) 91–132.

[39] Hewitt Kenyon, "Problem 5077," *Amer. Math. Monthly* 70 (1963) 216.

[40] Jussi Ketonen, "On the existence of P-points," *Fund. Math.* 92 (1976) 91–99.

[41] Eugene Kleinberg, "Strong partition properties for infinite cardinals," *J. Symbolic Logic* 35 (1970) 410–428.

[42] ———, *Infinitary Combinatorics and the Axiom of Determinateness,* Springer-Verlag, Lecture Notes in Math. 612 (1977).

[43] Kenneth Kunen, "Some points in βN," *Proc. Cambridge Phil. Soc.* 78 (1980) 385–398.

[44] ———, "Weak P-points in N^*," *Topology, Vol. II (Proc. Fourth Colloq. Budapest, 1978),* ed. Á. Császár, North-Holland, Colloq. Math. Soc. János Bolyai 23, (1980) 741–749.

[45] Claude Laflamme, "Forcing with filters and complete combinatorics," *Ann. Pure Appl. Logic* 42 (1989) 125–163.

[46] F. William Lawvere, "Functorial concepts of complexity for finite automata" *Theory and Applications of Categories* 13 (2004) 164–168.

[47] Donald A. Martin, "The axiom of determinateness and reduction principles in the analytical hierarchy," *Bull. Amer. Math. Soc.* 74 (1968) 687–689.

[48] Adrian Mathias, "Happy families," *Ann. Math. Logic* 12 (1977) 59–111.

[49] Heike Mildenberger, "Groupwise dense families," *Arch. Math. Logic* 40 (2001) 93–112.

[50] ———, "On the groupwise density number for filters," *Acta Univ. Carolin. Math. Phys.* 46 (2005) 55–63.

[51] Heike Mildenberger and Saharon Shelah, "The near coherence of filters principle does not imply the filter dichotomy principle," *Trans. Amer. Math. Soc.* 361 (2009) 2305–2317.

[52] Arnold Miller, "There are no Q-points in Laver's model for the Borel conjecture," *Proc. Amer. Math. Soc.* 78 (1980) 103–106.

[53] Jan Mycielski, "On the axiom of determinateness," *Fund. Math.* 53 (1963/64) 205–224.

[54] Jan Mycielski and Stanisław Świerczkowski, "On the Lebesgue measurability and the axiom of determinateness," *Fund. Math.* 54 (1964) 67–71.

[55] Peter Nyikos, "Special ultrafilters and cofinal subsets of ${}^{\omega}\omega$" (preprint)

[56] John Oxtoby, *Measure and Category,* Springer-Verlag, Graduate Texts in Mathematics 2 (1971).

[57] Bedřich Pospíšil, "Remark on bicompact spaces," *Ann. of Math. (2),* 38 (1937) 845–846.

[58] Judith Roitman, "Non-isomorphic H-fields from non-isomorphic ultrapowers," *Math. Z.* 181 (1982) 93–96.

[59] Ned Rosen, "Weakly Ramsey P-points," *Trans. Amer. Math. Soc.* 269 (1982) 415–427.

[60] Mary Ellen Rudin, "Types of ultrafilters," *1966 Topology Seminar (Wisconsin, 1965)*, Princeton Univ. Press, Ann. of Math. Studies 60, 147–151.

[61] Walter Rudin, "Homogeneity problems in the theory of Čech compactifications," *Duke Math. J.* 23 (1956) 409–419.

[62] Saharon Shelah, *Proper Forcing,* Springer-Verlag, Lecture Notes in Mathematics 940 (1982).

[63] ———, "Can you take Solovay's inaccessible away?" *Israel J. Math.* 48 (1984) 1–47.

[64] Wacław Sierpiński, "Fonctions additives non complètement additives et fonctions non mesurables," *Fund. Math.* 30 (1938) 96–99.

[65] Jack Silver, "Every analytic set is Ramsey," *J. Symbolic Logic* 35 (1970) 60–64.

[66] R. C. Solomon, "Families of sets and functions," *Czechoslovak Math. J.* 27 (1977) 556–559.

[67] Robert Solovay, "Measurable cardinals and the axiom of determinateness," Lecture notes in connection with the Summer Insitute on Axiomatic Set Theory, UCLA (1967).

[68] Petr Vopěnka, "General theory of ∇-models," *Comment. Math. Univ. Carolinae* 8 (1967) 145–170.

[69] Petr Vopěnka and Petr Hájek, *The Theory of Semisets,* Academia, Publishing House of the Czechoslovak Academy of Sciences (1972).

[70] Edward Wimmers, "The Shelah P-point independence theorem," *Israel J. Math.* 43 (1982) 28–48.

Mathematics Department, University of Michigan, Ann Arbor, MI 48109–1043, U.S.A.
E-mail address: ablass@umich.edu

Contemporary Mathematics
Volume **530**, 2010

Measure-centering ultrafilters

D.H.Fremlin

Like pure mathematicians in general, measure theorists in the last hundred years have often used ultrafilters as a tool. I suppose that the first person to notice that ultrafilters have intrinsic properties expressible in terms of measure theory was Sierpiński ([25]), who showed that if $\langle E_n\rangle_{n\in\mathbb{N}}$ is a stochastically independent sequence of measurable subsets of $[0,1]$, and $\mathcal{F}$ is a non-principal ultrafilter on $\mathbb{N}$, then $\lim_{n\to\mathcal{F}} E_n$ has inner measure 0 and outer measure 1. But if you are starting from an interest in ultrafilters rather than an interest in measure theory, your attention will be directed to ways in which measure theory can display differences between different classes of ultrafilter. In §538 of my book [14], I looked at p-point filters, Ramsey ultrafilters, rapid filters, 'measure-converging' filters (an idea due to Matt Foreman), and filters with what I call the 'Fatou property'. Rather than try to cover such a range here, however, I will concentrate on a single class, the 'measure-centering' or 'property M' ultrafilters. The most interesting results are due to Michael Benedikt.

The plan of this note is to begin with statements of the principal definitions and results, with some discussion (§1). Proofs are given in §§3-6, after a preliminary section §2 examining 'dependently selective' filters. Finally I comment on some open questions in §7.

1 Definitions and results

1A I start by defining the class of ultrafilters I mean to study. Its nature will perhaps be clearer if I move to a slightly more general context than is strictly necessary for the main theorems to follow. If $\mathfrak{A}$ is a Boolean algebra, a functional $\nu:\mathfrak{A}\to[0,1]$ is **additive** if $\nu(a\cup b)=\nu a+\nu b$ whenever a, $b\in\mathfrak{A}$ and $a\cap b=0$. In this language, we can define measure-centering ultrafilters ('property M ultrafilters') as follows.

> **Definition** An ultrafilter $\mathcal{F}$ on a set I is **measure-centering** if whenever $\mathfrak{A}$ is a Boolean algebra, $\nu:\mathfrak{A}\to[0,1]$ is an additive functional such that $\nu(1_{\mathfrak{A}})=1$, and $\langle a_i\rangle_{i\in I}$ is a family in $\mathfrak{A}$ such that $\inf_{i\in I}\nu a_i>0$, then there is a $J\in\mathcal{F}$ such that $\nu(\inf_{i\in K}a_i)>0$ for every finite $K\subseteq J$.

1991 *Mathematics Subject Classification.* Primary 28E15, secondary 03E02.

Key words and phrases. arrow ultrafilter, dependently selective filter, Hausdorff ultrafilter, measure-centering, measure-linking, nowhere dense ultrafilter, Ramsey ultrafilter, reduced product of probability algebras, ultrafilter, weakly Ramsey.

1B Principal ultrafilters are obviously measure-centering. There do not have to be any others (see Theorem 1Mj below). Subject to appropriate special axioms (in particular, the continuum hypothesis), we have a variety of types of measure-centering ultrafilter, which it is the purpose of this note to examine. To begin with, we have the following.

Theorem (a)(see [17]) A Ramsey ultrafilter is measure-centering.

(b) If κ is an infinite cardinal and $\operatorname{cov}\mathcal{N}_\kappa = 2^\kappa$, then there is a uniform measure-centering ultrafilter on κ.

(c) If $\operatorname{cov}\mathcal{N}_{\text{Leb}} = \mathfrak{c}$, there is a measure-centering ultrafilter on $\mathbb{N}$ which contains no set of zero asymptotic density.

(For the proof, see 3E-3G.) As you see, we are going to need rather a lot of definitions. Most of them are to be found in [14], but it will I expect help if I repeat some here. In particular:

Definitions (α)([7]) If I is an infinite set, an ultrafilter $\mathcal{F}$ on I is **Ramsey** (or 'selective') if it is uniform and for every $S \subseteq [I]^2$ there is a $J \in \mathcal{F}$ such that either $[J]^2 \subseteq S$ or $[J]^2 \cap S \neq \emptyset$.

(β) If (X, Σ, μ) is a measure space, set $\mu^* A = \inf\{\mu E : E \in \Sigma, E \supseteq A\}$ for every $A \subseteq X$. The **null ideal** of μ is $\mathcal{N} = \{A : \mu^* A = 0\}$. $\mathcal{N}_{\text{Leb}}$ will be the null ideal of Lebesgue measure on $[0, 1]$.

(γ) For any set I, I will write ν_I for the usual probability measure on $\{0,1\}^I$, the completed product measure if each copy of $\{0,1\}$ is given the uniform probability in which each point has measure $\frac{1}{2}$; $\mathcal{N}_I$ will be its null ideal.

(δ) If X is a set and $\mathcal{I}$ is an ideal of subsets of X such that $X = \bigcup \mathcal{I}$, then its **covering number** $\operatorname{cov}\mathcal{I}$ will be the least cardinal of any set $\mathcal{A} \subseteq \mathcal{I}$ such that $X = \bigcup \mathcal{A}$.

(ϵ) If $A \subseteq \mathbb{N}$ then the **upper asymptotic density** of A is $d^*(A) = \limsup_{n\to\infty} \frac{1}{n}\#(A \cap n)$. A has **zero asymptotic density** if $d^*(A) = 0$.

Recall that the continuum hypothesis is sufficient to ensure that there are Ramsey ultrafilters on $\mathbb{N}$; in fact it is sufficient to suppose that $\operatorname{cov}\mathcal{M} = \mathfrak{c}$, where $\mathcal{M}$ is the ideal of meager subsets of $\mathbb{R}$ ([14], 538Fg). For an uncountable cardinal κ, there is a Ramsey ultrafilter on κ iff κ is two-valued-measurable, and in this case an ultrafilter on κ is Ramsey iff it is isomorphic to a normal ultrafilter ([7], 9.6). There appears to be no bar to the number of such cardinals κ, but of course they must all be enormous. Note that if there is a Ramsey ultrafilter $\mathcal{F}$ on κ, then κ is regular and $\mathcal{F}$ is κ-complete.

To get a notion of the scope of (b) in this theorem, note that $\operatorname{cov}\mathcal{N}_\kappa \leq \operatorname{cov}\mathcal{N}_{\text{Leb}} \leq \mathfrak{c}$ for every infinite cardinal κ ([14], 523F), with equalities if Martin's axiom is true ([14], 524Na or [9], 32C); moreover, Martin's axiom implies that $2^\kappa = \mathfrak{c}$ whenever $\omega \leq \kappa < \mathfrak{c}$ ([14], 517Rb or [9], 21C). So we see that we can have many cardinals less than $\mathfrak{c}$ with uniform measure-centering ultrafilters.

Martin's axiom is sufficient to ensure that there are Ramsey ultrafilters on $\mathbb{N}$ (because it implies that $\operatorname{cov}\mathcal{M} = \mathfrak{c}$, or otherwise). But there is another important context in which (b) can be applied in the absence of any Ramsey ultrafilters at all. If we start with a model of ZFC and an uncountable regular cardinal λ such that $2^\kappa \le \lambda$ for every $\kappa < \lambda$, and add λ random reals, then in the resulting forcing language we shall have

$$\Vdash \operatorname{cov} \mathcal{N}_\kappa = 2^\kappa = \mathfrak{c} \text{ for every infinite } \kappa < \mathfrak{c},$$
$$\text{but there are no Ramsey ultrafilters on } \mathbb{N}$$

([14], 552B, 552G and 553H). The position is similar in any model in which $\mathfrak{c}$ is real-valued-measurable ([10], 5E, 6B and 5G).

1C In 1A I gave a definition of 'measure-centering' ultrafilter in a context well removed from the ordinary concerns of elementary measure theory. The original conception derived, as you would expect, from ideas closer to home, looking at filters on $\mathbb{N}$ and Lebesgue measure. To relate Lebesgue measure to the next result, recall that the usual measure on $\{0,1\}^{\mathbb{N}}$ is isomorphic to Lebesgue measure on $[0,1]$ ([11], 254K).

Proposition Let I be a set, and $\mathcal{F}$ an ultrafilter on I. Then the following are equiveridical, that is, if one is true so are the others:

(i) $\mathcal{F}$ is measure-centering;

(ii) whenever $\langle a_i \rangle_{i \in I}$ is a family in $\mathfrak{B}_I$ such that $\inf_{i \in I} \bar{\nu}_I a_i > 0$, there is an $A \in \mathcal{F}$ such that $\{a_i : i \in A\}$ is centered in $\mathfrak{B}_I$;

(iii) whenever $\langle E_i \rangle_{i \in I}$ is a family of measurable subsets of $\{0,1\}^I$ such that $\inf_{i \in I} \nu_I E_i > 0$, there is an $A \in \mathcal{F}$ such that $\bigcap_{i \in A} E_i \neq \emptyset$;

(iv) whenever (X, Σ, μ) is a compact probability space and $\langle E_i \rangle_{i \in I}$ is a family in Σ, then $\mu^*(\lim_{i \to \mathcal{F}} E_i) \ge \lim_{i \to \mathcal{F}} \mu E_i$.

(For the proof, see 3C.) Of course this calls for some more definitions from modern abstract measure theory.

Definitions (α) If $\mathfrak{A}$ is a Boolean algebra, a family $\langle a_i \rangle_{i \in I}$ in $\mathfrak{A}$ is **centered** if $\inf_{i \in K} a_i \neq 0$ for every finite $K \subseteq I$.

(β) If (X, Σ, μ) is a measure space, and $\mathcal{N}$ the null ideal of μ, the **measure algebra** of μ is the quotient Boolean algebra $\mathfrak{A} = \Sigma / \Sigma \cap \mathcal{N}$ together with the functional $\bar{\mu} : \mathfrak{A} \to [0, \infty]$ defined by setting $\bar{\mu} E^\bullet = \mu E$ for every $E \in \Sigma$. For any cardinal κ, I will write $(\mathfrak{B}_I, \bar{\nu}_I)$ for the measure algebra of ν_I, the usual measure on $\{0,1\}^I$.

(γ) A family $\mathcal{K}$ of sets is a **compact class** if $\bigcap \mathcal{L}$ is non-empty whenever $\mathcal{L} \subseteq \mathcal{K}$ has the finite intersection property, that is, $\bigcap \mathcal{L}' \neq \emptyset$ for every finite $\mathcal{L}' \subseteq \mathcal{L}$. If (X, Σ, μ) is a measure space, μ is **inner regular** with respect to a family $\mathcal{K}$ of sets if whenever $E \in \Sigma$ and $0 \le \gamma < \mu E$ there is a $K \in \mathcal{K} \cap \Sigma$ such that $K \subseteq E$ and $\mu K \ge \gamma$. A measure space (X, Σ, μ) is **compact** if μ is inner regular with respect to some compact class of sets.

(δ) If $\langle E_i \rangle_{i \in I}$ is a family of sets and $\mathcal{F}$ is an ultrafilter on I, I write $\lim_{i \to \mathcal{F}} E_i$ for

$$\bigcup_{A \in \mathcal{F}} \bigcap_{i \in A} E_i = \bigcap_{A \in \mathcal{F}} \bigcup_{i \in A} E_i = \{x : \{i : x \in E_i\} \in \mathcal{F}\},$$

the limit of $\langle E_i \rangle_{i \in I}$ along $\mathcal{F}$ in $\mathcal{P}X$ if X is any set including $\bigcup_{i \in I} E_i$ and $\mathcal{P}X \cong \{0,1\}^X$ is given its usual compact Hausdorff topology.

For basic results on compact measure spaces, see [12], §342. I remark here that a family $\mathcal{K}$ of subsets of a set X is a compact class iff there is a compact (not necessarily Hausdorff) topology on X such that every member of $\mathcal{K}$ is closed ([12], 342D); thus all Radon measures, and in particular Lebesgue measure, are compact measures in this sense.

1D It is natural to seek to explore the relationship of the class of measure-centering ultrafilters with the ordinary operations of the theory of ultrafilters. At an elementary level, we have the following.

Proposition (a) Let I and J be sets, $f : I \to J$ a function, and $\mathcal{F}$ a measure-centering ultrafilter on I. Then the image ultrafilter $f[[\mathcal{F}]]$ is a measure-centering ultrafilter on J.

(b)([3]) If $\mathcal{F}$ is a non-principal ultrafilter, then $\mathcal{F} \ltimes \mathcal{F}$ is not measure-centering.

(Proof in 3H.) Some relevant definitions are as follows.

Definitions (α) If I and J are sets, $f : I \to J$ is a function and $\mathcal{F}$ is a filter on I, then the **image filter** $f[[\mathcal{F}]]$ is $\{B : B \subseteq J,\ f^{-1}[B] \in \mathcal{F}\}$, that is, the filter on J generated by $\{f[A] : A \in \mathcal{F}\}$.

(β) If $\mathcal{F}$, $\mathcal{G}$ are filters on sets I, J respectively, then I write $\mathcal{F} \ltimes \mathcal{G}$ for the filter

$$\{A : A \subseteq I \times J,\ \{i : i \in I,\ A[\{i\}] \in \mathcal{G}\} \in \mathcal{F}\};$$

here $A[\{i\}] = \{j : (i,j) \in A\}$.

1E Extension of measures The original impulse to study measure-centering ultrafilters arose because they give an interesting expression of an ultrapower construction which I will describe shortly. The first result is a theorem on extension of probability measures.

Theorem Let (X, Σ, μ) be a compact probability space, and $\mathcal{F}$ a measure-centering ultrafilter on a set I. Let $\mathcal{A}$ be the family of all sets of the form $\lim_{i\to\mathcal{F}} E_i$ where $\langle E_i\rangle_{i\in I}$ is a family in Σ. Then there is a unique complete probability measure λ on X such that λ is inner regular with respect to $\mathcal{A}$ and $\lambda(\lim_{i\to\mathcal{F}} E_i) = \lim_{i\to\mathcal{F}} \mu E_i$ for every family $\langle E_i\rangle_{i\in I}$ in Σ.

(Proof in 3I. I ought perhaps to note that a measure λ is **complete** if λA is defined whenever $\lambda^* A = 0$; thus Lebesgue measure is complete.) Note that in the context of this theorem, λ must extend μ, because we can apply the defining formula to constant families $\langle E_i\rangle_{i\in I}$.

1F Reduced products of probability algebras We now need an abstract construction from the theory of measure algebras.

(a) First, let me define measure algebras in the abstract, as opposed to those constructed from measure spaces as in Definition 1C above. A **measure algebra** is a pair $(\mathfrak{A}, \bar\mu)$ where

$\mathfrak{A}$ is a Boolean algebra,

$\mathfrak{A}$ is **Dedekind σ-complete**, that is, every countable subset of $\mathfrak{A}$ has a least upper bound in $\mathfrak{A}$,

$\bar\mu : \mathfrak{A} \to [0,\infty]$ is **countably additive**, that is, $\bar\mu 0 = 0$ and $\bar\mu(\sup_{n\in\mathbb{N}} a_n)$ $= \sum_{n=0}^{\infty} \bar\mu a_n$ whenever $\langle a_n\rangle_{n\in\mathbb{N}}$ is a disjoint sequence in $\mathfrak{A}$.

It is straightforward to check that the measure algebras of Definition 1Cb are measure algebras in this sense. Conversely, any measure algebra as defined here is isomorphic to the measure algebra of some measure space ([12], 321J). A **probability algebra** is a measure algebra $(\mathfrak{A}, \bar{\mu})$ such that $\bar{\mu}1 = 1$, that is, $(\mathfrak{A}, \bar{\mu})$ is isomorphic to the measure algebra of a probability space.

(b) Let $\langle(\mathfrak{A}_i, \bar{\mu}_i)\rangle_{i \in I}$ be a family of probability algebras and $\mathcal{F}$ an ultrafilter on I. Write $\mathfrak{B}$ for the product Boolean algebra $\prod_{i \in I} \mathfrak{A}_i$, so that if $\boldsymbol{a} = \langle a_i \rangle_{i \in I}$ and $\boldsymbol{b} = \langle b_i \rangle_{i \in I}$ belong to $\mathfrak{B}$, then $\boldsymbol{a} * \boldsymbol{b} = \langle a_i * b_i \rangle_{i \in I}$ for all the Boolean operations $* = \triangle$, $\cap$, $\cup$ and $\setminus$. Define $\nu : \mathfrak{B} \to [0, 1]$ by setting $\nu(\langle a_i \rangle_{i \in I}) = \lim_{i \to \mathcal{F}} \bar{\mu} a_i$ whenever $\langle a_i \rangle_{i \in I} \in \mathfrak{B}$. Then ν is additive. Set $\mathcal{I} = \{\boldsymbol{a} : \boldsymbol{a} \in \mathfrak{B},\ \nu \boldsymbol{a} = 0\}$; then $\mathcal{I} \lhd \mathfrak{B}$. Let $\mathfrak{C}$ be the quotient Boolean algebra $\mathfrak{B}/\mathcal{I}$. Then we have a functional $\bar{\nu} : \mathfrak{C} \to [0, 1]$ defined by saying that $\bar{\nu}(\boldsymbol{a}^\bullet) = \nu \boldsymbol{a}$ for every $\boldsymbol{a} \in \mathfrak{B}$; and it turns out that $(\mathfrak{C}, \bar{\nu})$ is a probability algebra. I will call it the **reduced product** $\prod_{i \in I}(\mathfrak{A}_i, \bar{\mu}_i)|\mathcal{F}$. (See 4A below.)

1G Note that the construction in 1Fb does not depend on any property of the ultrafilter $\mathcal{F}$. When $\mathcal{F}$ is a measure-centering ultrafilter, however, we have the following result.

> **Theorem** Let (X, Σ, μ) be a compact probability space and $(\mathfrak{A}, \bar{\mu})$ its measure algebra. Let I be a set and $\mathcal{F}$ a measure-centering ultrafilter on I; write λ for the corresponding extension of μ as described in Theorem 1E, and $(\mathfrak{C}, \bar{\nu})$ for the reduced power $(\mathfrak{A}, \bar{\mu})^I|\mathcal{F}$ as described in 1Fb. Then we have a natural isomorphism between $(\mathfrak{C}, \bar{\nu})$ and the measure algebra $(\mathfrak{D}, \bar{\lambda})$ of λ defined by saying that $\langle E_i^\bullet \rangle_{i \in I}^\bullet \in \mathfrak{C}$ is matched with $(\lim_{i \to \mathcal{F}} E_i)^\bullet \in \mathfrak{D}$ for every family $\langle E_i \rangle_{i \in I}$ in Σ.

(Proof in 4C.)

1H Products of filters As will I hope become clear when we come to the proofs in §§3-4, all the results so far are more or less elementary, though some of them, naturally enough, demand graduate-level measure theory – in particular, Maharam's theorem and the lifting theorem – for their full strength. I want now to explain an astonishing theorem from [3]. This will depend on a construction of iterated products of filters, which may be of independent interest. I look at finite products of filters first.

> **Definition** For $n \in \mathbb{N}$ and filters $\mathcal{F}_0, \dots, \mathcal{F}_n$, define the product $\mathcal{F}_0 \ltimes \dots \ltimes \mathcal{F}_n$ inductively by saying that it is $\mathcal{F}_0$ when $n = 0$ and $(\mathcal{F}_0 \ltimes \dots \ltimes \mathcal{F}_{n-1}) \ltimes \mathcal{F}_n$ when $n \geq 1$.

> **Proposition** If $0 \leq m < n$ and $\mathcal{F}_0, \dots, \mathcal{F}_n$ are filters on $I_0, \dots, I_n$ respectively, then the natural bijection between $((\dots(I_0 \times I_1) \times \dots) \times I_m) \times ((\dots(I_{m+1} \times I_{m+2}) \times \dots) \times I_n)$ and $((\dots(I_0 \times I_1) \times \dots) \times I_n)$ identifies $\mathcal{F}_0 \ltimes \dots \ltimes \mathcal{F}_n$ with $(\mathcal{F}_0 \ltimes \dots \ltimes \mathcal{F}_m) \ltimes (\mathcal{F}_{m+1} \ltimes \dots \ltimes \mathcal{F}_n)$.

(The proof is a simple induction on n.)

1I Iterated products of filters The next bit works best for filters on $\mathbb{N}$ and countable iterations, but something can be done in a more general context.

(a) First, a scrap of notation. Let I be a set. If $m,n\in\mathbb{N}$, $\sigma\in I^m$ and $\tau\in I^n$, define the concatenation $\sigma^\frown\tau\in I^{m+n}$ by setting

$$\begin{aligned}(\sigma^\frown\tau)(k)&=\sigma(k)\text{ if }k<m,\\&=\tau(k-m)\text{ if }m\le k<m+n.\end{aligned}$$

For $i\in I$ write $<i>$ for the member of I^1 with value i.

(b) Now suppose that $\zeta>0$ is an ordinal, $\langle I_\xi\rangle_{1\le\xi\le\zeta}$ a family of sets, and $\mathcal{F}_\xi$ a filter on I_ξ for $1\le\xi\le\zeta$. Set $I=\bigcup_{1\le\xi\le\zeta}I_\xi$ and $S^*=\bigcup_{i\in\mathbb{N}}I^i$. Fix a function θ such that $\theta(\xi,i)<\xi$ for $1\le\xi\le\zeta$ and $i\in I_\xi$. For $\xi\le\zeta$, define $\mathcal{G}_\xi\subseteq\mathcal{P}S^*$ inductively, as follows. Start by taking $\mathcal{G}_0$ to be the principal filter generated by $\{\emptyset\}$. For $1\le\xi\le\zeta$, given that $\mathcal{G}_\eta$ has been defined for every $\eta<\xi$, set

$$\mathcal{G}_\xi=\{A:A\subseteq S^*,\ \{i:i\in I_\xi,\ \{\tau:<i>^\frown\tau\in A\}\in\mathcal{G}_{\theta(\xi,i)}\}\in\mathcal{F}_\xi\}.$$

It is elementary to check that every $\mathcal{G}_\xi$ is a filter. Moreover, if every $\mathcal{F}_\xi$ is an ultrafilter, so is every $\mathcal{G}_\xi$.

1J We are now ready for the statements of the main theorems.

Theorem In the construction of 1Ib above, suppose that ζ is countable, $I_\xi=I$ whenever $1\le\xi\le\zeta$, $\langle\mathcal{F}_\xi\rangle_{1\le\xi\le\zeta}$ is a family of Ramsey ultrafilters on I, no two isomorphic, and $\{i:i\in I,\ \theta(\xi,i)\ge\eta\}\in\mathcal{F}_\xi$ whenever $\eta<\xi\le\zeta$. Then $\mathcal{G}_\zeta$ is measure-centering.

(Proof in 5B. You will lose very little by restricting yourself to the case in which every sequence $\langle\theta(\xi,i)\rangle_{i\in\mathbb{N}}$ is non-decreasing, and is constant with value η when $\xi=\eta+1$ is a successor ordinal.) The point of this theorem is that there will be for each $\xi\in[1,\zeta]$ a function $f:S^*\to I$ such that $f[[\mathcal{G}_\zeta]]=\mathcal{F}_\xi$; starting from the family $\langle\mathcal{F}_\xi\rangle_{1\le\xi\le\zeta}$ of Ramsey ultrafilters on I, we can find a single measure-centering ultrafilter on S^* from which they can all be derived. I ought to point out straight away that if ζ is infinite, then the Ramsey ultrafilter $\mathcal{F}_\omega$ contains all the sets $\{i:\theta(\omega,i)\ge n\}$, for $n<\omega$, but not their intersection, so is not ω_1-additive. In this case, of course, $I=I_\omega$ cannot be uncountable. Thus we have either a finite iteration in which $\mathcal{G}_\zeta$ is the extension to $\bigcup_{n\in\mathbb{N}}I^n$ of the filter $\mathcal{F}_m\ltimes\mathcal{F}_{m-1}\ltimes\ldots\ltimes\mathcal{F}_1$ on I^m, or a countably infinite iteration in which I can be identified with $\mathbb{N}$. The finite-iteration case is in fact the hard part of a more general result: the skew product of finitely many non-isomorphic Ramsey ultrafilters is always measure-centering (Proposition 5E).

1K The second theorem is a universal extension theorem for Ramsey ultrafilters on $\mathbb{N}$.

Theorem Let (X,Σ,μ) be a compact probability space. Then there is a measure λ on X, extending μ, such that $\lambda(\lim_{i\to\mathcal{F}}E_i)$ is defined and equal to $\lim_{i\to\mathcal{F}}\mu E_i$ whenever $\mathcal{F}$ is a Ramsey ultrafilter on $\mathbb{N}$ and $\langle E_i\rangle_{i\in\mathbb{N}}$ is a sequence in Σ.

(Proof in 5J.) The idea here will be that for each countable family $\mathfrak{F}$ of Ramsey ultrafilters on $\mathbb{N}$, we can define a measure-centering ultrafilter $\mathcal{G}_{\mathfrak{F}}$ on S^* dominating every member of $\mathfrak{F}$, and that this can be done in such a way that the measures defined from the $\mathcal{G}_{\mathfrak{F}}$ by the process of Theorem 1E will have a common extension.

1L Perfect measure spaces Readers familiar with [3] and [4] may have noted that I speak of 'compact' measures where Benedikt deals with 'perfect' measures. The latter form a larger class, so it is not obvious that the results in this note really cover Benedikt's. The point is that a probability space (X, Σ, μ) is perfect iff $(X, \mathrm{T}, \mu\restriction\mathrm{T})$ is compact for every countably generated σ-subalgebra T of Σ ([23], or [13], 451F). Using this, it is easy to check that we have a variant on condition (iv) of Proposition 1C for filters on $\mathbb{N}$:

if $\mathcal{F}$ is an ultrafilter on $\mathbb{N}$, it is measure-centering iff whenever (X, Σ, μ) is a perfect probability space and $\langle E_n\rangle_{n\in\mathbb{N}}$ is a sequence in Σ, then $\mu^*(\lim_{n\to\mathcal{F}} E_n) \geq \lim_{n\to\mathcal{F}} \mu E_n$.

Versions of Theorems 1E and 1G for perfect probability spaces and ultrafilters on $\mathbb{N}$ are now easy to deduce. With a little more trouble – it is probably easiest to check that the proof in 5J applies essentially unchanged – we can confirm that Theorem 1K is true for all perfect probability spaces (X, Σ, μ).

1M Of course there are many classes of ultrafilters, associated with those considered above, which have been studied over the years. Six of them are the following.

Definitions Let $\mathcal{F}$ be an ultrafilter on a set I.

(α)([8]) An ultrafilter $\mathcal{F}$ on a set I is **Hausdorff** (or has 'property C') if whenever J is a set and $f : I \to J$, $g : I \to J$ are functions such that $\{i : f(i) \neq g(i)\} \in \mathcal{F}$, then $f[[\mathcal{F}]] \neq g[[\mathcal{F}]]$.

(β) $\mathcal{F}$ is **nowhere dense** if for every function $f : I \to \mathbb{R}$ the image filter $f[[\mathcal{F}]]$ contains a nowhere dense subset of $\mathbb{R}$.

(γ)([5]) $\mathcal{F}$ is **weakly Ramsey** if whenever S_0, S_1, S_2 are disjoint subsets of $[I]^2$ there is a $J \in \mathcal{F}$ such that $[J]^2$ is disjoint from at least one of S_0, S_1, S_2.

(δ)([2]) $\mathcal{F}$ is an **arrow ultrafilter** if whenever $S \subseteq [I]^2$ and $k \in \mathbb{N}$ then either there is a $K \in [I]^k$ such that $[K]^2 \cap S = \emptyset$ or there is a $J \in \mathcal{F}$ such that $[J]^2 \subseteq S$.

(ϵ)([4]) $\mathcal{F}$ is **measure-linking** if whenever $\mathfrak{A}$ is a Boolean algebra, $\nu : \mathfrak{A} \to [0,1]$ is an additive functional such that $\nu(1_{\mathfrak{A}}) = 1$, and $\langle a_i\rangle_{i\in I}$ is a family in $\mathfrak{A}$ such that $\inf_{i\in I} \nu a_i > 0$, then there is a $J \in \mathcal{F}$ such that $\nu(a_i \cap a_j) > 0$ for all $i, j \in J$.

(ζ)([1]) $\mathcal{F}$ is **closed Lebesgue null** if for every function $f : I \to [0,1]$ the image filter $f[[\mathcal{F}]]$ contains a closed Lebesgue negligible set.

An obvious strengthening of (ϵ) is

(η) $\mathcal{F}$ is **strongly measure-linking** if whenever $\mathfrak{A}$ is a Boolean algebra, $\nu : \mathfrak{A} \to [0,1]$ is an additive functional such that $\nu(1_{\mathfrak{A}}) = 1$, and $\langle a_i\rangle_{i\in I}$ is a family in $\mathfrak{A}$ such that $\inf_{i\in I} \nu a_i > 0$, then there is a $J \in \mathcal{F}$ such that $\inf_{i,j\in J} \nu(a_i \cap a_j) > 0$.

(I ought to remark that in all of [8], [5], [2], [1] and [4] only non-principal ultrafilters on countable sets are considered; that what I call 'measure-linking' is what [4] calls 'property M_2'; and that what I call 'closed Lebesgue null' is what [1] calls 'measure zero'.) Evidently Ramsey ultrafilters are weakly Ramsey, measure-centering ultrafilters are measure-linking, strongly measure-linking ultrafilters are measure-linking, and closed Lebesgue null filters are nowhere dense; it is also the case that p-point ultrafilters on $\mathbb{N}$ are closed Lebesgue null ([1]). Subject to the continuum hypothesis, there are non-principal weakly Ramsey ultrafilters on $\mathbb{N}$ which are not Ramsey ([5]). The results I wish to present here are the following:

Theorem (a)([3]) A measure-linking ultrafilter is Hausdorff.

(b)([24]) A measure-centering ultrafilter is nowhere dense.

(c)(see [2], Corollary 2.5) A weakly Ramsey ultrafilter is an arrow ultrafilter.

(d)(i)(see [4], p. 214, Proposition 3) An arrow ultrafilter is strongly measure-linking.

(ii) An arrow ultrafilter on $\mathbb{N}$ is nowhere dense.

(e) A strongly measure-linking ultrafilter on $\mathbb{N}$ contains a set of zero asymptotic density.

(f) A closed Lebesgue null ultrafilter on $\mathbb{N}$ contains a set of zero asymptotic density.

(g) If $\operatorname{cov}\mathcal{N}_{\text{Leb}} = \mathfrak{c}$, there is a measure-centering ultrafilter on $\mathbb{N}$ which is neither strongly measure-linking nor closed Lebesgue null.

(h) If $\mathfrak{c} = \omega_1$, there is a strongly measure-linking ultrafilter on $\mathbb{N}$ which is not nowhere dense, so is neither measure-centering nor an arrow ultrafilter.

(i) If $\mathfrak{p} = \mathfrak{c}$, there is a Hausdorff p-point ultrafilter which is not measure-centering.

(j)(see [24]) It is relatively consistent with ZFC to suppose that every measure-centering ultrafilter is a principal ultrafilter.

(Proof in 6A, 6C, 6E, 6H, 6L and 6O. Recall that $\mathfrak{p}$ is the least cardinal of any family $\mathcal{A}$ of infinite subsets of $\mathbb{N}$ such that $\bigcap \mathcal{A}_0$ is infinite for any finite $\mathcal{A}_0 \subseteq \mathcal{A}$, but there is no infinite $B \subseteq \mathbb{N}$ such that $B \setminus A$ is finite for every $A \in \mathcal{A}$.)

2 Dependently selective filters

A particularly important property of Ramsey ultrafilters is preserved under certain intersections of such ultrafilters, and it is in this form that it will be used in §5. I therefore isolate it in the next definition. The results which will be needed in the proofs of Theorems 1Ba, 1J and 1K are special cases of Propositions 2D and 2E, but I think it is worth while expressing the intermediate lemmas 2B and 2C in their full natural strength.

2A Definition Let $\mathcal{F}$ be a filter on a set I. I will say that $\mathcal{F}$ is **dependently selective** if it has the following property:

whenever $\mathcal{S} \subseteq [I]^{<\omega}$ is such that $\emptyset \in \mathcal{S}$ and $\{i : K \cup \{i\} \in \mathcal{S}\} \in \mathcal{F}$ for every $K \in \mathcal{S}$, then there is a $J \in \mathcal{F}$ such that $[J]^{<\omega} \subseteq \mathcal{S}$.

In the present paper I will give only those results which are necessary for the applications in §5; for a fuller account of this class of filters, see my note [15].

2B Lemma A uniform dependently selective ultrafilter is a Ramsey ultrafilter.

Proof. Let $\mathcal{F}$ be a uniform dependently selective ultrafilter on a set I. It will be enough to consider the case in which $I = \kappa$ is a cardinal. Take any $S \subseteq [\kappa]^2$. For $\xi < \kappa$ set $A_\xi = \{\eta : \{\xi, \eta\} \in S\}$. Let $\mathcal{S}$ be the family of finite subsets K of κ such that for all ξ, $\eta \in K$ such that $\xi < \eta$, $\{\xi, \eta\} \in S$ iff $A_\xi \in \mathcal{F}$. If $K \in \mathcal{S}$, then (because $\mathcal{F}$ is an ultrafilter) there is a $J \in \mathcal{F}$ such that, for every $\xi \in K$, J is either included in A_ξ or disjoint from A_ξ. Now $K \cup \{\eta\} \in \mathcal{S}$ whenever $\eta \in J$ and $\eta > \xi$ for every $\xi \in K$. So $\mathcal{S}$ satisfies the condition of 2A. Let $J \in \mathcal{F}$ be such that $[J]^{<\omega} \subseteq \mathcal{S}$. In this case, if ξ, $\eta \in J$ and $\xi < \eta$, $\{\xi, \eta\} \in S$ iff $A_\xi \in \mathcal{F}$. Now

$$J_1 = \{\xi : \xi \in J,\, A_\xi \in \mathcal{F}\}, \quad J_0 = \{\xi : \xi \in J,\, A_\xi \notin \mathcal{F}\}$$

have union J and one of them must belong to $\mathcal{F}$; while $[J_0]^2 \cap S = \emptyset$ and $[J_1]^2 \subseteq S$. As S is arbitrary, $\mathcal{F}$ is a Ramsey ultrafilter.

2C Lemma (a) Let I be an infinite set, $\mathcal{F}$ a Ramsey ultrafilter on I, and $\mathcal{A} \subseteq \mathcal{F}$ a set of size at most $\#(I)$. Then there is a $C \in \mathcal{F}$ such that $\#(C \setminus A) < \#(I)$ for every $A \in \mathcal{A}$.

(b) Let κ be an infinite cardinal, $\lambda \leq \kappa$ another cardinal, and $\langle \mathcal{F}_\alpha \rangle_{\alpha<\lambda}$ a family of distinct Ramsey ultrafilters on κ. Then there is a disjoint family $\langle A_\alpha \rangle_{\alpha<\lambda}$ of subsets of κ such that $A_\alpha \in \mathcal{F}_\alpha$ for every $\alpha < \lambda$.

PROOF. **(a)** Set $A^* = \kappa \cap \bigcap \mathcal{A}$. If $A^* \in \mathcal{F}$, we can set $C = A^*$ and stop. Otherwise, enumerate $\mathcal{A}$ as $\langle A_\alpha \rangle_{\alpha<\lambda}$, where $\lambda \leq \kappa$. For $i \in I$, set $f(i) = \min\{\alpha : \alpha < \lambda,\, i \notin A_\alpha \setminus A^*\}$. Then there is a $C \in \mathcal{F}$ such that $f \restriction C$ is either constant or injective ([7], 9.6). The former is impossible, because $\{i : f(i) = \alpha\} \subseteq A^* \cup (\kappa \setminus A_\alpha)$ never belongs to $\mathcal{F}$. So $f \restriction C$ is injective and $C \setminus A_\alpha \subseteq \{i : i \in C,\, f(i) \leq \alpha\}$ has cardinal less than κ for every $\alpha < \lambda$.

(b) For $\alpha < \beta < \lambda$, take $A_{\alpha\beta} \in \mathcal{F}_\beta \setminus \mathcal{F}_\alpha$. For each $\alpha < \kappa$, there is a $B_\alpha \in \mathcal{F}_\alpha$ such that $\#(B_\alpha \cap A_{\alpha\beta}) < \kappa$ for every $\beta > \alpha$ (apply (a) to $\{I \setminus A_{\alpha\beta} : \alpha < \beta < \lambda\} \subseteq \mathcal{F}_\alpha$)). Set

$$A_\beta = B_\beta \setminus \textstyle\bigcup_{\alpha<\beta} B_\alpha$$

for $\beta < \lambda$. Of course $\langle A_\beta \rangle_{\beta<\lambda}$ is disjoint. On the other hand, for each $\beta < \lambda$, $A'_\beta = B_\beta \cap \bigcap_{\alpha<\beta} A_{\alpha\beta}$ belongs to $\mathcal{F}$ because $\mathcal{F}$ is κ-complete; and $A'_\beta \setminus A_\beta \subseteq \bigcup_{\alpha<\beta} A_{\alpha\beta} \cap B_\alpha$ has cardinal less than κ, so A_β also belongs to $\mathcal{F}$.

2D Proposition Let I be an infinite set, and $\mathfrak{F}$ a non-empty family of non-isomorphic Ramsey ultrafilters on I with $\#(\mathfrak{F}) \leq \#(I)$. Then $\mathcal{H} = \bigcap \mathfrak{F}$ is a dependently selective filter on I.

PROOF. **(a)** It is enough to consider the case in which $I = \kappa$ is a cardinal. Let $\langle \mathcal{F}_\alpha \rangle_{\alpha<\lambda}$ be an enumeration of $\mathfrak{F}$.

(b) If $\langle A_\alpha \rangle_{\alpha<\lambda}$ is such that $A_\alpha \in \mathcal{F}_\alpha$ for $\alpha < \lambda$, then there is a family $\langle D_\alpha \rangle_{\alpha<\lambda}$ such that $D_\alpha \in \mathcal{F}_\alpha$ and $D_\alpha \subseteq A_\alpha$ for every $\alpha < \lambda$, and whenever $\xi < \eta < \kappa$, α, $\beta < \lambda$ are such that $\xi \in D_\alpha$ and $\eta \in D_\beta$, there is a $\zeta \in A_\beta$ such that $\xi \leq \zeta < \eta$.
P By 2Cb, we may suppose that $\langle A_\alpha \rangle_{\alpha<\lambda}$ is disjoint. For any $\zeta < \kappa$, $\{\alpha : \alpha < \lambda,\, A_\alpha \cap \zeta \neq \emptyset\}$ has cardinal less than κ; so there is a closed cofinal set $F \subseteq \kappa$, containing 0, such that $A_\alpha \cap \zeta' \setminus \zeta \neq \emptyset$ whenever $\zeta < \zeta'$ in F, $\alpha < \lambda$ and $A_\alpha \cap \zeta \neq \emptyset$. Set $f(\xi) = \max\{\zeta : \zeta \in F,\, \zeta \leq \xi\}$ for $\xi < \kappa$. Then $\langle f[[\mathcal{F}_\alpha]] \rangle_{\alpha<\lambda}$ is a family of κ-complete uniform ultrafilters on F, so there must be a cofinal set $V \subseteq F$ not belonging to any of them. (We can easily build inductively a family $\langle V_\xi \rangle_{\xi<\kappa^+}$ of cofinal subsets of F such that $\#(V_\xi \cap V_\eta) < \kappa$ whenever $\xi < \eta < \kappa^+$, and now each $f[[\mathcal{F}_\alpha]]$ can contain V_ξ for at most one ξ, so there is a ξ left over for which we can set $V = V_\xi$.) Set $M = f^{-1}[V]$; then $A_\alpha \setminus M \in \mathcal{F}_\alpha$ for each α.

Define $g:\kappa\to\kappa$ by setting $g(\xi)=\min\{\zeta:\xi\le\zeta\in V\}$ for $\xi<\kappa$. For each $\alpha<\lambda$, there is a $B_\alpha\in\mathcal{F}_\alpha$ on which g is injective ([7], 9.6), so that $g[[\mathcal{F}_\alpha]]$ is a Ramsey ultrafilter on κ isomorphic to $\mathcal{F}_\alpha$. Because the $\mathcal{F}_\alpha$ are non-isomorphic, all the $g[[\mathcal{F}_\alpha]]$ are different. By 2Cb again, there is a disjoint family $\langle G_\alpha\rangle_{\alpha<\lambda}$ of sets such that $G_\alpha\in g[[\mathcal{F}_\alpha]]$ for every α.

Set

$$C_\alpha=A_\alpha\cap B_\alpha\cap g^{-1}[G_\alpha]\setminus M,\qquad D_\alpha=C_\alpha\setminus\{\min C_\alpha\}\in\mathcal{F}_\alpha$$

for each $\alpha<\lambda$. Suppose that $\xi\in D_\alpha$, $\eta\in D_\beta$ and $\xi<\eta$. Then $g(\xi)<g(\eta)$. (If $\alpha=\beta$, this is because $g\restriction B_\alpha$ is injective; otherwise, it is because $G_\alpha\cap G_\beta$ is empty.) It follows that $g(\xi)<\eta$. Let η_0 be the least member of C_β. We have $\eta_0<\eta$. If $\xi\le\eta_0$, then η_0 is a member of $A_\beta\cap\eta\setminus\xi$. Otherwise, $A_\beta\cap g(\xi)\ne\emptyset$, so there is a $\zeta\in A_\beta\cap\gamma\setminus g(\xi)$, where γ is the next member of F above $g(\xi)$. Now $\gamma\setminus g(\xi)=f^{-1}[\{g(\xi)\}]\subseteq M$ is disjoint from D_β, so $\gamma\le\eta$ and $\zeta\in A_\beta\cap\eta\setminus\xi$.

Thus $\langle D_\alpha\rangle_{\alpha<\lambda}$ is a suitable family. **Q**[1]

(c) Now suppose that $\mathcal{S}$ is a family of finite subsets of κ such that $\emptyset\in\mathcal{S}$ and $\{\xi:K\cup\{\xi\}\in\mathcal{S}\}\in\mathcal{H}$ for every $K\in\mathcal{S}$. Set

$$S=\{\{\xi,\eta\}:\xi<\eta<\kappa,\ K\cup\{\eta\}\in\mathcal{S}\text{ whenever }K\in\mathcal{S}\text{ and }K\subseteq\xi+1\}.$$

For each $\alpha<\lambda$, there is an $A_\alpha\in\mathcal{F}_\alpha$ such that $[A_\alpha]^2$ is either included in or disjoint from S. But taking $\xi=\min A_\alpha$, we see that $\{\eta:\eta>\xi,\ K\cup\{\eta\}\in\mathcal{S}\}$ belongs to $\mathcal{H}\subseteq\mathcal{F}_\alpha$ for every $K\in\mathcal{S}$; because $\mathcal{F}_\alpha$ is κ-complete, there must be an $\eta\in A_\alpha$ such that $\eta>\xi$ and $K\cup\{\eta\}\in\mathcal{S}$ whenever $K\in\mathcal{S}$ and $K\subseteq\xi+1$, in which case $\{\xi,\eta\}\in S$. So we must have $[A_\alpha]^2\subseteq S$. Set $A'_\alpha=\{\xi:\xi\in A_\alpha,\ \{\xi\}\in\mathcal{S}\}$; then $A'_\alpha\in\mathcal{F}_\alpha$ because $\{\xi:\{\xi\}\in\mathcal{S}\}\in\mathcal{H}\subseteq\mathcal{F}_\alpha$.

By (b), we have a family $\langle D_\alpha\rangle_{\alpha<\lambda}$ of sets such that $D_\alpha\in\mathcal{F}_\alpha$ and $D_\alpha\subseteq A'_\alpha$ for every $\alpha<\lambda$, and whenever $\xi<\eta<\kappa$, α, $\beta<\lambda$ are such that $\xi\in D_\alpha$ and $\eta\in D_\beta$, there is a $\zeta\in A'_\beta$ such that $\xi\le\zeta<\eta$. Set $A=\bigcup_{\alpha<\lambda}D_\alpha\in\mathcal{H}$. Then $[A]^n\subseteq\mathcal{S}$ for every n. **P** Induce on n. The case $n=0$ is trivial, and the case $n=1$ has been dealt with when defining A'_α. For the inductive step to $n+1\ge2$, suppose that $I\in[A]^{n+1}$. Let $\xi<\eta$ be the two greatest points of I; suppose that $\eta\in D_\beta$. Then there is a $\zeta\in A'_\beta$ such that $\xi\le\zeta<\eta$. In this case, $K=I\setminus\{\eta\}$ belongs to $[A]^n\subseteq\mathcal{S}$ and $K\subseteq\zeta+1$. Also $\{\zeta,\eta\}\in[A_\beta]^2\subseteq S$, so $I=K\cup\{\eta\}\in\mathcal{S}$. Thus the induction continues. **Q**

So $[A]^{<\omega}\subseteq\mathcal{S}$. As $\mathcal{S}$ is arbitrary, $\mathcal{F}$ is dependently selective.

Remark In particular, every Ramsey ultrafilter is dependently selective. Compare the 'weak T-ideals' of [16], and also §4 of [6].

2E Proposition Let I be a set, and $\mathfrak{F}$ a non-empty countable family of non-isomorphic dependently selective ultrafilters on I. Then

(a) there is a disjoint family $\langle A_\mathcal{F}\rangle_{\mathcal{F}\in\mathfrak{F}}$ of sets such that $A_\mathcal{F}\in\mathcal{F}$ for every $\mathcal{F}\in\mathfrak{F}$,

(b) $\mathcal{H}=\bigcap\mathfrak{F}$ is dependently selective.

Proof. **(a)** For each $\mathcal{F}\in\mathfrak{F}$, let $I_\mathcal{F}\in\mathcal{F}$ be a set of minimal size. Let K be the countable set $\{\#(I_\mathcal{F}):\mathcal{F}\in\mathfrak{F}\}$; for $\kappa\in$ K, set $\mathfrak{F}_\kappa=\{\mathcal{F}:\mathcal{F}\in\mathfrak{F},\ \#(I_\mathcal{F})=\kappa\}$ and $J_\kappa=\bigcup_{\mathcal{F}\in\mathfrak{F}_\kappa}I_\mathcal{F}$, so that $\#(J_\kappa)=\kappa$. (For if $\kappa=1$, any member of $\mathfrak{F}_\kappa$ is a principal

[1]Each pair of symbols **P**, **Q** in this paper is used to enclose an argument justifying the statement immediately preceding the **P**.

ultrafilter, and there can be at most one such.) Set $J'_\kappa = J_\kappa \setminus \bigcup_{\lambda\in\mathrm{K},\lambda<\kappa} J_\lambda$ for $\kappa\in\mathrm{K}$; then $\langle J'_\kappa\rangle_{\kappa\in\mathrm{K}}$ is disjoint and $J'_\kappa\in\mathcal{F}$ whenever $\kappa\in\mathrm{K}$ and $\mathcal{F}\in\mathfrak{F}_\kappa$.

For $\mathcal{F}\in\mathfrak{F}$, let $\mathcal{F}'=\mathcal{F}\cap\mathcal{P}J'_\kappa$ be the trace of $\mathcal{F}$ on J'_κ, where $\kappa\in\mathrm{K}$ is such that $\mathcal{F}\in\mathfrak{F}_\kappa$. It is easy to check that $\mathcal{F}'$ is dependently selective, so is either principal (if $\kappa=1$) or a Ramsey ultrafilter (Lemma 2B). Moreover, $\mathcal{F}'$ and $\mathcal{G}'$ must be non-isomorphic whenever $\mathcal{F}$, $\mathcal{G}$ are distinct members of the same $\mathfrak{F}_\kappa$. So 2Cb tells us that we have for each $\kappa\in\mathrm{K}$ a disjoint family $\langle A_\mathcal{F}\rangle_{\mathcal{F}\in\mathfrak{F}_\kappa}$ of subsets of J'_κ such that $A_\mathcal{F}\in\mathcal{F}'$ for every $\mathcal{F}\in\mathfrak{F}_\kappa$, and 2D tells us that $\mathcal{H}_\kappa=\bigcap\{\mathcal{F}':\mathcal{F}\in\mathfrak{F}_\kappa\}$ is dependently selective for every $\kappa\in\mathrm{K}$. Assembling the families $\langle A_\mathcal{F}\rangle_{\mathcal{F}\in\mathfrak{F}_\kappa}$, we have a disjoint family $\langle A_\mathcal{F}\rangle_{\mathcal{F}\in\mathfrak{F}}$ such that $A_\mathcal{F}\in\mathcal{F}$ for every $\mathcal{F}\in\mathfrak{F}$.

(b) Evidently

$$\mathcal{H}=\{A:A\subseteq I,\ A\cap J'_\kappa\in\mathcal{H}_\kappa \text{ for every } \kappa\in\mathrm{K}\}.$$

Now suppose that $\mathcal{S}\subseteq[I]^{<\omega}$ is such that $\emptyset\in\mathcal{S}$ and $\{i:K\cup\{i\}\in\mathcal{S}\}\in\mathcal{H}$ for every $K\in\mathcal{S}$. Choose $\langle B_\kappa\rangle_{\kappa\in\mathrm{K}}$ inductively, as follows. Given that $\kappa\in\mathrm{K}$, that $B_\lambda\in\mathcal{H}_\lambda$ has been defined for $\lambda\in\mathrm{K}\cap\kappa$ and that $[\bigcup_{\lambda\in\mathrm{K}\cap\kappa}B_\lambda]^{<\omega}\subseteq\mathcal{S}$, note that $\#(\bigcup_{\lambda\in\mathrm{K}\cap\kappa}J'_\lambda)<\kappa$, because if $\kappa>\omega$ then κ is two-valued-measurable and certainly has uncountable cofinality. So $C_\kappa=\bigcup_{\lambda\in\mathrm{K}\cap\kappa}B_\lambda$ and $[C_\kappa]^{<\omega}$ have cardinal less than κ.

Set

$$\mathcal{S}_\kappa=\{K:K\in[J'_\kappa]^{<\omega},\ K\cup L\in\mathcal{S}\text{ for every }L\in[C_\kappa]^{<\omega}\}.$$

Then $\emptyset\in\mathcal{S}_\kappa$, by the hypothesis on C_κ. If $K\in\mathcal{S}_\kappa$, then for each $L\in[C_\kappa]^{<\omega}$ the set $C_L=\{i:i\in J'_\kappa,\ K\cup L\cup\{i\}\in\mathcal{S}\}$ belongs to $\mathcal{H}_\kappa$; but $\mathcal{H}_\kappa$, being an intersection of κ-complete filters, is again κ-complete, so $C=\bigcap\{C_L:L\in[C_\kappa]^{<\omega}\}\in\mathcal{H}_\kappa$, and $K\cup\{i\}\in\mathcal{S}_\kappa$ for every $i\in C$. As $\mathcal{H}_\kappa$ is dependently selective, there is a $B_\kappa\in\mathcal{H}_\kappa$ such that $[B_\kappa]^{<\omega}\subseteq\mathcal{S}_\kappa$ and $[B_\kappa\cup C_\kappa]^{<\omega}\subseteq\mathcal{S}$.

The inductive hypothesis

$$[\textstyle\bigcup_{\lambda\in\mathrm{K}\cap\kappa}B_\lambda]^{<\omega}\subseteq\mathcal{S}$$

gives no difficulty when $\kappa\in\mathrm{K}$ is a limit in K, so the induction proceeds to the end. Setting $A=\bigcup_{\kappa\in\mathrm{K}}B_\kappa$, we have $A\in\mathcal{H}$ and $[A]^{<\omega}\subseteq\mathcal{S}$. As $\mathcal{S}$ is arbitrary, $\mathcal{H}$ is dependently selective.

3 Proofs of Theorems 1B-1E

The main work of this paper begins with proofs of the relatively elementary results up to Theorem 1E. These will not be done in exactly the order in which they were presented in §1; in the hope of minimising confusion, I will restate the results as I come to prove them. I begin with a review of basic fragments of measure theory which will be used later.

3A Measure spaces: definitions and facts

(a)(i) Let (X,Σ,μ) be a measure space. Its **completion** is the measure space $(X,\hat\Sigma,\hat\mu)$, where $\hat\Sigma=\{E\triangle F:E\in\Sigma,\ F\text{ belongs to the null ideal of }\mu\}$ and $\hat\mu$ is the unique monotonic extension of μ to $\hat\Sigma$ ([11], 212C).

(ii) If a measure is inner regular with respect to a class $\mathcal{K}$ of sets, so is its completion ([13], 412H).

(iii) If X is a set and μ_1, μ_2 are two complete probability measures on X such that μ_1 is inner regular with respect to $\{K : K \in \operatorname{dom}\mu_1 \cap \operatorname{dom}\mu_2,\ \mu_1 K = \mu_2 K\}$, then μ_2 extends μ_1 ([13], 412K).

(iv) If X is a set and μ_1, μ_2 are two complete probability measures on X both inner regular with respect to $\{K : K \in \operatorname{dom}\mu_1 \cap \operatorname{dom}\mu_2,\ \mu_1 K = \mu_2 K\}$, they are equal ([13], 412L).

(b) Let X be a set, and Λ a family of probability measures on X such that (α) for all λ_0, $\lambda_1 \in \Lambda$ there is a $\lambda \in \Lambda$ which extends both λ_0 and λ_1 (β) for every countable $\Lambda_0 \subseteq \Lambda$ there is a probability measure on X (not necessarily belonging to Λ) extending every measure in Λ_0. Then there is a probability measure on X extending every measure in Λ. ([13], 457G.)

(c) A **Radon probability space** is a quadruple $(X, \mathfrak{T}, \Sigma, \mu)$ where $(X, \mathfrak{T})$ is a Hausdorff topological space, μ is a complete probability measure on X with domain Σ, $\mathfrak{T} \subseteq \Sigma$ (so that μ measures every Borel subset of X), and μ is inner regular with respect to the family of compact subsets of X. The usual measure ν_I on $\{0,1\}^I$ is always a Radon probability measure ([13], 416Ub).

(d) Suppose that Z is a zero-dimensional compact Hausdorff space and $\mathfrak{A}$ is the Boolean algebra of open-and-closed subsets of Z. If $\nu : \mathfrak{A} \to [0,1]$ is an additive functional such that $\nu Z = 1$, there is a unique Radon probability measure on Z extending ν ([13], 416Qa).

(e) Let I be an infinite set.

(i) $\#(\mathfrak{B}_I)$ is the cardinal power $\#(I)^\omega$ ([14], 524Ma).

(ii) Let T_I be the domain of ν_I. If $F \in \mathrm{T}_I$ and $\nu_I F > 0$, set $\Sigma = \mathrm{T}_I \cap \mathcal{P}F$ and $\mu E = \frac{1}{\nu_I F}\nu_I E$ for $E \in \Sigma$; then (F, Σ, μ) is isomorphic to $(\{0,1\}^I, \mathrm{T}_I, \nu_I)$ ([12], 344L[2]). Writing $\mathcal{N}_I$ for the null ideal of ν_I, so that $\mathcal{N}_I \cap \mathcal{P}F$ is the null ideal of μ, we see that $\mathcal{N}_I$ and $\mathcal{N}_I \cap \mathcal{P}F$ are isomorphic, and have the same covering number.

(iii) Suppose that $\mathcal{E} \subseteq \mathrm{T}_I$ is a non-empty downwards-directed family of measurable sets such that $\#(\mathcal{E}) < \operatorname{cov}\mathcal{N}_I$ and $\gamma = \inf_{E\in\mathcal{E}} \nu_I E > 0$. Then $\bigcap \mathcal{E} \neq \emptyset$. **P** For $n \in \mathbb{N}$ we can find $F_n \in \mathcal{E}$ such that $\nu_I F_n \leq \gamma + 2^{-n}$; because $\mathcal{E}$ is downwards-directed, we can suppose that $\langle F_n\rangle_{n\in\mathbb{N}}$ is non-decreasing, so that $F = \bigcap_{n\in\mathbb{N}} F_n$ has measure $\lim_{n\to\infty} \nu_I F_n = \gamma$. If $E \in \mathcal{E}$, then $\nu_I(E \cap F_n) \geq \gamma$ for each n, so $\nu_I(E \cap F) = \gamma$ and $\nu_I(F \setminus E) = 0$. As $\#(\mathcal{E})$ is less than $\operatorname{cov}\mathcal{N}_I$, which by (ii) is the covering number of the subspace measure on F, F cannot be covered by $\{F \setminus E : E \in \mathcal{E}\}$, and $\bigcap \mathcal{E}$ is non-empty. **Q**

(f) Let X be a set, Σ an algebra of subsets of X, and $\nu : \Sigma \to [0,1]$ an additive functional such that $\nu X = 1$ and $\lim_{n\to\infty} \nu E_n = 0$ whenever $\langle E_n\rangle_{n\in\mathbb{N}}$ is a non-increasing sequence in Σ with empty intersection. Then ν has a unique extension to a complete probability measure on X which is inner regular with respect to the family Σ_δ of intersections of sequences in Σ ([13], 413K).

3B Measure algebras: definitions and facts

(a) If $(\mathfrak{A}, \bar\mu)$ is a probability algebra, then $\mathfrak{A}$ is ccc ([12], 322G). So if $A \subseteq \mathfrak{A}$ is any set, there is a countable $B \subseteq A$ with the same upper bounds as A ([12],

[2] Later editions only; obtainable through `http://www.essex.ac.uk/maths/staff/fremlin/mtcont.htm`.

316E), and A has a least upper bound. (Thus $\mathfrak{A}$ is Dedekind complete.) $\bar{\mu}$ is order-continuous in the sense that

— if $A \subseteq \mathfrak{A}$ is non-empty and upwards-directed, with supremum c, then $\bar{\mu}c = \sup_{a\in A} \bar{\mu}a$,

— if $A \subseteq \mathfrak{A}$ is non-empty and downwards-directed, with infimum c, then $\bar{\mu}c = \inf_{a\in A} \bar{\mu}a$.

([12], 321C and 321F).

(b) If $\mathfrak{A}$ is a Boolean algebra and $A \subseteq \mathfrak{A}$, I say that A **τ-generates** $\mathfrak{A}$ if $\mathfrak{A}$ is the smallest order-closed subalgebra of itself including A. The **Maharam type** $\tau(\mathfrak{A})$ is the least cardinal of any set $A \subseteq \mathfrak{A}$ which τ-generates $\mathfrak{A}$. If $(\mathfrak{A}, \bar{\mu})$ is a probability algebra, I is an infinite set and $\tau(\mathfrak{A}) \leq \#(I)$, then $(\mathfrak{A}, \bar{\mu})$ can be embedded in $(\mathfrak{B}_I, \bar{\nu}_I)$ in the sense that there is an injective Boolean homomorphism $\pi : \mathfrak{A} \to \mathfrak{B}_I$ such that $\bar{\nu}_I(\pi a) = \bar{\mu}a$ for every $a \in \mathfrak{A}$ ([12], 332N).

(c) If (X, Σ, μ) is a measure space with measure algebra $(\mathfrak{A}, \bar{\mu})$, and $\mathcal{E} \subseteq \Sigma$ σ-generates Σ in the sense that Σ is the σ-algebra of subsets of X generated by $\mathcal{E}$, then $\{E^\bullet : E \in \mathcal{E}\}$ τ-generates $\mathfrak{A}$. (If $\mathfrak{A}'$ is an order-closed subalgebra of $\mathfrak{A}$ including $\{E^\bullet : E \in \mathcal{E}\}$, then $\{F : F^\bullet \in \mathfrak{A}'\}$ must be a σ-subalgebra of Σ including $\mathcal{E}$, so is the whole of Σ.)

(d) Let (X, Σ, μ) be a measure space with measure algebra $(\mathfrak{A}, \bar{\mu})$. A **lifting** for μ is a Boolean homomorphism $\theta : \mathfrak{A} \to \Sigma$ such that $(\theta a)^\bullet = a$ for every $a \in \mathfrak{A}$. Every complete probability measure has a lifting ([12], 341K).

If (X, Σ, μ) is a probability space with measure algebra $(\mathfrak{A}, \bar{\mu})$, $\theta : \mathfrak{A} \to \Sigma$ is a lifting, and $A \subseteq \mathfrak{A}$ is a non-empty set with supremum c in $\mathfrak{A}$, then $\theta c \setminus \bigcup_{a\in A} \theta a$ is negligible. **P** There is a sequence $\langle a_n \rangle_{n\in\mathbb{N}}$ in A such that $c = \sup_{n\in\mathbb{N}} a_n$ ((a) above). Now

$$(\textstyle\bigcup_{n\in\mathbb{N}} \theta(a_n))^\bullet = \sup_{n\in\mathbb{N}} \theta(a_n)^\bullet = \sup_{n\in\mathbb{N}} a_n = c = (\theta c)^\bullet,$$

so $\theta c \setminus \bigcup_{a\in A} \theta a \subseteq \theta c \setminus \bigcup_{n\in\mathbb{N}} \theta(a_n)$ is negligible. **Q**

(e) If I is an infinite set, I will say that the **standard generating family** in the probability algebra $\mathfrak{B}_I$ is the family $\langle e_i \rangle_{i\in I} = \langle E_i^\bullet \rangle_{i\in I}$ where $E_i = \{x : x \in \{0,1\}^I, x(i) = 1\}$ for each $i \in I$. Note that $\bar{\nu}_I e_i = \frac{1}{2}$ and $\bar{\nu}_I(e_i \setminus e_j) = \frac{1}{4}$ for all distinct i, $j \in I$.

3C We are ready to begin work on the proofs of results announced in §1.

Proof of Proposition 1C Let I be a set, and $\mathcal{F}$ an ultrafilter on I. Then the following are equiveridical:

(i) $\mathcal{F}$ is measure-centering;

(ii) whenever $\langle a_i \rangle_{i\in I}$ is a family in $\mathfrak{B}_I$ such that $\inf_{i\in I} \bar{\nu}_I a_i > 0$, there is an $A \in \mathcal{F}$ such that $\{a_i : i \in A\}$ is centered in $\mathfrak{B}_I$;

(iii) whenever $\langle E_i \rangle_{i\in I}$ is a family of measurable subsets of $\{0,1\}^I$ such that $\inf_{i\in I} \nu_I E_i > 0$, there is an $A \in \mathcal{F}$ such that $\bigcap_{i\in A} E_i \neq \emptyset$;

(iv) whenever (X, Σ, μ) is a compact probability space and $\langle E_i \rangle_{i\in I}$ is a family in Σ, then $\mu^*(\lim_{i\to\mathcal{F}} E_i) \geq \lim_{i\to\mathcal{F}} \mu E_i$.

PROOF. The case in which $\mathcal{F}$ is a principal ultrafilter is trivial, so I shall assume henceforth that $\mathcal{F}$ is non-principal; in particular, that I is infinite.

(i)⇒(ii) is trivial.

not-(iv)⇒not-(ii) Suppose there are a compact probability space (X, Σ, μ) and a family $\langle E_i \rangle_{i \in I}$ in Σ such that $\lim_{i \to \mathcal{F}} \mu E_i > \mu^*(\lim_{i \to \mathcal{F}} E_i)$. Let $F \in \Sigma$ be such that $\lim_{i \to \mathcal{F}} \mu E_i \subseteq F$ and $\mu F < \lim_{i \to \mathcal{F}} \mu E_i$; let $\gamma > 0$ be such that $\lim_{i \to \mathcal{F}} \mu E_i > \mu F + \gamma$, and set $C = \{i : \mu E_i > \gamma + \mu F\}$, so that $C \in \mathcal{F}$ and $\mu(E_i \setminus F) > \gamma$ for every $i \in C$.

Let $\mathcal{K}$ be a compact class such that μ is inner regular with respect to $\mathcal{K}$. For $i \in C$, let $K_i \in \mathcal{K} \cap \Sigma$ be such that $K_i \subseteq E_i \setminus F$ and $\mu K_i \geq \gamma$. For $i \in I \setminus C$, set $K_i = X$. Observe that

$$\lim_{i \to \mathcal{F}} K_i \subseteq \lim_{i \to \mathcal{F}} (E_i \setminus F) = (\lim_{i \to \mathcal{F}} E_i) \setminus F = \emptyset.$$

Let T be the σ-subalgebra of Σ generated by $\{K_i : i \in I\}$, and $\nu = \mu \restriction \mathrm{T}$. Then the measure algebra $(\mathfrak{B}, \bar{\nu})$ of (X, T, ν) is a probability algebra of Maharam type at most $\#(I)$, by 3Bc. By 3Bb, there is a measure-preserving Boolean homomorphism $\pi : \mathfrak{B} \to \mathfrak{B}_I$. Set $a_i = \pi(K_i^\bullet)$ for $i \in C$; then $\bar{\nu}_I a_i = \mu K_i \geq \gamma$ for every $i \in I$.

If $A \in \mathcal{F}$, then $A \cap C \in \mathcal{F}$ so

$$\bigcap_{i \in A} K_i \subseteq \bigcap_{i \in A \cap C} E_i \setminus F \subseteq (\lim_{i \to \mathcal{F}} E_i) \setminus F = \emptyset.$$

As $\mathcal{K}$ is a compact class, there must be a finite set $J \subseteq A \cap C$ such that $\bigcap_{i \in J} K_i = \emptyset$. But this means that

$$\begin{aligned} \inf_{i \in J} a_i &= \inf_{i \in J} \pi(K_i^\bullet) = \pi(\inf_{i \in J} K_i^\bullet) = \pi((\bigcap_{i \in J} K_i)^\bullet) \\ &= \pi(\emptyset^\bullet) = \pi 0 = 0 \end{aligned}$$

in $\mathfrak{B}_I$. This shows that $\{a_i : i \in A\}$ is not centered. As A is arbitrary, $\langle a_i \rangle_{i \in I}$ witnesses that (ii) is false.

(iv)⇒(i) Suppose that (iv) is true. Take a Boolean algebra $\mathfrak{A}$, an additive functional $\nu : \mathfrak{A} \to [0, \infty[$ such that $\nu 1 = 1$, and a family $\langle a_i \rangle_{i \in I}$ in $\mathfrak{A}$ such that $\inf_{i \in I} \nu a_i > 0$. Let Z be the Stone space of $\mathfrak{A}$, so that Z is a compact Hausdorff space and $\mathfrak{A}$ can be identified with the algebra of open-and-closed subsets of Z. Then there is an extension of ν to a Radon probability measure μ on Z (Fact 3Ad). Since μ is inner regular with respect to the compact class of compact subsets of Z, (Z, μ) is a compact probability space.

Let $\mathcal{G}$ be the family of μ-negligible open subsets of Z, and H its union. Then H is an open set, so μH is defined. If $K \subseteq H$ is compact, K is covered by finitely many members of $\mathcal{G}$, so $\mu K = 0$; as μ is inner regular with respect to the compact sets, $\mu H = 0$.

By (iv),

$$\begin{aligned} \mu^*(\lim_{i \to \mathcal{F}} (a_i \setminus H)) &\geq \lim_{i \to \mathcal{F}} \mu(a_i \setminus H) \geq \inf_{i \in I} \mu(a_i \setminus H) \\ &= \inf_{i \in I} \mu a_i = \inf_{i \in I} \nu a_i > 0. \end{aligned}$$

So $\lim_{i \to \mathcal{F}} (a_i \setminus H)$ is non-empty and there is a $z \in Z \setminus H$ such that $A = \{i : z \in a_i\} \in \mathcal{F}$. If $J \subseteq A$ is finite and not empty, then $\bigcap_{i \in J} a_i$ is an open set containing z, so is not included in H and does not belong to $\mathcal{G}$, and

$$0 < \mu(\bigcap_{i\in J} a_i) = \nu(\inf_{i\in J} a_i).$$

Thus $A \in \mathcal{F}$ has the property demanded in Definition 1A; as $\mathfrak{A}$, ν and $\langle a_i\rangle_{i\in I}$ are arbitrary, $\mathcal{F}$ is measure-centering.

(iv)⇒(iii) is elementary, once we know that every ν_I is a Radon measure (Fact 3Ac).

(iii)⇒(ii) Suppose that (iii) is true, and that $\langle a_i\rangle_{i\in I}$ is a family in $\mathfrak{B}_I$ such that $\epsilon = \inf_{i\in I} \bar\nu_I a_i$ is greater than 0. Write T_I for the domain of ν_I, and let $\theta : \mathfrak{B}_I \to \mathrm{T}_I$ be a lifting (Fact 3Bd); set $E_i = \theta a_i$ for each $i \in I$. Then

$$\nu_I E_i = \bar\nu_I(E_i^\bullet) = \bar\nu_I a_i \geq \epsilon$$

for every $i \in I$. By (iii), there is an $A \in \mathcal{F}$ such that $\bigcap_{i\in A} E_i \neq \emptyset$. If $J \subseteq A$ is finite and not empty, then

$$\theta(\inf_{i\in J} a_i) = \bigcap_{i\in J} \theta a_i = \bigcap_{i\in J} E_i \neq \emptyset,$$

so $\inf_{i\in J} a_i \neq 0$; as J is arbitrary, $\{a_i : i \in A\}$ is centered in $\mathfrak{B}_I$; as $\langle a_i\rangle_{i\in I}$ is arbitrary, (ii) is true.

3D In Definition 1A, it is clear that we can expect to have more difficulty in finding a centering set in $\mathcal{F}$ if $\inf_{i\in I} \nu a_i$ is nearly 0, and it is natural to focus on that case as the essence of the definition. As it happens, however, it makes no difference.

Proposition Let $\mathcal{F}$ be an ultrafilter on a set I, and suppose that $\gamma < 1$ is such that whenever $\mathfrak{A}$ is a Boolean algebra, $\nu : \mathfrak{A} \to [0,1]$ is an additive functional such that $\nu 1 = 1$, and $\langle a_i\rangle_{i\in I}$ is a family in $\mathfrak{A}$ such that $\nu a_i \geq \gamma$ for every i, then there is a $J \in \mathcal{F}$ such that $\nu(\inf_{i\in K} a_i) > 0$ for every finite $K \subseteq J$. Then $\mathcal{F}$ is measure-centering.

PROOF. Suppose that $\mathfrak{A}$ is a Boolean algebra, $\nu : \mathfrak{A} \to [0,1]$ is an additive functional such that $\nu 1 = 1$, and $\langle a_i\rangle_{i\in I}$ is a family in $\mathfrak{A}$ such that $\inf_{i\in I} \nu a_i = \epsilon > 0$. Let $m \in \mathbb{N}$ be such that $(1-\epsilon)^m \leq 1-\gamma$, and let $\mathfrak{C} = \bigotimes_m \mathfrak{A}$ be the free product of m copies of $\mathfrak{A}$, so that we have Boolean homomorphisms $\varepsilon_k : \mathfrak{A} \to \mathfrak{C}$, for $k < m$, such that $\inf_{k<m} \varepsilon_k b_k \neq 0$ whenever $b_k \in \mathfrak{A} \setminus \{0\}$ for every $k < m$ ([12], §315). Let $\lambda : \mathfrak{C} \to [0,1]$ be the additive functional such that $\lambda(\inf_{k<m} \varepsilon_k b_k) = \prod_{k>m} \nu_k b_k$ whenever $\langle b_k\rangle_{k<m} \in \mathfrak{A}^m$ ([12], 326Q). For $i \in I$, set $c_i = \sup_{k<m} \varepsilon_k a_i$; then

$$\lambda(1\setminus c_i) = (\nu(1\setminus a_i))^m \leq (1-\epsilon)^m \leq 1-\gamma$$

and $\lambda c_i \geq \gamma$ for every $i \in I$. By hypothesis, there is a $J \in \mathcal{F}$ such that $\lambda(\inf_{i\in K} c_i) > 0$ for every finite $K \subseteq J$. Let $D \subseteq \mathfrak{C}$ be a maximal set such that $c_i \in D$ for every $i \in J$ and $\lambda(\inf D_0) > 0$ for every finite $D_0 \subseteq D$. Then for every $i \in J$ there is a $k_i < m$ such that $\varepsilon_{k_i}(a_i) \in D$. Because $\mathcal{F}$ is an ultrafilter, there is a $k < m$ such that $J' = \{i : i \in J,\ k_i = k\}$ belongs to $\mathcal{F}$; and now $\nu(\inf_{i\in K} a_i) = \lambda(\inf_{i\in K} \varepsilon_k a_i) > 0$ for every finite $K \subseteq J'$. As $\mathfrak{A}$, ν and $\langle a_i\rangle_{i\in I}$ are arbitrary, $\mathcal{F}$ is measure-centering.

3E Proof of Theorem 1Ba A Ramsey ultrafilter is measure-centering.

PROOF. Let $\mathcal{F}$ be a Ramsey ultrafilter on a set I, and $\langle a_i\rangle_{i\in I}$ a family in $\mathfrak{B}_I$ such that $\epsilon = \inf_{i\in I} \bar\nu_I a_i$ is greater than 0. For $C \subseteq I$ set $b_C = \sup_{i\in C} a_i$; then $\bar\nu_I b_C \geq \epsilon$ for every $C \in \mathcal{F}$. Set $b = \inf_{C\in\mathcal{F}} b_C$; because $\mathcal{F}$ is downwards-directed, so is $\{b_C : C \in \mathcal{F}\}$, and $\bar\nu_I b \geq \epsilon$ (Fact 3Ba). In particular, $b \neq 0$.

Let $\mathcal{S}$ be the set of those finite subsets K of I such that $b \cap \inf_{i\in K} a_i$ is non-zero, counting $\inf\emptyset$ as 1, so that $\emptyset\in\mathcal{S}$. If $K\in\mathcal{S}$, then $C=\{i : K\cup\{i\}\in\mathcal{S}\}$ belongs to $\mathcal{F}$. **P?** Otherwise, $I\setminus C\in\mathcal{F}$ and $b\subseteq b_{I\setminus C}$. Set

$$d = b\cap \inf_{k\in K} a_k \subseteq b_{I\setminus C} = \sup_{i\in I\setminus C} a_i;$$

as $d\neq 0$, there is a $j\in I\setminus C$ such that $d\cap a_j\neq 0$. But $d\cap a_j = b\cap\inf_{i\in K\cup\{j\}} a_i$, so $K\cup\{j\}\in\mathcal{S}$ and $j\in C$, which is absurd. **X**[3]**Q**

So $\mathcal{S}$ satisfies the conditions of 2A. Since $\mathcal{F}$ is dependently selective (Proposition 2D), there is a $J\in\mathcal{F}$ such that $[J]^{<\omega}\subseteq\mathcal{S}$, that is, $\{a_i : i\in J\}$ is centered. As $\langle a_i\rangle_{i\in I}$ is arbitrary, $\mathcal{F}$ is measure-centering.

3F Proof of Theorem 1Bb If κ is an infinite cardinal and $\operatorname{cov}\mathcal{N}_\kappa = 2^\kappa$, then there is a uniform measure-centering ultrafilter on κ.

PROOF. **(a)** $\#(\mathfrak{B}_\kappa)=\kappa^\omega$ (Fact 3A(e-i)), so we can enumerate $\mathfrak{B}_\kappa^\kappa$ as $\langle \boldsymbol{a}_\zeta\rangle_{\zeta<2^\kappa}$. Let T_κ be the domain of ν_κ and $\theta:\mathfrak{B}_\kappa\to\mathrm{T}_\kappa$ a lifting. Set $\lambda=\operatorname{cf}\kappa$. If $\lambda=\kappa$, set $I_\alpha=\{\alpha\}$ for every $\alpha<\kappa$; otherwise, let $\langle\kappa_\alpha\rangle_{\alpha<\lambda}$ be a strictly increasing family of cardinals less than κ with supremum κ, and set $I_\alpha=\kappa_\alpha^+\setminus\bigcup_{\beta<\alpha}I_\beta$ for $\alpha<\lambda$, so that $\langle I_\alpha\rangle_{\alpha<\lambda}$ is a partition of κ and $\operatorname{otp}(I_\alpha)=\kappa_\alpha^+$ for every $\alpha<\lambda$.

(b) Construct families $\langle C_{\alpha\zeta}\rangle_{\alpha<\lambda,\zeta\le 2^\kappa}$, $\langle D_\zeta\rangle_{\zeta<2^\kappa}$, $\langle\mathcal{C}_{\alpha\zeta}\rangle_{\alpha<\lambda,\zeta\le 2^\kappa}$ and $\langle\mathcal{D}_\zeta\rangle_{\zeta<2^\kappa}$ inductively, as follows. Start by setting $\mathcal{C}_{\alpha 0}=\{I_\alpha\setminus\xi : \xi\in I_\alpha\}$, so that $\mathcal{C}_{\alpha 0}$ is a filter base of subsets of I_α of cardinal less than κ; let $\mathcal{D}_0$ be $\{\lambda\}$. Given that $\zeta<2^\kappa$, that $\mathcal{D}_\zeta$ is a filter base of subsets of λ of cardinal at most $\max(\omega,\#(\zeta))$, and $\mathcal{C}_{\alpha\zeta}$ is a filter base of subsets of I_α of cardinal at most $\max(\kappa,\#(\zeta))$ for each $\alpha<\lambda$, consider $\boldsymbol{a}_\zeta=\langle a_\xi\rangle_{\xi<\kappa}$ say. Set $\epsilon=\inf_{\xi<\kappa}\bar\nu_\kappa a_\xi$. If $\epsilon=0$, set $D_{\zeta+1}=\lambda$ and $C_{\alpha,\zeta+1}=I_\alpha$ for every $\alpha<\lambda$. Otherwise, set $b_C=\sup_{\xi\in C}a_\xi$ for $C\subseteq\kappa$, and $c_\alpha=\inf_{C\in\mathcal{C}_{\alpha\zeta}}b_C$ for $\alpha<\lambda$; as in 3E above, $\bar\nu_\kappa c_\alpha=\inf_{C\in\mathcal{C}_{\alpha\zeta}}\bar\nu_\kappa b_C\ge\epsilon$. Set $d_D=\sup_{\alpha\in D}c_\alpha$ for $D\in\mathcal{D}_\zeta$ and $e=\inf_{D\in\mathcal{D}_\zeta}d_D$; then the same arguments show that $\bar\nu_\kappa e\ge\epsilon$.

For each $C\subseteq\kappa$, $\theta(b_C)\setminus\bigcup_{\xi\in C}\theta(a_\xi)$ is negligible (Fact 3Bd). So $E_{\alpha C}=\theta(c_\alpha)\setminus\bigcup_{\xi\in C}\theta(a_\xi)$ is negligible whenever $\alpha<\lambda$ and $C\in\mathcal{C}_\alpha$. Similarly, $E_D=\theta(e)\setminus\bigcup_{\alpha\in D}\theta(c_\alpha)$ is negligible for every $D\in\mathcal{D}$. Now $\mathcal{D}_\zeta\cup\{(\alpha,C):\alpha<\lambda,\,C\in\mathcal{C}_{\alpha\zeta}\}$ has cardinal at most $\max(\omega,\#(\zeta),\#(\kappa))<2^\kappa$, so $\{E_D : D\in\mathcal{D}_\zeta\}\cup\{E_{\alpha C}:\alpha<\lambda,$ $C\in\mathcal{C}_{\alpha\zeta}\}$ cannot cover the non-negligible measurable set $\theta(e)$ (Fact 3A(e-iii)), and there must be an $x_\zeta\in\theta(e)$ such that $x_\zeta\notin\bigcup_{D\in\mathcal{D}_\zeta}E_D\cup\bigcup_{\alpha<\lambda,C\in\mathcal{C}_\alpha}E_{\alpha C}$. Set $D_\zeta=\{\alpha:\alpha<\lambda,\,x_\zeta\in\theta(c_\alpha)\}$; then D_ζ meets every member of $\mathcal{D}_\zeta$. For $\alpha\in D_\zeta$, set $C_{\alpha\zeta}=\{\xi:\xi\in I_\alpha,\,x_\zeta\in\theta(a_\xi)\}$; for $\alpha\in\kappa\setminus D_\zeta$, set $C_{\alpha\zeta}=I_\alpha$; then $C_{\alpha\zeta}$ meets every member of $\mathcal{C}_{\alpha\zeta}$.

Now define $\mathcal{D}_{\zeta+1}$, $\mathcal{C}_{\alpha,\zeta+1}$ by setting

$$\mathcal{D}_{\zeta+1}=\mathcal{D}_\zeta\cup\{D\cap D_\zeta : D\in\mathcal{D}_\zeta\},$$

$$\mathcal{C}_{\alpha,\zeta+1}=\mathcal{C}_{\alpha\zeta}\cup\{C\cap C_{\alpha\zeta}: C\in\mathcal{C}_{\alpha\zeta}\}$$

for $\alpha<\lambda$. Then $\mathcal{D}_{\zeta+1}$ is a filter base of subsets of λ of cardinal at most $\max(\omega,\#(\zeta+1))$, and $\mathcal{C}_{\alpha,\zeta+1}$ is a filter base of subsets of I_α of cardinal at most $\max(\kappa,\#(\zeta+1))$ for every $\alpha<\lambda$.

---

[3]Each pair of symbols **?**, **X** in this paper is used to enclose an argument from contradiction refuting the hypothesis advanced immediately after the **?**.

For non-zero limit ordinals $\zeta \le 2^\kappa$, set $\mathcal{D}_\zeta = \bigcup_{\xi<\zeta} \mathcal{D}_\xi$ and $\mathcal{C}_{\alpha\zeta} = \bigcup_{\xi<\zeta} \mathcal{C}_{\alpha\xi}$; once again, $\mathcal{D}_\zeta$ will be a filter base of subsets of λ of cardinal at most $\max(\omega, \#(\zeta))$, and $\mathcal{C}_{\alpha\zeta}$ will a filter base of subsets of I_α of cardinal at most $\max(\kappa, \#(\zeta))$ for each $\alpha < \lambda$.

(c) At the end of the induction, let $\mathcal{F}$ be an ultrafilter on κ containing all sets of the form $\bigcup_{\alpha\in D} C_\alpha$ where $D \in \mathcal{D}_{2^\kappa}$ and $C_\alpha \in \mathcal{C}_{\alpha,2^\kappa}$ for every $\alpha \in D$. Then $\mathcal{F}$ is measure-centering. **P** Let $\boldsymbol{a} = \langle a_\xi\rangle_{\xi<\kappa} \in \mathfrak{B}_\kappa^\kappa$ be such that $\inf_{\xi<\kappa} \bar\nu_\kappa a_\xi > 0$. Then there is a $\zeta < 2^\kappa$ such that $\boldsymbol{a} = \boldsymbol{a}_\zeta$. In this case, x_ζ is defined and $J = \bigcup_{\alpha\in D_\zeta} C_{\alpha\zeta}$ belongs to $\mathcal{F}$. If $\alpha \in D_\zeta$ and $\xi \in C_{\alpha\zeta}$, then $x_\zeta \in \theta(a_\xi)$. But this means that if $K \subseteq J$ is finite and not empty,

$$x_\zeta \in \bigcap_{\xi\in K} \theta(a_\xi) = \theta(\inf_{\xi\in K} a_\xi),$$

and $\inf_{\xi\in K} a_\xi \ne 0$. Thus $\{a_\xi : \xi \in J\}$ is centered. As $\boldsymbol{a}$ is arbitrary, $\mathcal{F}$ is measure-centering, by Proposition 1C(ii). **Q**

(d) Of course I should note that $\mathcal{F}$ is uniform because if $D \in \mathcal{D}_{2^\kappa}$ and $C_\alpha \in \mathcal{C}_{\alpha,2^\kappa}$ for every $\alpha \in D$, then D meets every member of $\mathcal{D}_0$, so has cardinal λ. Set $B = \bigcup_{\alpha\in D} C_\alpha$. Since C_α meets every member of $\mathcal{C}_{\alpha 0}$, $\#(C_\alpha) = \#(I_\alpha)$ for every $\alpha \in D$, and $\#(B) = \#(\bigcup_{\alpha\in D} I_\alpha) = \kappa$.

3G Proof of Theorem 1Bc If $\operatorname{cov}\mathcal{N}_{\text{Leb}} = \mathfrak{c}$, there is a measure-centering ultrafilter on $\mathbb{N}$ which contains no set of zero asymptotic density.

Proof. **(a)** I start with a general fact about upper asymptotic density $d^* : \mathcal{P}\mathbb{N} \to [0,1]$. Let (X, Σ, μ) be a probability space, $I \in \mathcal{P}\mathbb{N} \setminus \mathcal{Z}$, and $\langle E_n\rangle_{n\in\mathbb{N}}$ a sequence in Σ with $\inf_{n\in\mathbb{N}} \mu E_n = \gamma > 0$. For $x \in X$ set $J_x = \{n : n \in I,\ x \in E_n\}$. Then $\mu\{x : d^*(J_x) > 0\} \ge \gamma$.

P Set $\delta = d^*(I)$; then there is a disjoint sequence $\langle K_n\rangle_{n\in\mathbb{N}}$ of non-empty finite subsets of $\mathbb{N}$, all of the form $\{i : k \le i < 2k\}$, such that $\#(I \cap K_n) \ge \frac{1}{3}\delta\#(K_n)$ for every n. If $n \in \mathbb{N}$ then

$$\gamma\#(I\cap K_n) \le \sum_{i\in I\cap K_n} \mu E_i = \int \sum_{i\in I\cap K_n} \chi E_i(x)\mu(dx) = \int \#(J_x \cap K_n)\mu(dx),$$

so if $\eta > 0$

$$\mu\{x : \#(J_x \cap K_n) \ge \eta\#(I\cap K_n)\} \ge \gamma - \eta.$$

Consequently

$$\begin{aligned}
&\mu\{x : d^*(J_x) \ge \frac{1}{6}\delta\eta\}\\
&\qquad \ge \mu\{x : \text{for infinitely many } n,\ \#(J_x\cap K_n) \ge \frac{1}{3}\delta\eta\#(K_n)\}\\
&\qquad \ge \mu\{x : \text{for infinitely many } n,\ \#(J_x\cap K_n) \ge \eta\#(I\cap K_n)\}\\
&\qquad \ge \gamma - \eta.
\end{aligned}$$

As η is arbitrary, $\mu\{x : d^*(J_x) > 0\} \ge \gamma$. **Q**

(b) Now, given that $\operatorname{cov}\mathcal{N}_{\text{Leb}} = \mathfrak{c}$, enumerate the family of all sequences $\langle E_n\rangle_{n\in\mathbb{N}}$ of Borel subsets of $\{0,1\}^\omega$ such that $\inf_{n\in\mathbb{N}} \nu E_n > 0$ as $\langle\langle E_{\xi n}\rangle_{n\in\mathbb{N}}\rangle_{\xi<\mathfrak{c}}$.

Build filter bases $\mathcal{E}_\xi \subseteq \mathcal{P}\mathbb{N}$, for $\xi \le \mathfrak{c}$, as follows. Start with $\mathcal{E}_0 = \{\mathbb{N}\}$. The inductive hypothesis will be that $\#(\mathcal{E}_\xi) \le \max(\omega, \#(\xi))$ and $\mathcal{E}_\xi \cap \mathcal{Z} = \emptyset$. For the inductive step to $\xi + 1$, set $\epsilon = \inf_{n\in\mathbb{N}} \nu E_{\xi n}$. For each $I \in \mathcal{E}_\xi$, set

$$F_I = \{x : d^*(J_x \cap I) > 0\},$$

where $J_x = \{n : x \in E_{\xi n}\}$. By (a), $\nu F_I \ge \epsilon$; because $\langle F_I\rangle_{I\in\mathcal{E}_\xi}$ is downwards-directed, and $\#(\mathcal{E}_\xi) < \operatorname{cov}\mathcal{N}_{\text{Leb}} = \operatorname{cov}\mathcal{N}_\omega$, there is a point $x \in \bigcap_{I\in\mathcal{E}_\xi} F_I$ (Fact 3A(e-iii)). Set $I_\xi = J_x$,

$$\mathcal{E}_{\xi+1} = \mathcal{E}_\xi \cup \{I \cap I_\xi : I \in \mathcal{E}_\xi\}.$$

Then $\mathcal{E}_{\xi+1}$ is a filter base, including $\mathcal{E}_\xi$, disjoint from $\mathcal{Z}$, of cardinal at most $\max(\omega, \#(\xi+1))$, and containing a set I_ξ such that $\bigcap_{n\in I_\xi} E_{\xi n}$ is non-empty. At non-zero limit ordinals $\xi \le \mathfrak{c}$, set $\mathcal{E}_\xi = \bigcup_{\eta<\xi} \mathcal{E}_\eta$.

Let $\mathcal{G}$ be the filter on $\mathbb{N}$ generated by $\mathcal{E}_\mathfrak{c}$. Then $\mathcal{G} \cap \mathcal{Z} = \emptyset$, so there is an ultrafilter $\mathcal{F}$ on $\mathbb{N}$, including $\mathcal{G}$, and still disjoint from $\mathcal{Z}$. If $\langle F_n\rangle_{n\in\mathbb{N}}$ is any sequence of measurable subsets of $\{0,1\}^{\mathbb{N}}$ such that $\inf_{n\in\mathbb{N}} \nu F_n > 0$, there is a $\xi < \mathfrak{c}$ such that $E_{\xi n} \subseteq F_n$ for every n; now $I_\xi \in \mathcal{F}$ and $\bigcap_{n\in I_\xi} F_n \ne \emptyset$. By Proposition 1C(iii), $\mathcal{F}$ is measure-centering.

3H Proof of Proposition 1D (a) Let I and J be sets, $f : I \to J$ a function, and $\mathcal{F}$ a measure-centering ultrafilter on I. Then the image ultrafilter $f[[\mathcal{F}]]$ is a measure-centering ultrafilter on J.

(b) If $\mathcal{F}$ is a non-principal ultrafilter, then $\mathcal{F} \ltimes \mathcal{F}$ is not measure-centering.

Proof. **(a)** Let $\mathfrak{A}$ be a Boolean algebra, $\nu : \mathfrak{A} \to [0,1]$ an additive functional such that $\nu 1 = 1$, and $\langle a_j\rangle_{j\in J}$ a family in $\mathfrak{A}$ such that $\inf_{j\in J} \nu a_j > 0$. Set $b_i = a_{f(i)}$ for $i \in I$; then $\inf_{i\in I} \nu b_i > 0$, so there is an $A \in \mathcal{F}$ such that $\nu(\inf_{i\in K} b_i) > 0$ for every $K \in [A]^{<\omega}$. Now $f[A] \in f[[\mathcal{F}]]$, and if $L \subseteq f[A]$ is finite there is a finite $K \subseteq A$ such that $L = f[K]$, so that

$$\nu(\inf_{j\in L} a_j) = \nu(\inf_{i\in K} b_i) > 0.$$

This shows that $f[[\mathcal{F}]]$ has the property of Definition 1A and is measure-centering.

(b) Let I be a set and $\mathcal{F}$ a non-principal ultrafilter on I. Let $\langle e_i\rangle_{i\in I}$ be the standard generating family in $\mathfrak{B}_I$, and for $i, j \in I$ set

$$\begin{aligned} a_{ij} &= e_i \setminus e_j \text{ if } i \ne j, \\ &= 1 \text{ if } i = j. \end{aligned}$$

Then $\bar\nu_I a_{ij} \ge \frac{1}{4}$ for all $i, j \in I$. If $A \in \mathcal{F} \ltimes \mathcal{F}$, then $B = \{i : A[\{i\}] \in \mathcal{F}\}$ belongs to $\mathcal{F}$. Take any $i \in B$; then B must meet $A[\{i\}]$ in more than one point, because $\mathcal{F}$ is non-principal; take $j \in B \cap A[\{i\}] \setminus \{i\}$ and $k \in A[\{j\}] \setminus \{j\}$. The points (i,j) and (j,k) are distinct points of A and $a_{jk} \cap a_{ij} \subseteq e_j \setminus e_j = 0$. As A is arbitrary, $\mathcal{F} \ltimes \mathcal{F}$ cannot be measure-centering.

Remark I include (b) here to show that in Theorem 1J we really need to have non-isomorphic filters. In fact rather more can be said; the argument here already shows that $\mathcal{F} \ltimes \mathcal{F}$ is not measure-linking, and in fact it is not Hausdorff ([8]).

3I Proof of Theorem 1E Let (X, Σ, μ) be a compact probability space, and $\mathcal{F}$ a measure-centering ultrafilter on a set I. Let $\mathcal{A}$ be the family of all sets of the form $\lim_{i\to\mathcal{F}} E_i$ where $\langle E_i\rangle_{i\in I}$ is a family in Σ. Then there is a unique complete probability measure λ on X such that λ is inner regular with respect to $\mathcal{A}$ and $\lambda(\lim_{i\to\mathcal{F}} E_i) = \lim_{i\to\mathcal{F}} \mu E_i$ for every family $\langle E_i\rangle_{i\in I}$ in Σ.

PROOF. **(a)** The key fact is that if $\langle E_i\rangle_{i\in I}$ and $\langle F_i\rangle_{i\in I}$ are two families in Σ such that $\lim_{i\to\mathcal{F}} E_i = \lim_{i\to\mathcal{F}} F_i$, then $\lim_{i\to\mathcal{F}} \mu E_i = \lim_{i\to\mathcal{F}} \mu F_i$. **P** We have

$$|\lim_{i\to\mathcal{F}} \mu E_i - \lim_{i\to\mathcal{F}} \mu F_i| = \lim_{i\to\mathcal{F}} |\mu E_i - \mu F_i| \le \lim_{i\to\mathcal{F}} \mu(E_i \triangle F_i)$$
$$\le \mu^*(\lim_{i\to\mathcal{F}} (E_i\triangle F_i))$$

(by Proposition 1C(iv))

$$\le \mu^*(\lim_{i\to\mathcal{F}} E_i \triangle \lim_{i\to\mathcal{F}} F_i) = \mu^*\emptyset = 0. \textbf{ Q}$$

(b) The formula $\phi(\lim_{i\to\mathcal{F}} E_i) = \lim_{i\to\mathcal{F}} \mu E_i$ therefore defines a functional $\phi : \mathcal{A} \to [0, 1]$. If $*$ is any of the Boolean operations $\setminus$, $\cap$ and $\cup$, then

$$\lim_{i\to\mathcal{F}} E_i * \lim_{i\to\mathcal{F}} F_i = \lim_{i\to\mathcal{F}}(E_i * F_i)$$

for all families $\langle E_i\rangle_{i\in I}$ and $\langle F_i\rangle_{i\in I}$ in Σ, so $\mathcal{A}$ is an algebra of subsets of X. As noted in 1E, ϕ extends μ. If $\langle E_i\rangle_{i\in I}$ and $\langle F_i\rangle_{i\in I}$ are two families in Σ,

$$\begin{aligned}
\phi(\lim_{i\to\mathcal{F}} E_i \cup \lim_{i\to\mathcal{F}} F_i) &+ \phi(\lim_{i\to\mathcal{F}} E_i \cap \lim_{i\to\mathcal{F}} F_i)\\
&= \phi(\lim_{i\to\mathcal{F}} (E_i \cup F_i)) + \phi(\lim_{i\to\mathcal{F}} (E_i \cap F_i))\\
&= \lim_{i\to\mathcal{F}} \mu(E_i \cup F_i) + \lim_{i\to\mathcal{F}} \mu(E_i \cap F_i)\\
&= \lim_{i\to\mathcal{F}} \mu(E_i \cup F_i) + \mu(E_i \cap F_i)\\
&= \lim_{i\to\mathcal{F}} \mu E_i + \mu F_i = \phi(\lim_{i\to\mathcal{F}} E_i) + \phi(\lim_{i\to\mathcal{F}} F_i));
\end{aligned}$$

so $\phi(A\cup B) + \phi(A\cap B) = \phi A + \phi B$ for all A, $B \in \mathcal{A}$. Because $\phi\emptyset = \mu\emptyset = 0$, this is enough to show that $\phi : \mathcal{A} \to [0, 1]$ is additive.

(c) If $\langle A_n\rangle_{n\in\mathbb{N}}$ is a non-increasing sequence in $\mathcal{A}$ and $0 \le \gamma < \inf_{n\in\mathbb{N}} \phi A_n$, then there is an $A \in \mathcal{A}$ such that $A \subseteq \bigcap_{n\in\mathbb{N}} A_n$ and $\phi A \ge \gamma$. **P** For each n, express A_n as $\lim_{i\to\mathcal{F}} E_{ni}$, where $E_{ni} \in \Sigma$ for each $i \in I$; replacing E_{ni} by $\bigcap_{m\le n} E_{mi}$ if necessary, we may suppose that $E_{ni} \subseteq E_{mi}$ whenever $m \le n$ and $i \in I$. For each $i \in I$, define F_i by saying that

$$\begin{aligned}
F_i &= X \text{ if } \mu E_{0i} < \gamma,\\
&= E_{ni} \text{ if } n \in \mathbb{N} \text{ and } \mu E_{ni} \ge \gamma > \mu E_{n+1,i},\\
&= \bigcap_{n\in\mathbb{N}} E_{ni} \text{ if } \mu E_{ni} \ge \gamma \text{ for every } n \in \mathbb{N}.
\end{aligned}$$

Then $\mu F_i \ge \gamma$ for every $i \in I$, and $F_i \subseteq E_{ni}$ whenever $n \in \mathbb{N}$, $i \in I$ and $\mu E_{ni} \ge \gamma$. For each $n \in \mathbb{N}$, therefore, $\{i : F_i \subseteq E_{ni}\}$ belongs to $\mathcal{F}$, and $A = \lim_{i\to\mathcal{F}} F_i \subseteq A_n$; while $\phi A \ge \gamma$. **Q**

(d) In particular, ϕ is countably additive in the sense that if $\langle A_n\rangle_{n\in\mathbb{N}}$ is a non-increasing sequence in $\mathcal{A}$ such that $\bigcap_{n\in\mathbb{N}} A_n$ is empty, then $\lim_{n\to\infty} \phi A_n = 0$. There is therefore an extension of ϕ to a complete probability measure λ on X which is inner regular with respect to $\mathcal{A}_\delta$, the family of subsets of X expressible as the intersection of a sequence in $\mathcal{A}$ (Fact 3Af). From (c) we see that if $B \in \mathcal{A}_\delta$ and $\gamma < \lambda B$, there must be an $A \in \mathcal{A}$ such that $A \subseteq B$ and $\phi A \ge \gamma$. So in fact λ is inner regular with respect to $\mathcal{A}$.

(e) Thus we have a suitable extension of μ. By 3A(a-iv), λ is uniquely defined.

Remark If you have seen a construction of Loeb measure ([21]), you will recognise the method above; the special properties of measure-centering ultrafilters and compact measure spaces mean, in effect, that the original set X has full outer measure in the Loeb measure space.

4 Reduced products of probability algebras

The construction offered in 1F is straightforward enough, but requires some support, in particular in the assertion that the reduced product $(\mathfrak{C}, \bar\lambda)$ there is a probability algebra in the full sense of the phrase as used here. In this section I fill in the details, with a proof of Theorem 1G, and also of a further result, Proposition 4B, which will be needed for the proof of Theorem 1J in the next section.

4A I start with a slightly expanded version of 1Fb.

Proposition Let $\langle(\mathfrak{A}_i, \bar\mu_i)\rangle_{i\in I}$ be a family of probability algebras and $\mathcal{F}$ an ultrafilter on I. Let $\mathfrak{B}$ be the product Boolean algebra $\prod_{i\in I}\mathfrak{A}_i$. Define $\nu : \mathfrak{B} \to [0,1]$ by setting $\nu(\langle a_i\rangle_{i\in I}) = \lim_{i\to\mathcal{F}} \bar\mu a_i$ whenever $\langle a_i\rangle_{i\in I} \in \mathfrak{B}$, and set $\mathcal{I} = \{\boldsymbol{a} : \boldsymbol{a} \in \mathfrak{B}, \nu\boldsymbol{a} = 0\}$.

(a) ν is additive; $\mathcal{I}$ is an ideal of $\mathfrak{B}$; and if $\mathfrak{C}$ is the quotient Boolean algebra $\mathfrak{B}/\mathcal{I}$, we have an additive functional $\bar\nu : \mathfrak{C} \to [0,1]$ defined by saying that $\bar\nu(\boldsymbol{a}^\bullet) = \nu\boldsymbol{a}$ for every $\boldsymbol{a} \in \mathfrak{B}$. If $\langle a_i\rangle_{i\in I}$, $\langle b_i\rangle_{i\in I} \in \mathfrak{B}$, then

— if $\{i : a_i \subseteq b_i\} \in \mathcal{F}$, then $\langle a_i\rangle_{i\in I}^\bullet \subseteq \langle b_i\rangle_{i\in I}^\bullet$;

— if $\{i : a_i = b_i\} \in \mathcal{F}$, then $\langle a_i\rangle_{i\in I}^\bullet = \langle b_i\rangle_{i\in I}^\bullet$.

(b) $(\mathfrak{C}, \bar\nu)$ is a probability algebra.

PROOF. **(a)** is entirely elementary.

(b)(i) Perhaps I should begin by remarking that

$$\bar\nu(1_{\mathfrak{C}}) = \bar\nu(1_{\mathfrak{B}}^\bullet) = \nu(1_{\mathfrak{B}}) = \nu(\langle 1_{\mathfrak{A}_i}\rangle_{i\in I}) = \lim_{i\to\mathcal{F}} \mu_i(1_{\mathfrak{A}_i}) = 1.$$

Next, if $c \in \mathfrak{C}\setminus\{0\}$, then $c = \boldsymbol{b}^\bullet$ for some $\boldsymbol{b} \in \mathfrak{B}\setminus\mathcal{I}$, so $\bar\nu c = \nu\boldsymbol{b}$ is non-zero.

(ii) If $\langle c_n\rangle_{n\in\mathbb{N}}$ is a non-decreasing sequence in $\mathfrak{C}$, there is a $c \in \mathfrak{C}$ such that $c_n \subseteq c$ for every $n \in \mathbb{N}$ and $\bar\nu c = \sup_{n\in\mathbb{N}} \bar\nu c_n$. **P** For each $n \in \mathbb{N}$, let $\langle a_{ni}\rangle_{i\in I} \in \mathfrak{B}$ be such that $c_n = \langle a_{ni}\rangle_{i\in I}^\bullet$. Set $a'_{ni} = \sup_{m\le n} a_{mi}$ for $n \in \mathbb{N}$ and $i \in I$; then $c_n = \langle a'_{ni}\rangle_{i\in I}^\bullet$ for each n, and $\langle a'_{ni}\rangle_{n\in\mathbb{N}}$ is non-decreasing for each $i \in I$. Let γ be $\sup_{n\in\mathbb{N}} \bar\nu c_n = \sup_{n\in\mathbb{N}} \lim_{i\to\mathcal{F}} \mu_i a'_{ni}$, and for $i \in I$ set

$$\begin{aligned} a_i &= a'_{ni} \text{ if } \mu_i a'_{ni} \le \gamma + 2^{-n} \text{ and } \mu_i a'_{n+1,i} > \gamma + 2^{-n-1}, \\ &= \sup_{n\in\mathbb{N}} a'_{ni} \text{ if } \mu_i a'_{ni} \le \gamma + 2^{-n} \text{ for every } n \in \mathbb{N}. \end{aligned}$$

Consider $c = \langle a_i\rangle^\bullet_{i\in I}$. For any $n \in \mathbb{N}$,

$$\begin{aligned} \{i : a'_{ni} \subseteq a_i,\ \mu_i a_i \le \gamma + 2^{-n}\} &\supseteq \{i : \mu_i a'_{ni} \le \gamma + 2^{-n}\} \\ &\supseteq \{i : \mu_i a'_{ni} \le \lim_{j\to\mathcal{F}} \mu_j a'_{nj} + 2^{-n}\} \in \mathcal{F}, \end{aligned}$$

so

$$c_n = \langle a'_{ni}\rangle^\bullet_{i\in I} \subseteq \langle a_i\rangle^\bullet_{i\in I} = c.$$

It follows at once that $\bar\nu c \ge \gamma$. At the same time, $\bar\nu c \le \gamma + 2^{-n}$ for every n, so $\bar\nu c \le \gamma$. **Q**

(iii) $\mathfrak{C}$ is Dedekind σ-complete. **P** If $C \subseteq \mathfrak{C}$ is a countable set, take a sequence $\langle c_n\rangle_{n\in\mathbb{N}}$ running over $C \cup \{0_\mathfrak{C}\}$. Set $c'_n = \sup_{m\le n} c_m$ for each n. By (ii), there is a $c \in \mathfrak{C}$ such that $c'_n \subseteq c$ for every n and $\bar\nu c = \sup_{n\in\mathbb{N}} \bar\nu c'_n$. Because $c_n \subseteq c'_n \subseteq c$ for every n, c is an upper bound of C. If c' is any other upper bound of C, then $c' \supseteq c'_n$ for every $n \in \mathbb{N}$, so

$$\bar\nu(c\setminus c') \le \inf_{n\in\mathbb{N}} \bar\nu(c\setminus c'_n) \le \inf_{n\in\mathbb{N}} \bar\nu c - \bar\nu c'_n = 0,$$

$c\setminus c' = 0$ and $c \subseteq c'$. Thus c is the least upper bound of C; as C is arbitrary, $\mathfrak{C}$ is Dedekind σ-complete. **Q**

(iv) $\bar\nu$ is countably additive. **P** This time, let $\langle c_n\rangle_{n\in\mathbb{N}}$ be a disjoint sequence in $\mathfrak{C}$. Again set $c'_n = \sup_{m\le n} c_m$ for each $n \in \mathbb{N}$. Re-running the argument of (iii), we see that if c is the least upper bound $\sup_{n\in\mathbb{N}} c'_n$ then

$$\bar\nu c = \sup_{n\in\mathbb{N}} \bar\nu c'_n = \sup_{n\in\mathbb{N}} \sum_{m=0}^n \bar\nu c_m = \sum_{n=0}^\infty \bar\nu c_n,$$

while c is also the least upper bound of $\{c_n : n \in \mathbb{N}\}$. **Q**

Putting these together, $(\mathfrak{C}, \bar\nu)$ is a probability algebra.

4B Directed families Some further phenomena appear if $\langle(\mathfrak{A}_i, \bar\mu_i)\rangle_{i\in I}$ is appropriately linked, as follows.

Proposition Let $(I, \le)$ be a non-empty pre-ordered set (that is, $\le$ is a reflexive transitive relation on I), and $\langle(\mathfrak{A}_i, \bar\mu_i)\rangle_{i\in I}$ a family of probability algebras. Suppose that for $i \le j$ in I we are given a measure-preserving Boolean homomorphism $\pi_{ji} : \mathfrak{A}_i \to \mathfrak{A}_j$, and that $\pi_{ki} = \pi_{kj}\pi_{ji}$ whenever $i \le j \le k$ in I. Let $\mathcal{F}$ be an ultrafilter on I such that $\{j : i \le j\}$ belongs to $\mathcal{F}$ for every $i \in I$, and let $(\mathfrak{C}, \bar\nu)$ be the reduced product $\prod_{i\in I}(\mathfrak{A}, \bar\mu_i)|\mathcal{F}$.

(a) For each $i \in I$ we have a measure-preserving Boolean homomorphism $\pi_i : \mathfrak{A}_i \to \mathfrak{C}$ defined by saying that $\pi_i a = \langle a_j\rangle^\bullet_{j\in I}$ whenever $\langle a_j\rangle_{j\in I} \in \prod_{j\in I} \mathfrak{A}_j$ is such that $a_j = \pi_{ji} a_i$ for every $j \ge i$.

(b) $\pi_i = \pi_j \pi_{ji}$ whenever $i \le j$ in I.

(c) $\langle a_i\rangle^\bullet_{i\in I} \subseteq \sup_{j\in A} \pi_j a_j$ whenever $\langle a_i\rangle_{i\in I} \in \prod_{i\in I} \mathfrak{A}_i$ and $A \in \mathcal{F}$.

PROOF. **(a)** π_i is well-defined because $\{j : j \geq i\} \in \mathcal{F}$; now it is a measure-preserving Boolean homomorphism because every π_{ji} is.

(b) If $a_k = \pi_{ki}a$ for $k \geq i$, then $a_k = \pi_{kj}\pi_{ji}a$ for $k \geq j$, so $\pi_j\pi_{ji}a = \langle a_k\rangle^\bullet_{k\in I} = \pi_i a$.

(c) Set $c = \sup_{j\in A}\pi_j a_j$ in $\mathfrak{C}$. For every $\epsilon > 0$, there is a finite $K \subseteq A$ such that $\bar{\nu}c \leq \epsilon + \bar{\nu}(\sup_{j\in K}\pi_j a_j)$, because $\bar{\nu}$ is order-continuous (Fact 3Ba). The set $B = \{k : k \in I,\ j \leq k \text{ for every } j \in K\}$ belongs to $\mathcal{F}$; fix $k \in B$, and set $b = \sup_{j\in K}\pi_{kj}a_j \in \mathfrak{A}_k$,

$$
\begin{aligned}
b_i &= \pi_{ik}b \text{ if } k \leq i,\\
&= 0_{\mathfrak{A}_i} \text{ for other } i \in I.
\end{aligned}
$$

Then

$$\langle b_i\rangle^\bullet_{i\in I} = \pi_k b = \pi_k(\sup_{j\in K}\pi_{kj}a_j) = \sup_{j\in K}\pi_k\pi_{kj}a_j = \sup_{j\in K}\pi_j a_j \subseteq c.$$

If $i \in A$ and $i \geq k$, then

$$
\begin{aligned}
\bar{\mu}_i(a_i \setminus b_i) &= \bar{\nu}(\pi_i a_i \setminus \pi_i b_i) = \bar{\nu}(\pi_i a_i \setminus \pi_i\pi_{ik}b)\\
&= \bar{\nu}(\pi_i a_i \setminus \pi_k b) = \bar{\nu}(\pi_i a_i \setminus \sup_{j\in K}\pi_j a_j) \leq \bar{\nu}(c \setminus \sup_{j\in K}\pi_j a_j) \leq \epsilon
\end{aligned}
$$

by the choice of K. Because $\{i : i \in A,\ i \geq k\} \in \mathcal{F}$,

$$
\begin{aligned}
\bar{\nu}(\langle a_i\rangle^\bullet_{i\in I} \setminus c) &\leq \bar{\nu}(\langle a_i\rangle^\bullet_{i\in I} \setminus \pi_k b) = \bar{\nu}(\langle a_i \setminus b_i\rangle^\bullet_{i\in I})\\
&= \lim_{i\to\mathcal{F}}\bar{\mu}_i(a_i \setminus b_i) \leq \sup_{i\in A, i\geq k}\bar{\mu}_i(a_i \setminus b_i) \leq \epsilon.
\end{aligned}
$$

As ϵ is arbitrary, $\bar{\nu}(\langle a_i\rangle^\bullet_{i\in I} \setminus c) = 0$ and $\langle a_i\rangle^\bullet_{i\in I} \subseteq c$.

4C Proof of Theorem 1G Let (X, Σ, μ) be a compact probability space and $(\mathfrak{A}, \bar{\mu})$ its measure algebra. Let I be a set and $\mathcal{F}$ a measure-centering ultrafilter on I; write λ for the corresponding extension of μ as described in Theorem 1E, and $(\mathfrak{C}, \bar{\nu})$ for the reduced power $(\mathfrak{A}, \bar{\mu})^I|\mathcal{F}$. Then we have a natural isomorphism between $(\mathfrak{C}, \bar{\nu})$ and the measure algebra $(\mathfrak{D}, \bar{\lambda})$ of λ defined by saying that $\langle E_i^\bullet\rangle^\bullet_{i\in I} \in \mathfrak{C}$ is matched with $(\lim_{i\to\mathcal{F}} E_i)^\bullet \in \mathfrak{D}$ for every family $\langle E_i\rangle_{i\in I}$ in Σ.

PROOF. If $\langle E_i\rangle_{i\in I}$ and $\langle F_i\rangle_{i\in I}$ are families in Σ, then $(\lim_{i\to\mathcal{F}} E_i)^\bullet = (\lim_{i\to\mathcal{F}} F_i)^\bullet$ in $\mathfrak{D}$ iff $\langle E_i^\bullet\rangle^\bullet_{i\in I} = \langle F_i^\bullet\rangle^\bullet_{i\in I}$ in $\mathfrak{C}$. **P**

$$
\begin{aligned}
(\lim_{i\to\mathcal{F}} E_i)^\bullet = (\lim_{i\to\mathcal{F}} F_i)^\bullet &\iff (\lim_{i\to\mathcal{F}} E_i)^\bullet \triangle (\lim_{i\to\mathcal{F}} F_i)^\bullet = 0_{\mathfrak{D}}\\
&\iff ((\lim_{i\to\mathcal{F}} E_i) \triangle (\lim_{i\to\mathcal{F}} F_i))^\bullet = 0_{\mathfrak{D}}\\
&\iff (\lim_{i\to\mathcal{F}} E_i \triangle F_i)^\bullet = 0_{\mathfrak{D}}\\
&\iff \lambda(\lim_{i\to\mathcal{F}} E_i \triangle F_i) = 0\\
&\iff \lim_{i\to\mathcal{F}} \mu(E_i \triangle F_i) = 0\\
&\iff \lim_{i\to\mathcal{F}} \bar\mu(E_i \triangle F_i)^\bullet = 0\\
&\iff \lim_{i\to\mathcal{F}} \bar\mu(E_i^\bullet \triangle F_i^\bullet) = 0\\
&\iff \bar\nu(\langle E_i^\bullet \triangle F_i^\bullet\rangle_{i\in I}^\bullet) = 0\\
&\iff \bar\nu(\langle E_i^\bullet\rangle_{i\in I}^\bullet \triangle \langle F_i^\bullet\rangle_{i\in I}^\bullet) = 0\\
&\iff \langle E_i^\bullet\rangle_{i\in I}^\bullet \triangle \langle F_i^\bullet\rangle_{i\in I}^\bullet = 0_{\mathfrak{C}}\\
&\iff \langle E_i^\bullet\rangle_{i\in I}^\bullet = \langle F_i^\bullet\rangle_{i\in I}^\bullet. \quad \mathbf{Q}
\end{aligned}
$$

We therefore have a function $\pi : \mathfrak{D} \to \mathfrak{C}$ defined by saying that $\pi(\langle E_i^\bullet\rangle_{i\in I}^\bullet) = (\lim_{i\to\mathcal{F}} E_i)^\bullet$ for every family $\langle E_i\rangle_{i\in I}$ in Σ. Following the formulae just above, we see that the same ideas tell us that π is a Boolean homomorphism and that $\bar\lambda\pi = \bar\nu$. Next, $\pi[\mathfrak{C}] = \{A^\bullet : A \in \mathcal{A}\}$, where $\mathcal{A}$ is the family of Theorem 1E. But this means that every $d \in \mathfrak{D}$ is the supremum of a non-decreasing sequence in $\pi[\mathfrak{C}]$ and is expressible as $\sup_{n\in\mathbb{N}} \pi c_n$ for some non-decreasing sequence $\langle c_n\rangle_{n\in\mathbb{N}}$ in $\mathfrak{C}$. In this case, if $c = \sup_{n\in\mathbb{N}} c_n$, we must have $d \subseteq \pi c$ and $\bar\lambda d = \bar\lambda(\pi c)$, so that $d = \pi c \in \pi[\mathfrak{C}]$. This shows that π is an isomorphism between $\mathfrak{C}$ and $\mathfrak{D}$, as required.

5 Products of filters

I come now to the proofs of Theorems 1J and 1K, based on the ideas of §§2-4. The first step is to find a way of simultaneously representing many reduced products inside a single probability algebra.

5A Lemma Let $\zeta \geq 1$ be an ordinal, and suppose that for $1 \leq \xi \leq \zeta$ we are given a set I_ξ, an ultrafilter $\mathcal{F}_\xi$ on I_ξ, and a function $i \mapsto \theta(\xi, i) : I_\xi \to \xi$ such that $\{i : i \in I_\xi,\ \theta(\xi,i) \geq \eta\} \in \mathcal{F}_\xi$ for every $\eta < \xi$. Set $I = \bigcup_{1\leq\xi\leq\zeta} I_\xi$, and let $\langle \mathcal{G}_\xi\rangle_{\xi\leq\zeta}$ be the family of ultrafilters on $S^* = \bigcup_{i\in\mathbb{N}} I^i$ constructed from $\langle\mathcal{F}_\xi\rangle_{1\leq\xi\leq\zeta}$ and θ as in 1I. Let $(\mathfrak{A}, \bar\mu)$ be a probability algebra. Then there are a probability algebra $(\mathfrak{C}, \bar\nu)$, a family $\langle\mathfrak{C}_\xi\rangle_{\xi\leq\zeta}$ of closed subalgebras of $\mathfrak{C}$, a measure-preserving Boolean homomorphism $\pi : \mathfrak{A} \to \mathfrak{C}$ such that $\pi[\mathfrak{A}] = \mathfrak{C}_0$, and a family $\langle\psi_\xi\rangle_{1\leq\xi\leq\zeta}$ such that, for $1 \leq \xi \leq \zeta$, $\psi_\xi : \prod_{i\in I_\xi} \mathfrak{C}_{\theta(\xi,i)} \to \mathfrak{C}_\xi$ is a Boolean homomorphism and

$$\bar\nu\psi_\xi(\langle c_i\rangle_{i\in I_\xi}) = \lim_{i\to\mathcal{F}_\xi} \bar\nu c_i, \qquad \psi_\xi(\langle c_i\rangle_{i\in I_\xi}) \subseteq \sup_{i\in A} c_i$$

whenever $\langle c_i\rangle_{i\in I_\xi} \in \prod_{i\in I_\xi} \mathfrak{C}_{\theta(\xi,i)}$ and $A \in \mathcal{F}_\xi$.

PROOF. Define families $\langle(\mathfrak{A}_\xi, \bar\mu_\xi)\rangle_{\xi\leq\zeta}$, $\langle\phi_{\xi\eta}\rangle_{\eta\leq\xi\leq\zeta}$ inductively, as follows. The inductive hypothesis will be that each $(\mathfrak{A}_\xi, \bar\mu_\xi)$ is a probability algebra and that each $\phi_{\xi\eta}$ is a measure-preserving Boolean homomorphism from $\mathfrak{A}_\eta$ to $\mathfrak{A}_\xi$ such that $\phi_{\xi\eta'}\phi_{\eta'\eta} = \phi_{\xi\eta}$ whenever $\eta \leq \eta' \leq \xi$.

Start with $(\mathfrak{A}_0, \bar\mu_0) = (\mathfrak{A}, \bar\mu)$ and $\phi_{00} : \mathfrak{A}_0 \to \mathfrak{A}_0$ the identity map. Given $\langle(\mathfrak{A}_\eta, \bar\mu_\eta)\rangle_{\eta<\xi}$ and $\langle\phi_{\eta'\eta}\rangle_{\eta\le\eta'<\xi}$, where $0 < \xi \le \zeta$, let $(\mathfrak{A}_\xi, \bar\mu_\xi)$ be the reduced power $\prod_{i\in I_\xi}(\mathfrak{A}_{\theta(\xi,i)}, \bar\mu_{\theta(\xi,i)})|\mathcal{F}_\xi$, as defined in 1Fb/4A. Let $\le_\xi$ be the pre-order on I_ξ defined by saying that $i \le_\xi j$ iff $\theta(\xi, i) \le \theta(\xi, j)$; then our hypothesis on θ ensures that $\{j : i \le_\xi j\} \in \mathcal{F}_\xi$ for every $i \in I_\xi$. For $i \le_\xi j \in I_\xi$, $\tilde\phi_{ji} = \phi_{\theta(\xi,j),\theta(\xi,i)} : \mathfrak{A}_{\theta(\xi,i)} \to \mathfrak{A}_{\theta(\xi,j)}$ is defined; and if $i \le_\xi j \le_\xi k$, then $\tilde\phi_{ki} = \tilde\phi_{kj}\tilde\phi_{ji}$. By 4B, we have measure-preserving Boolean homomorphisms $\tilde\phi_i : \mathfrak{A}_{\theta(\xi,i)} \to \mathfrak{A}_\xi$ such that $\tilde\phi_i = \tilde\phi_j\tilde\phi_{ji}$ for $i \le_\xi j$. If $i \le_\xi j$ and $\eta \le \theta(\xi, i)$, then

$$\tilde\phi_j\phi_{\theta(\xi,j),\eta} = \tilde\phi_j\phi_{\theta(\xi,j),\theta(\xi,i)}\phi_{\theta(\xi,i),\eta} = \tilde\phi_i\phi_{\theta(\xi,i),\eta},$$

so we can take this common value for $\phi_{\xi\eta} : \mathfrak{A}_\eta \to \mathfrak{A}_\xi$. If $\eta \le \eta' < \xi$, take $i \in I_\xi$ such that $\eta' \le \theta(\xi, i)$, and see that

$$\phi_{\xi\eta'}\phi_{\eta'\eta} = \tilde\phi_i\phi_{\theta(\xi,i),\eta'}\phi_{\eta'\eta} = \tilde\phi_i\phi_{\theta(\xi,i),\eta} = \phi_{\xi\eta},$$

so the induction proceeds.

At the end of the induction, set $\mathfrak{C} = \mathfrak{A}_\zeta$, $\bar\nu = \bar\mu_\zeta$, $\pi_\xi = \phi_{\zeta\xi} : \mathfrak{A}_\xi \to \mathfrak{C}$ and $\mathfrak{C}_\xi = \pi_\xi[\mathfrak{A}_\xi] \subseteq \mathfrak{C}$. If $\eta \le \xi \le \zeta$, then $\pi_\eta = \pi_\xi\phi_{\xi\eta}$, so $\mathfrak{C}_\eta \subseteq \mathfrak{C}_\xi$.

For each $\xi > 0$, we have a canonical map $\langle a_i\rangle_{i\in I_\xi} \mapsto \langle a_i\rangle^\bullet_{i\in I_\xi} : \prod_{i\in I_\xi}\mathfrak{A}_{\theta(\xi,i)} \to \mathfrak{A}_\xi$. Since $\pi_\eta : \mathfrak{A}_\eta \to \mathfrak{C}_\eta$ is always a measure-preserving isomorphism, we have corresponding maps $\psi_\xi : \prod_{i\in I_\xi}\mathfrak{C}_{\theta(\xi,i)} \to \mathfrak{C}_\xi$. Reading off the basic facts from 4A-4B, we see that

$$\bar\nu\psi_\xi(\langle c_i\rangle_{i\in I_\xi}) = \lim_{i\to\mathcal{F}_\xi}\bar\nu c_i \text{ whenever } \langle c_i\rangle_{i\in I_\xi} \in \prod_{i\in I_\xi}\mathfrak{C}_{\theta(\xi,i)},$$

$$\psi_\xi(\langle c_i\rangle_{i\in I_\xi}) \subseteq \sup_{i\in A} c_i \text{ whenever } \langle c_i\rangle_{i\in I_\xi} \in \prod_{i\in\mathbb{N}}\mathfrak{C}_{\theta(\xi,i)} \text{ and } A \in \mathcal{F}_\xi$$

(we can take the suprema in $\mathfrak{C}$ because $\mathfrak{C}_\xi$, being a closed subalgebra, is regularly embedded in $\mathfrak{C}$, as noted in [12], 314Ga).

5B Proof of Theorem 1J Let $\zeta \ge 1$ be a countable ordinal, and I an infinite set. Suppose that for $1 \le \xi \le \zeta$ we are given a Ramsey ultrafilter $\mathcal{F}_\xi$ on I, and a function $i \mapsto \theta(\xi, i) : I \to \xi$ such that $\{i : \theta(\xi, i) \ge \eta\}$ belongs to $\mathcal{F}_\xi$ for every $\eta < \xi$. Let $\langle\mathcal{G}_\xi\rangle_{\xi\le\zeta}$ be the family of ultrafilters on $S^* = \bigcup_{n\in\mathbb{N}} I^n$ constructed from $\langle\mathcal{F}_\xi\rangle_{1\le\xi\le\zeta}$ and θ as in 1I. Suppose further that all the $\mathcal{F}_\xi$ are non-isomorphic. Then $\mathcal{G}_\zeta$ is measure-centering.

PROOF. **(a)** It is enough to consider the case in which $I = \kappa$ is a cardinal. Let $(\mathfrak{A}, \bar\mu)$ be a probability algebra, and $\langle a_\tau\rangle_{\tau\in S^*}$ a family in $\mathfrak{A}$ such that $\epsilon = \inf_{\tau\in S^*}\bar\mu a_\tau$ is greater than 0. By 5A, there are a probability algebra $(\mathfrak{C}, \bar\nu)$, a family $\langle\mathfrak{C}_\xi\rangle_{\xi\le\zeta}$ of closed subalgebras of $\mathfrak{C}$, a measure-preserving Boolean homomorphism $\pi : \mathfrak{A} \to \mathfrak{C}$ such that $\pi[\mathfrak{A}] = \mathfrak{C}_0$, and a family $\langle\psi_\xi\rangle_{1\le\xi\le\zeta}$ such that, for $1 \le \xi \le \zeta$, $\psi_\xi : \prod_{\delta<\kappa}\mathfrak{C}_{\theta(\xi,\delta)} \to \mathfrak{C}_\xi$ is a Boolean homomorphism and

$$\bar\nu\psi_\xi(\langle c_\delta\rangle_{\delta<\kappa}) = \lim_{\delta\to\mathcal{F}_\xi}\bar\nu c_\delta, \quad \psi_\xi(\langle c_\delta\rangle_{\delta<\kappa}) \subseteq \sup_{\delta\in A} c_\delta$$

whenever $\langle c_\delta\rangle_{\delta<\kappa} \in \prod_{\delta<\kappa}\mathfrak{C}_{\theta(\xi,\delta)}$ and $A \in \mathcal{F}_\xi$.

(b) By 2Ea, there is a disjoint family $\langle A_\xi\rangle_{1\le\xi\le\zeta}$ of subsets of I such that $A_\xi \in \mathcal{F}_\xi$ for every ξ. (This is where we need to know that ζ is countable.) Define $T \subseteq S^*$ and $\alpha : T \to [0, \zeta]$ as follows. Start by saying that $\emptyset \in T$ and that $\alpha(\emptyset) = \zeta$. Having determined $T \cap \kappa^n$ and $\alpha : T \cap \kappa^n \to [0, \zeta]$, where $n \in \mathbb{N}$, then for $\tau \in \kappa^{n+1}$ say that $\tau \in T$ iff τ is of the form $\sigma^\frown{<}\delta{>}$ where

$$\sigma \in T \cap \kappa^n, \quad \alpha(\sigma) > 0, \quad \delta \in A_{\alpha(\sigma)}, \quad \sigma(m) < \delta \text{ for every } m < n,$$

and in this case set $\alpha(\tau) = \theta(\alpha(\sigma), \delta)$. Continue. Observe that every member of T is a strictly increasing finite sequence in κ. For $D \subseteq \kappa$, set $T_D = T \cap \bigcup_{n\in\mathbb{N}} D^n$.

(c) Set $\mathcal{H} = \bigcap_{1\le\xi\le\zeta} \mathcal{F}_\xi$. Then $T_D^* = \{\tau : \tau \in T_D,\ \alpha(\tau) = 0\}$ belongs to $\mathcal{G}_\zeta$ for every $D \in \mathcal{H}$. **P** I aim to show by induction on ξ that if $\tau \in T_D$ and $\alpha(\tau) = \xi$ then $\{\sigma : \tau^\frown\sigma \in T_D^*\} \in \mathcal{G}_\xi$. The induction starts with $\alpha(\tau) = 0$ and $\{\sigma : \tau^\frown\sigma \in T_D^*\} = \{\emptyset\} \in \mathcal{G}_0$. For the inductive step to $\alpha(\tau) = \xi > 0$,

$$\{\delta : \{\sigma : \tau^\frown{<}\delta{>}^\frown\sigma \in T_D^*\} \in \mathcal{G}_{\theta(\xi,\delta)}\}$$
$$\supseteq \{\delta : \delta \in D,\ \tau^\frown{<}\delta{>} \in T,\ \alpha(\tau^\frown{<}\delta{>}) = \theta(\xi,\delta)\}$$

(by the inductive hypothesis)

$$= \{\delta : \delta \in A_\xi \cap D,\ \tau(m) < \delta \text{ for every } m < \operatorname{dom}\tau\} \in \mathcal{F}_\xi,$$

so $\{\sigma : \tau^\frown\sigma \in T_D^*\} \in \mathcal{G}_\xi$. At the end of the induction, we can apply this to $\tau = \emptyset$ and $\xi = \zeta$. **Q**

(d) Set $c_\tau = \pi a_\tau$ for $\tau \in S^* \setminus T$. For $\tau \in T$ define $c_\tau \in \mathfrak{C}_{\alpha(\tau)}$ by induction on $\alpha(\tau)$, as follows. If $\alpha(\tau) = 0$, set $c_\tau = \pi a_\tau$. For the inductive step to $\alpha(\tau) = \xi > 0$, $\alpha(\tau^\frown{<}\delta{>}) = \theta(\alpha(\tau), \delta) < \alpha(\tau)$ whenever $\tau^\frown{<}\delta{>} \in T$. So $c_{\tau^\frown<\delta>}$ is defined and belongs to $\mathfrak{C}_{\theta(\xi,\delta)}$ for every $\delta < \kappa$. We can therefore set $c_\tau = \psi_\xi(\langle c_{\tau^\frown<\delta>}\rangle_{\delta<\kappa})$, and continue. Inducing on $\alpha(\tau)$ when $\tau \in T$, we see that $\bar\nu c_\tau \ge \epsilon$ for every $\tau \in S^*$.

(e) For $K \subseteq I$, set $e_K = \inf_{\tau\in T_K} c_\tau$; let $\mathcal{S}$ be the family of those finite sets $K \subseteq I$ such that $e_K \ne 0$. Interpreting $\inf\emptyset$ as $1_{\mathfrak{C}}$, $\emptyset \in \mathcal{S}$. Moreover, if $K \in \mathcal{S}$, then $\{\delta : K \cup \{\delta\} \in \mathcal{S}\}$ belongs to $\mathcal{H}$. **P** Set $\gamma = \sup K$. Take any ξ such that $1 \le \xi \le \zeta$. Set

$$d_\delta = \inf\{c_{\tau^\frown<\delta>} : \tau \in T_K,\ \alpha(\tau) = \xi\}$$

for $\delta < \kappa$,

$$B = \{\delta : \delta \in A_\xi,\ \delta > \gamma,\ d_\delta \cap e_K \ne 0\}.$$

If $\delta \in B$, then

$$T_{K\cup\{\delta\}} = T_K \cup \{\tau^\frown{<}\delta{>} : \tau \in T_K,\ \alpha(\tau) = \xi\},$$

because every member of T is strictly increasing and $\tau^\frown{<}\delta{>}$ can belong to T only when $\delta \in A_{\alpha(\tau)}$, that is, when $\alpha(\tau) = \xi$. So $e_{K\cup\{\delta\}} = d_\delta \cap e_K \ne 0$ and $K \cup \{\delta\} \in \mathcal{S}$.

? If $B \notin \mathcal{F}_\xi$, then $B' = \{\delta : \delta \in A_\xi,\ \delta > \gamma,\ d_\delta \cap e_K = 0\}$ belongs to $\mathcal{F}_\xi$. So

$$e_K \subseteq \inf\{c_\tau : \tau \in T_K,\ \alpha(\tau) = \xi\}$$
$$= \inf_{\substack{\tau\in T_K\\ \alpha(\tau)=\xi}} \psi_\xi(\langle c_{\tau^\frown<\delta>}\rangle_{\delta<\kappa}) = \psi_\xi(\langle \inf_{\substack{\tau\in T_K\\ \alpha(\tau)=\xi}} c_{\tau^\frown<\delta>}\rangle_{\delta<\kappa})$$

(because ψ_ξ is a Boolean homomorphism and T_K is finite)

$$\subseteq \sup_{\delta\in B'} \inf_{\substack{\tau\in T_K\\ \alpha(\tau)=\xi}} c_{\tau^\frown<\delta>}$$

(as noted in (a) above)

$$= \sup_{\delta\in B'} d_\delta.$$

But $e_K \cap d_\delta = 0$ for every $\delta \in B'$ and $e_K \neq 0$. **X**

Thus $\{\delta : K \cup \{\delta\} \in \mathcal{S}\} \supseteq B$ belongs to $\mathcal{F}_\xi$. As ξ is arbitrary, $\{\delta : K \cup \{\delta\} \in \mathcal{S}\} \in \mathcal{H}$. **Q**

(f) At this point, recall that $\mathcal{H}$ is dependently selective, by 2D. So there is a $D \in \mathcal{H}$ such that $[D]^{<\omega} \subseteq \mathcal{S}$, that is, $e_K \neq 0$ for every $K \in [D]^{<\omega}$, that is, $\{c_\tau : \tau \in T_D\}$ is centered in $\mathfrak{C}$. It follows that $\{c_\tau : \tau \in T_D^*\}$ is centered; but for $\tau \in T_D^*$, $c_\tau = \pi a_\tau$, so $\{a_\tau : \tau \in T_D^*\}$ is centered, while $T_D^* \in \mathcal{G}_\zeta$, by (c). As $(\mathfrak{A}, \bar\mu)$ and $\langle a_\tau\rangle_{\tau\in S^*}$ are arbitrary, $\mathcal{G}_\zeta$ is measure-centering, by Proposition 1C(ii).

5C Lemma Suppose that I, J are sets and that $\mathcal{F}$, $\mathcal{G}$ are measure-centering ultrafilters on I, J respectively such that $\mathcal{F}$ is $\#(J)^+$-complete. If $(\mathfrak{A}, \bar\mu)$ is a probability algebra and $\langle a_{ij}\rangle_{i\in I, j\in J}$ is a family in $\mathfrak{A}$ such that $\inf_{i\in I, j\in J} \bar\mu a_{ij} > 0$, there are $A \in \mathcal{F}$, $B \in \mathcal{G}$ such that $\{a_{ij} : i \in A, j \in B\}$ is centered.

Proof. If J is finite then $\mathcal{G}$ is principal; taking $B = \{j\}$ to be the singleton belonging to $\mathcal{G}$, there is an $A \in \mathcal{F}$ such that $\{a_{ij} : i \in A\}$ is centered, and we're done. Otherwise, let κ be the greatest cardinal such that $\mathcal{F}$ is κ-complete. Then κ is two-valued-measurable and greater than $\#(J)$, so must be greater than $2^{\#(J)}$. For $K \in [J]^{<\omega}$ and $i \in J$ set $b_{iK} = \inf_{j\in K} a_{ij}$. Because $\mathcal{F}$ is κ-complete, there is a family $\langle \gamma_K\rangle_{K\in[J]^{<\omega}}$ in $[0,1]$ such that $A^* = \{i : \bar\mu b_{iK} = \gamma_K$ for every finite $K \subseteq I\}$ belongs to $\mathcal{F}$. Take any $i_0 \in A^*$; then there must be a $B \in \mathcal{G}$ such that $\{a_{i_0 j} : j \in B\}$ is centered, so that $\gamma_K > 0$ for every $K \in [B]^{<\omega}$. In this case, for each $K \in [B]^{<\omega}$, there is an $A_K \in \mathcal{F}$ such that $\{b_{iK} : i \in A_K\}$ is centered. Set $A = A^* \cap \bigcap_{K\subseteq B \text{ is finite}} A_K$; then $A \in \mathcal{F}$ and $\{a_{ij} : i \in A, j \in B\}$ is centered.

5D Lemma Suppose that $\mathcal{F}_0, \dots, \mathcal{F}_n$ are filters on sets $I_0, \dots, I_n$. Let (K, L) be a non-trivial partition of $\{0, \dots, n\}$, and $(i_0, \dots, i_k)$, $(j_0, \dots, j_l)$ the increasing enumerations of K, L respectively. For $A \subseteq I_{i_0} \times \dots \times I_{i_k}$ and $B \subseteq I_{j_0} \times \dots \times I_{j_l}$ set

$$A\#B = \{(x_0, \dots, x_n) : (x_{i_0}, \dots, x_{i_k}) \in A,\ (x_{j_0}, \dots, x_{j_l}) \in B\} \subseteq I_0 \times \dots \times I_n.$$

If $A \in \mathcal{F}_{i_0} \ltimes \dots \ltimes \mathcal{F}_{i_k}$ and $B \in \mathcal{F}_{j_0} \ltimes \dots \ltimes \mathcal{F}_{j_l}$ then $A\#B \in \mathcal{F}_0 \ltimes \dots \ltimes \mathcal{F}_n$.

Proof. Induce on n. The induction starts with $n = 1$ and $k = l = 0$ and either $A\#B = A \times B$ with $A \in \mathcal{F}_0$ and $B \in \mathcal{F}_1$, or $A\#B = B \times A$ with $B \in \mathcal{F}_0$ and $A \in \mathcal{F}_1$; in either case the result is trivial.

For the inductive step to $n > 1$, suppose to begin with that $0 \in K$. If $K = \{0\}$, then we have $A \in \mathcal{F}_0$ and $B \in \mathcal{F}_1 \ltimes \dots \ltimes \mathcal{F}_n$, so that $A\#B$ can be identified with $A \times B \in \mathcal{F}_0 \ltimes (\mathcal{F}_1 \ltimes \dots \ltimes \mathcal{F}_n)$, identified with $\mathcal{F}_0 \ltimes \dots \ltimes \mathcal{F}_n$. Otherwise, set $K' = K \setminus \{0\}$, so that (K', L) is a non-trivial partition of $\{1, \dots, n\}$. For $x \in I_0$, it is easy to see that

$$(A\#B)[\{x\}] = \{(x_1, \dots, x_n) : (x, x_1, \dots, x_n) \in A\#B\}$$

and

$$A[\{x\}]\#B = \{(x_{i_1}, \dots, x_{i_k}) : (x, x_{i_1}, \dots, x_{i_k}) \in A\}\#B$$

are equal, where the interleaving $A[\{x\}]\#B$ is computed with regard to the partition (K', L) and the increasing enumerations $(i_1, \dots, i_k)$ and $(j_0, \dots, j_l)$. Now

$$\{x : (A\#B)[\{x\}] \in \mathcal{F}_1 \ltimes \dots \ltimes \mathcal{F}_n\} = \{x : A[\{x\}]\#B \in \mathcal{F}_1 \ltimes \dots \ltimes \mathcal{F}_n\}$$
$$\supseteq \{x : A[\{x\}] \in \mathcal{F}_{i_1} \ltimes \dots \ltimes \mathcal{F}_{i_k}\}$$

(by the inductive hypothesis, because $B \in \mathcal{F}_{j_0} \ltimes \dots \ltimes \mathcal{F}_{j_l}$)

$$\in \mathcal{F}_{i_0} = \mathcal{F}_0$$

because $A \in \mathcal{F}_{i_0} \ltimes \dots \ltimes \mathcal{F}_{i_k}$. So $A\#B \in \mathcal{F}_0 \ltimes \dots \ltimes \mathcal{F}_n$.

If $0 \in L$, the same argument applies, but looking at sections $B[\{x\}]$ rather than $A[\{x\}]$ and the partition $(K, L \setminus \{0\})$ of $\{1, \dots, n\}$. So the induction proceeds.

5E Proposition Suppose that $\mathcal{F}_0, \dots, \mathcal{F}_n$ are non-isomorphic Ramsey ultrafilters. Then $\mathcal{F}_0 \ltimes \dots \ltimes \mathcal{F}_n$ is measure-centering.

PROOF. It is enough to consider the case in which every $\mathcal{F}_i$ is based on a cardinal λ_i. Induce on n. If $n = 0$ we just use Theorem 1Ba. For the inductive step, set $\kappa = \max_{i \le n} \lambda_i$, $K = \{i : i \le n,\ \lambda_i = \kappa\}$, $L = \{i : i \le n,\ \lambda_i < \kappa\}$. If $K = \{0, \dots, n\}$ then we can use Theorem 1J with $I = \kappa$ and $\zeta = n + 1$. Otherwise, (K, L) is a proper partition of $\{0, \dots, n\}$. Let $(i_0, \dots, i_k)$ and $(j_0, \dots, j_l)$ be the increasing enumerations of K, L respectively. Then the inductive hypothesis tells us that $\mathcal{G} = \mathcal{F}_{i_0} \ltimes \dots \ltimes \mathcal{F}_{i_k}$ and $\mathcal{H} = \mathcal{F}_{j_0} \ltimes \dots \ltimes \mathcal{F}_{j_l}$ are measure-centering. Observe next that as every $\mathcal{F}_{i_m}$ is κ-complete, so is $\mathcal{G}$, while the base set of $\mathcal{H}$ is $\lambda_{j_0} \times \dots \times \lambda_{j_l}$, which has cardinal less than κ.

Let $(\mathfrak{A}, \bar{\mu})$ be a probability measure, and $\langle a_x \rangle_{x \in I}$ a family in $\mathfrak{A}$ such that $\inf_{x \in I} \bar{\mu} a_x > 0$, where $I = \lambda_0 \times \dots \times \lambda_n$. For $x \in I$, set $x' = (x_{i_0}, \dots, x_{i_k})$ and $x'' = \{x_{j_0}, \dots, x_{j_l}\}$, so that $x \mapsto (x', x'')$ is a bijection between I and $(I_{i_0} \times \dots \times I_{i_k}) \times (I_{j_0} \times \dots \times I_{j_l})$. By Lemma 5C, there are sets $A \in \mathcal{G}$, $B \in \mathcal{H}$ such that $\{a_x : x' \in A,\ x'' \in B\}$ is centered; that is, in the language of Lemma 5D, $\{a_x : x \in A\#B\}$ is centered. But 5D tells us that $A\#B \in \mathcal{F}_0 \ltimes \dots \ltimes \mathcal{F}_n$. As $\langle a_x \rangle_{x \in I}$ is arbitrary, $\mathcal{F}_0 \ltimes \dots \ltimes \mathcal{F}_n$ is measure-centering, and the induction continues.

5F To prove Theorem 1K, we need to know a little more both about the extensions of measures described in 1E and about the iterated products of 1I. It will be convenient to have a name for a relation extending the Rudin-Keisler pre-ordering of the ultrafilters on a given set.

Definition If $\mathcal{F}$ and $\mathcal{G}$ are filters on sets I, J respectively, I will say that $\mathcal{F} \le_{\mathrm{RK}} \mathcal{G}$ if there is a function $f : J \to I$ such that $\mathcal{F} = f[[\mathcal{G}]]$. Observe that $\le_{\mathrm{RK}}$ is a reflexive transitive relation on the class of all filters.

5G Proposition Let $\mathcal{F}$ and $\mathcal{G}$ be measure-centering ultrafilters such that $\mathcal{F} \le_{\mathrm{RK}} \mathcal{G}$. Let (X, Σ, μ) be a compact probability space, and $\lambda_{\mathcal{F}}$, $\lambda_{\mathcal{G}}$ the extensions of μ defined from $\mathcal{F}$ and $\mathcal{G}$ as in Theorem 1E. Then $\lambda_{\mathcal{G}}$ extends $\lambda_{\mathcal{F}}$.

PROOF. Set $I = \bigcup \mathcal{F}$ and $J = \bigcup \mathcal{G}$, and let $f : J \to I$ be such that $\mathcal{F} = f[[\mathcal{G}]]$. Defining $\mathcal{A}_{\mathcal{F}}$ and $\mathcal{A}_{\mathcal{G}}$ from $\mathcal{F}$ and $\mathcal{G}$ as in 1E, $\mathcal{A}_{\mathcal{F}} \subseteq \mathcal{A}_{\mathcal{G}}$ and $\lambda_{\mathcal{G}} A = \lambda_{\mathcal{F}} A$ for every

$A \in \mathcal{A}_{\mathcal{F}}$. **P** Express A as $\lim_{i\to\mathcal{F}} E_i$ where $\langle E_i\rangle_{i\in I}$ is a family in Σ. For $j \in J$, set $F_j = E_{f(j)}$; then $A = \lim_{j\to\mathcal{G}} F_j \in \mathcal{A}_{\mathcal{G}}$ and

$$\lambda_{\mathcal{F}} A = \lim_{i\to\mathcal{F}} \mu E_i = \lim_{j\to\mathcal{G}} \mu F_j = \lambda_{\mathcal{G}} A. \textbf{ Q}$$

Since $\lambda_{\mathcal{F}}$ and $\lambda_{\mathcal{G}}$ are complete probability measures and $\lambda_{\mathcal{F}}$ is inner regular with respect to $\mathcal{A}_{\mathcal{F}}$, $\lambda_{\mathcal{G}}$ extends $\lambda_{\mathcal{F}}$, by 3A(a-iii).

5H Proposition Let $\mathcal{F}_0, \dots, \mathcal{F}_n$ be filters.

(a) $\mathcal{F}_0 \le_{\mathrm{RK}} \mathcal{F}_0 \ltimes \mathcal{F}_1$ and $\mathcal{F}_1 \le_{\mathrm{RK}} \mathcal{F}_0 \ltimes \mathcal{F}_1$.

(b) If $\mathcal{G}_0, \dots, \mathcal{G}_n$ are filters such that $\mathcal{G}_k \le_{\mathrm{RK}} \mathcal{F}_k$ for every $k \le n$, then $\mathcal{G}_0 \ltimes \dots \ltimes \mathcal{G}_n \le_{\mathrm{RK}} \mathcal{F}_0 \ltimes \dots \ltimes \mathcal{F}_n$.

(c) If $0 \le k_0 < k_1 < \dots < k_m \le n$, then $\mathcal{F}_{k_0} \ltimes \dots \ltimes \mathcal{F}_{k_m} \le_{\mathrm{RK}} \mathcal{F}_0 \ltimes \dots \ltimes \mathcal{F}_n$.

PROOF. Let $I_k = \bigcup \mathcal{F}_k$ be the base set of $\mathcal{F}_k$ for each k.

(a) If $f_0 : I_0 \times I_1 \to I_0$ and $f_1 : I_0 \times I_1 \to I_1$ are the canonical projections, it is easy to see that $\mathcal{F}_0 = f_0[[\mathcal{F}_0 \ltimes \mathcal{F}_1]]$ and $\mathcal{F}_1 = f_1[[\mathcal{F}_0 \ltimes \mathcal{F}_1]]$.

(b)(i) If $n = 1$, set $J_k = \bigcup \mathcal{G}_k$ and let $f_k : I_k \to J_k$ be such that $\mathcal{G}_k = f_k[[\mathcal{F}_k]]$ for $k = 0$ and $k = 1$. Setting $h(i, j) = (f_0(i), f_1(j))$ for $i \in I_0$ and $j \in J_0$, it is easy to check that $\mathcal{G}_0 \ltimes \mathcal{G}_1 = h[[\mathcal{F}_0 \ltimes \mathcal{F}_1]] \le_{\mathrm{RK}} \mathcal{F}_0 \ltimes \mathcal{F}_1$.

(ii) For $n \ge 2$ the result now follows by induction.

(c)(i) Induce on n to see that $\mathcal{F}_k \le_{\mathrm{RK}} \mathcal{F}_0 \ltimes \dots \ltimes \mathcal{F}_n$ whenever $0 \le k \le n$; for the case $n = k > 0$, apply (a) to the product $(\mathcal{F}_0 \ltimes \dots \ltimes \mathcal{F}_{k-1}) \ltimes \mathcal{F}_k$; for the inductive step to $n + 1 > k$, observe that

$$\mathcal{F}_k \le_{\mathrm{RK}} \mathcal{F}_0 \ltimes \dots \ltimes \mathcal{F}_n \le_{\mathrm{RK}} (\mathcal{F}_0 \ltimes \dots \ltimes \mathcal{F}_n) \ltimes \mathcal{F}_{n+1}$$

by the other half of (a).

(ii) Now induce on m; the inductive step to $m + 1$ is

$$\begin{aligned} &(\mathcal{F}_{k_0} \ltimes \mathcal{F}_{k_1} \ltimes \dots \ltimes \mathcal{F}_{k_m}) \ltimes \mathcal{F}_{k_{m+1}} \\ &\qquad \le_{\mathrm{RK}} (\mathcal{F}_0 \ltimes \mathcal{F}_1 \ltimes \dots \ltimes \mathcal{F}_{k_m}) \ltimes (\mathcal{F}_{k_m+1} \ltimes \dots \ltimes \mathcal{F}_n) \\ &\qquad \cong \mathcal{F}_0 \ltimes \dots \ltimes \mathcal{F}_n, \end{aligned}$$

using the inductive hypothesis, (i) here and (b) above for the first step.

5I These have been easy. For the next result we have to think a little harder, but the principles are the same.

Proposition Let ζ, $\langle I_\xi\rangle_{1\le\xi\le\zeta}$, $\langle \mathcal{F}_\xi\rangle_{1\le\xi\le\zeta}$, I, S^*, θ and $\langle \mathcal{G}_\xi\rangle_{\xi\le\zeta}$ be as in 1Ib, and suppose that $\{i : i \in I_\xi,\ \theta(\xi, i) \ge \eta\} \in \mathcal{F}_\xi$ whenever $\eta < \xi \le \zeta$. Then $\mathcal{F}_{\xi_n} \ltimes \dots \ltimes \mathcal{F}_{\xi_0} \le_{\mathrm{RK}} \mathcal{G}_\zeta$ whenever $1 \le \xi_0 < \xi_1 < \dots < \xi_n \le \zeta$.

PROOF. **(a)** The first step is to show that $\mathcal{F}_\xi \le_{\mathrm{RK}} \mathcal{G}_\xi$ whenever $1 \le \xi \le \zeta$. **P** Let $f : S^* \to I_\xi$ be such that $f(\tau) = \tau(0)$ whenever $\tau \ne \emptyset$ and $\tau(0) \in I_\xi$. For $A \subseteq I_\xi$,

$$\begin{aligned} f^{-1}[A] \in \mathcal{G}_\xi &\iff \{i : i \in I_\xi,\, \{\tau : {<}i{>}^\frown\tau \in f^{-1}[A]\} \in \mathcal{G}_{\theta(\xi,i)}\} \in \mathcal{F}_\xi \\ &\iff \{i : i \in I_\xi,\, \{\tau : f({<}i{>}^\frown\tau) \in A\} \in \mathcal{G}_{\theta(\xi,i)}\} \in \mathcal{F}_\xi \\ &\iff \{i : i \in A,\, S^* \in \mathcal{G}_{\theta(\xi,i)}\} \in \mathcal{F}_\xi \iff A \in \mathcal{F}_\xi. \end{aligned}$$

So $\mathcal{F}_\xi = f[[\mathcal{G}_\xi]] \leq_{\mathrm{RK}} \mathcal{G}_\xi$. **Q**

(b) Next, $\mathcal{G}_\eta \leq_{\mathrm{RK}} \mathcal{G}_\xi$ whenever $\eta \leq \xi \leq \zeta$. **P** Induce on ξ. If $\xi = \eta$, the result is trivial. For the inductive step to $\xi > \eta$, set $J = \{i : i \in I_\xi,\, \theta(\xi,i) \geq \eta\}$; then $J \in \mathcal{F}_\xi$, by hypothesis. For $i \in J$, the inductive hypothesis tells us that $\mathcal{G}_\eta \leq_{\mathrm{RK}} \mathcal{G}_{\theta(\xi,i)}$; let $g_i : S^* \to S^*$ be such that $g_i[[\mathcal{G}_{\theta(\xi,i)}]] = \mathcal{G}_\eta$. Now let $g : S^* \to S^*$ be such that $g({<}i{>}^\frown\sigma) = g_i(\sigma)$ whenever $\sigma \in S^*$ and $i \in J$. For $A \subseteq S^*$,

$$\begin{aligned} g^{-1}[A] \in \mathcal{G}_\xi &\iff \{i : i \in I_\xi,\, \{\tau : g({<}i{>}^\frown\tau) \in A\} \in \mathcal{G}_{\theta(\xi,i)}\} \in \mathcal{F}_\xi \\ &\iff \{i : i \in J,\, \{\tau : g_i(\tau) \in A\} \in \mathcal{G}_{\theta(\xi,i)}\} \in \mathcal{F}_\xi \\ &\iff \{i : i \in J,\, g_i^{-1}[A] \in \mathcal{G}_{\theta(\xi,i)}\} \in \mathcal{F}_\xi \\ &\iff \{i : i \in J,\, A \in \mathcal{G}_\eta\} \in \mathcal{F}_\xi \iff A \in \mathcal{G}_\eta, \end{aligned}$$

so $G_\eta = g[[\mathcal{G}_\xi]] \leq_{\mathrm{RK}} g[[\mathcal{G}_\xi]]$. **Q**

(c) It follows that if $1 \leq \xi_0 < \ldots < \xi_n \leq \zeta$ then $\mathcal{F}_{\xi_n} \ltimes \ldots \ltimes \mathcal{F}_{\xi_0} \leq \mathcal{G}_{\xi_n}$. **P** Induce on the pair (n, ξ_n). If $\xi_n = 1$ then $n = 0$ and we just have to know that $\mathcal{F}_1 \leq_{\mathrm{RK}} \mathcal{G}_1$, as in (a). For the inductive step to $\xi_n > 1$, if $n = 0$ we again have only to know that $\mathcal{F}_{\xi_0} \leq_{\mathrm{RK}} \mathcal{G}_{\xi_0} \leq_{\mathrm{RK}} \mathcal{G}_{\xi_n}$, this time using (b). If $n > 0$, let J be $\{i : i \in I_{\xi_n},\, \theta(\xi_n, i) \geq \xi_{n-1}\} \in \mathcal{F}_{\xi_n}$. For $i \in J$,

$$\mathcal{F}_{\xi_{n-1}} \ltimes \ldots \ltimes \mathcal{F}_{\xi_0} \leq_{\mathrm{RK}} \mathcal{G}_{\xi_{n-1}} \leq_{\mathrm{RK}} \mathcal{G}_{\theta(\xi_n,i)}$$

by the inductive hypothesis; let $h_i : S^* \to I_{\xi_{n-1}} \times \ldots \times I_{\xi_0}$ be a function witnessing this. Now let $h : S^* \to I_{\xi_n} \times I_{\xi_{n-1}} \times \ldots \times I_{\xi_0}$ be such that $h({<}i{>}^\frown\sigma) = (i, h_i(\sigma))$ whenever $i \in J$ and $\sigma \in S^*$. For $A \subseteq I_{\xi_n} \times \ldots \times I_{\xi_0}$,

$$\begin{aligned} h^{-1}[A] \in \mathcal{G}_{\xi_n} &\iff \{i : i \in I_{\xi_n},\, \{\tau : h({<}i{>}^\frown\tau) \in A\} \in \mathcal{G}_{\theta(\xi_n,i)}\} \in \mathcal{F}_{\xi_n} \\ &\iff \{i : i \in J,\, \{\tau : h({<}i{>}^\frown\tau) \in A\} \in \mathcal{G}_{\theta(\xi_n,i)}\} \in \mathcal{F}_{\xi_n} \\ &\iff \{i : i \in J,\, \{\tau : h_i(\tau) \in A[\{i\}]\} \in \mathcal{G}_{\theta(\xi_n,i)}\} \in \mathcal{F}_{\xi_n} \\ &\iff \{i : i \in J,\, A[\{i\}] \in h_i[[\mathcal{G}_{\theta(\xi_n,i)}\}]] \in \mathcal{F}_{\xi_n} \\ &\iff \{i : i \in J,\, A[\{i\}] \in \mathcal{F}_{\xi_{n-1}} \ltimes \ldots \ltimes \mathcal{F}_{\xi_0}\} \in \mathcal{F}_{\xi_n} \\ &\iff A \in \mathcal{F}_{\xi_n} \ltimes (\mathcal{F}_{\xi_{n-1}} \ltimes \ldots \ltimes \mathcal{F}_{\xi_0}) \cong \mathcal{F}_{\xi_n} \ltimes \ldots \ltimes \mathcal{F}_{\xi_0}. \end{aligned}$$

So h witnesses that $\mathcal{F}_{\xi_n} \ltimes \ldots \ltimes \mathcal{F}_{\xi_0} \leq_{\mathrm{RK}} \mathcal{G}_{\xi_n}$, and the induction proceeds. **Q**

(d) Finally, if $1 \leq \xi_0 < \ldots < \xi_n \leq \zeta$ then $\mathcal{F}_{\xi_n} \ltimes \ldots \ltimes \mathcal{F}_{\xi_0} \leq \mathcal{G}_{\xi_n} \leq_{\mathrm{RK}} \mathcal{G}_\zeta$.

5J Proof of Theorem 1K Let (X, Σ, μ) be a compact probability space. Then there is a measure λ on X, extending μ, such that $\lambda(\lim_{i\to\mathcal{F}} E_i)$ is defined and equal to $\lim_{i\to\mathcal{F}} \mu E_i$ whenever $\mathcal{F}$ is a Ramsey ultrafilter on $\mathbb{N}$ and $\langle E_i\rangle_{i\in\mathbb{N}}$ is a sequence in Σ.

PROOF. **(a)** If there are no Ramsey ultrafilters on $\mathbb{N}$, we can set $\lambda=\mu$ and stop; so let us suppose that there is at least one Ramsey ultrafilter. Let $\mathfrak{F}$ be a family of Ramsey ultrafilters on $\mathbb{N}$ consisting of just one member of each isomorphism class. Fix a well-ordering $\preccurlyeq$ of $\mathfrak{F}$ with greatest member $\mathcal{F}^*$ and a ladder system $\langle\theta(\xi,i)\rangle_{1\le\xi<\omega_1,i\in\mathbb{N}}$ such that $\langle\theta(\xi,i)\rangle_{i\in\mathbb{N}}$ is a non-decreasing sequence in ξ, and $\{\theta(\xi,i):i\in\mathbb{N}\}$ is cofinal with ξ, whenever $1\le\xi<\omega_1$.

(b)(i) For any non-empty finite set $V\subseteq\mathfrak{F}$, list it in $\preccurlyeq$-increasing order as $\mathcal{F}_0\prec\mathcal{F}_1\prec\ldots\prec\mathcal{F}_n$, and set $\mathcal{H}_V=\mathcal{F}_n\ltimes\ldots\ltimes\mathcal{F}_0$.

(ii) For any non-empty countable set $W\subseteq\mathfrak{F}$ containing $\mathcal{F}^*$, list it in $\preccurlyeq$-increasing order as $\langle\mathcal{F}_\xi\rangle_{1\le\xi\le\zeta}$, where $\zeta\ge1$ is a countable ordinal, and let $\mathcal{G}_W$ be the final ultrafilter on $S^*=\bigcup_{n\in\mathbb{N}}\mathbb{N}^n$ defined from $\langle\mathcal{F}_\xi\rangle_{1\le\xi\le\zeta}$ and $\langle\theta(\xi,i)\rangle_{1\le\xi\le\zeta,i\in\mathbb{N}}$ by the process of 1Ib. By Theorem 1J, $\mathcal{G}_W$ is measure-centering.

(iii) If $V\subseteq W\subseteq\mathfrak{F}$, W is countable and contains $\mathcal{F}^*$, and V is finite and not empty, then $\mathcal{H}_V\le_{\mathrm{RK}}\mathcal{G}_W$, by 5I; so $\mathcal{H}_V$ is measure-centering (Proposition 1Da). Let λ_V be the corresponding extension of μ as described in Theorem 1E.

(c) Consider the family $\Lambda=\{\lambda_V:V\in[\mathfrak{F}]^{<\omega}\setminus\{\emptyset\}\}$. This is a collection of probability measures on X. It is upwards-directed in the sense that if we have any two members of Λ they have a common extension belonging to Λ. **P** If V_0, V_1 are non-empty finite subsets of $\mathfrak{F}$ with union V, then $\mathcal{H}_{V_0}\le_{\mathrm{RK}}\mathcal{H}_V$, by 5Hc, so λ_V extends λ_{V_0}, by 5G; and similarly λ_V extends λ_{V_1}. **Q** Next, if $\Lambda_0\subseteq\Lambda$ is countable, then there is a measure on X extending every member of Λ_0. **P** Let $W\subseteq\mathfrak{F}$ be a non-empty countable set, containing $\mathcal{F}^*$, such that $\Lambda_0\subseteq\{\lambda_V:V\in[W]^{<\omega}\setminus\{\emptyset\}\}$. Let $\lambda^\#$ be the extension of μ corresponding to the measure-centering ultrafilter $\mathcal{G}_W$. If $V\subseteq W$ is finite and non-empty, then $\mathcal{H}_V\le_{\mathrm{RK}}\mathcal{G}_W$, so $\lambda^\#$ extends λ_V; thus $\lambda^\#$ extends every member of Λ_0. **Q**

By 3Ab, there is a measure λ on X extending every member of Λ.

(d) Suppose that $\langle E_i\rangle_{i\in\mathbb{N}}$ is a sequence in Σ and that $\mathcal{F}$ is a Ramsey ultrafilter on $\mathbb{N}$. Then there is a $\mathcal{F}'\in\mathfrak{F}$ such that $\mathcal{F}'=\mathcal{H}_{\{\mathcal{F}'\}}$ is isomorphic to $\mathcal{F}$. Let $f:\mathbb{N}\to\mathbb{N}$ be such that $f[[\mathcal{F}']]=\mathcal{F}$, and set $F_j=E_{f(j)}$, as in the proof of 5G; then $\lim_{i\to\mathcal{F}}E_i=\lim_{j\to\mathcal{F}'}F_j$ and

$$\lambda(\lim_{i\to\mathcal{F}}E_i)=\lambda(\lim_{j\to\mathcal{F}'}F_j)=\lambda_{\{\mathcal{F}'\}}(\lim_{j\to\mathcal{F}'}F_j)$$

is defined and equal to $\lim_{j\to\mathcal{F}'}\mu F_j=\lim_{i\to\mathcal{F}}\mu E_i$. So λ has the required property.

6 Other kinds of ultrafilter

I return to the relatively concrete context of a proof of Theorem 1M, describing the relationships between various classes of ultrafilters.

6A Proof of Theorem 1M, parts (a), (c) and (j) As usual, I begin by repeating the statements of the results in question.

(a) A measure-linking ultrafilter is Hausdorff.

(b) A measure-centering ultrafilter is nowhere dense.

(j) It is relatively consistent with ZFC to suppose that every measure-centering ultrafilter is principal.

PROOF. **(a)** Let $\mathcal{F}$ be a measure-linking ultrafilter on a set I, and $f : I \to J$, $g : I \to J$ two functions such that $A = \{i : f(i) \neq g(i)\} \in \mathcal{F}\}$. Let $\langle e_j \rangle_{j \in J}$ be the standard generating family in $\mathfrak{B}_J$ (Definition 3Be), and for $i \in I$ set

$$a_i = e_{f(i)} \setminus e_{g(i)} \text{ if } i \in A,$$
$$= 1 \text{ otherwise.}$$

Then $\inf_{i \in I} \bar{\nu}_J a_i \geq \frac{1}{4}$, so there is a $B \in \mathcal{F}$ such that $a_i \cap a_j \neq 0$ for all $i, j \in B$; we can suppose that $B \subseteq A$. Now $f[B] \cap g[B] = \emptyset$. **P?** Otherwise, there are $i, j \in B$ such that $f(i) = g(j)$. In this case,

$$0 \neq a_i \cap a_j \subseteq e_{f(i)} \setminus e_{g(j)} = 0. \ \mathbf{XQ}$$

Since $f[B] \in f[[\mathcal{F}]]$ and $g[B] \in g[[\mathcal{F}]]$, $f[[\mathcal{F}]] \neq g[[\mathcal{F}]]$; as f and g are arbitrary, $\mathcal{F}$ is a Hausdorff ultrafilter.

(b) Let I be a set, $\mathcal{F}$ a measure-centering ultrafilter on I and $f : I \to \mathbb{R}$ a function. Let μ be Lebesgue measure on $\mathbb{R}$. Let $G \subseteq \mathbb{R}$ be an open set, including $\mathbb{Q}$, of measure at most $\frac{1}{2}$; for $x \in \mathbb{R}$, write $<x> = x - \lfloor x \rfloor$ for the fractional part of x. For each $i \in I$, set

$$K_i = [0, 1] \setminus (G + <f(i)>).$$

Then $\mu K_i \geq \frac{1}{2}$. Applying the definition 1A with $\mathfrak{A}$ the algebra of measurable subsets of $[0, 1]$, we see that there is an $A \in \mathcal{F}$ such that $\mu(\bigcap_{i \in L} K_i) > 0$ for every non-empty finite $L \subseteq A$. Because all the sets K_i are closed subsets of the compact set $[0, 1]$, there must be a point x of $\bigcap_{i \in A} K_i$. In this case, $x \notin G + <f(i)>$ for every $i \in A$, that is, $<f(i)> \notin x - G$ for every $i \in A$. Now $x - G$ is a dense open subset of $\mathbb{R}$, so $H = (x - G) \cap]0, 1[$ is a dense open subset of $]0, 1[$, and $H + \mathbb{Z}$ is a dense open subset of $\mathbb{R}$; while $f(i) \notin H + \mathbb{Z}$ for every $i \in A$. Thus $f[A]$ is nowhere dense in $\mathbb{R}$; as f is arbitrary, $\mathcal{F}$ is nowhere dense.

(j)(α) The point is that if there is a non-principal measure-centering ultrafilter $\mathcal{F}$ on a set I, then either there is a non-principal nowhere dense ultrafilter on $\mathbb{N}$, or there is a two-valued-measurable cardinal. **P** If $\mathcal{F}$ is not closed under countable intersections, there is a partition of I into a sequence $\langle A_n \rangle_{n \in \mathbb{N}}$ of sets not belonging to $\mathcal{F}$. Setting $f(i) = n$ for $i \in A_n$, we get a function $f : I \to \mathbb{N}$ such that $f[[\mathcal{F}]]$ is a non-principal ultrafilter on $\mathbb{N}$. By Proposition 1Da $f[[\mathcal{F}]]$ is measure-centering, and by (b) above it is nowhere dense. On the other hand, if $\mathcal{F}$ is closed under countable intersections, then there is a two-valued-measurable cardinal ([7], 8.31; [19], 10.2). **Q**

(β) In [24], Theorem 3.1, Shelah proved that if $\mathfrak{c} = \omega_1$ and $\diamondsuit_{\omega_2}(\{\gamma : \gamma < \omega_2, \operatorname{cf}\gamma = \omega_1\})$ is true, then there is a proper forcing notion $\mathbb{P}$ of cardinality ω_2 such that

$$\Vdash_{\mathbb{P}} \text{there are no non-principal nowhere dense ultrafilters on } \mathbb{N}.$$

In particular, this is so if we start from a model of $V = L$ ([19], 13.20 & Exercise 27.4). Now, if we begin with such a model, so that there are no two-valued-measurable cardinals ([19], 17.1), then we shall certainly have

$$\Vdash_{\mathbb{P}} \text{there are no two-valued-measurable cardinals}$$

([19], 21.2). So (α) tells us that

$$\Vdash_{\mathbb{P}} \text{all measure-centering ultrafilters are principal.}$$

6B Lemma (a) (cf. [5], Theorem 5) Let $\mathcal{F}$ be a uniform κ-complete weakly Ramsey ultrafilter on a regular infinite cardinal κ. If $\langle A_\xi\rangle_{\xi<\kappa}$ is any family in $\mathcal{F}$, there is an $A \in \mathcal{F}$ such that $\#(A \setminus A_\xi) < \kappa$ for every $\xi < \kappa$.

(b) Let $\mathcal{F}$ be a weakly Ramsey ultrafilter on a set I, and $\mathcal{D}$ a disjoint family of subsets of κ, none belonging to $\mathcal{F}$. Set $Q = \bigcup_{D\in\mathcal{D}}[D]^2$. Then for any $S \subseteq [\kappa]^2$ there is an $A \in \mathcal{F}$ such that $Q \cap [A]^2$ is either included in S or disjoint from S.

(c) Let $\mathcal{F}$ be a weakly Ramsey ultrafilter on a set I. If $A \in \mathcal{F}$, then $\mathcal{F}\restriction A$ is weakly Ramsey, where $\mathcal{F}\restriction A = \mathcal{F} \cap \mathcal{P}A$.

(d) Let $\mathcal{F}$ be an ultrafilter on a set I, and suppose that there is an $A \in \mathcal{F}$ such that $\mathcal{F}\restriction A$ is an arrow ultrafilter. Then $\mathcal{F}$ is an arrow ultrafilter.

PROOF. **(a)** For $\alpha < \kappa$, set $h(\alpha) = \min\{\xi : \xi < \kappa,\ \alpha \notin A_\xi \setminus \xi\}$. Note that $h^{-1}[\xi] \notin \mathcal{F}$ for every $\xi < \kappa$, because $\mathcal{F}$ is uniform and κ-complete; so $\sup h[A] = \kappa$ for every $A \in \mathcal{F}$. Set

$$S_0 = \{\{\alpha,\beta\} : \alpha < \beta < \kappa \text{ and } h(\alpha) > h(\beta)\},$$

$$S_1 = \{\{\alpha,\beta\} : \alpha < \beta < \kappa \text{ and } h(\alpha) = h(\beta)\},$$

$$S_2 = \{\{\alpha,\beta\} : \alpha < \beta < \kappa \text{ and } h(\alpha) < h(\beta)\},$$

If $A \in \mathcal{F}$ then $[A]^2 \cap S_2 \neq \emptyset$. **P?** Otherwise, $h(\alpha) \geq h(\beta)$ whenever $\alpha, \beta \in A$ and $\alpha < \beta$. Set $\xi = \min h[A]$ and let $\alpha \in A$ be such that $h(\alpha) = \xi$; then $h(\beta) = \xi$ whenever $\beta \in A \setminus \alpha$. But this means that $A \setminus \alpha$ and $A_\xi \setminus \xi$ are disjoint sets both belonging to $\mathcal{F}$. **XQ**

Because $\mathcal{F}$ is weakly Ramsey, there is an $A \in \mathcal{F}$ such that $[A]^2$ is disjoint from at least one of S_0 and S_1.

case 0 If $[A]^2 \cap S_0 = \emptyset$, then $h(\alpha) \leq h(\beta)$ whenever $\alpha, \beta \in A$ and $\alpha < \beta$. If $\xi < \kappa$, take $\alpha \in A$ such that $h(\alpha) > \xi$. If $\beta \in A \setminus \alpha$, then $h(\beta) \geq h(\alpha) > \xi$ so $\beta \in A_\xi \setminus \xi$; thus $A \setminus A_\xi \subseteq A \cap \alpha$ has cardinal less than κ.

case 1 If $[A]^2 \cap S_1 = \emptyset$, then $h\restriction A$ is injective. If $\xi < \kappa$, take α such that $h(\beta) > \xi$ for every $\beta \in A \setminus \alpha$; then again $A \setminus A_\xi \subseteq A \cap \alpha$ has cardinal less than κ, as required.

(b) There is an $A \in \mathcal{F}$ such that $[A]^2$ is disjoint from at least one of $Q \cap S$, $Q \setminus S$ and $[I]^2 \setminus Q$. Since $A \not\subseteq D$ for any $D \in \mathcal{D}$, $[A]^2 \setminus Q \neq \emptyset$. So $[A]^2 \cap Q$ is either disjoint from S or included in S.

(c) If S_0, S_1, S_2 are disjoint subsets of $[A]^2$, then there is a $B \in \mathcal{F}$ such that $[B]^2$ is disjoint from some S_j, and now $A \cap B \in \mathcal{F}\restriction A$ and $[A \cap B]^2 \cap S_j = \emptyset$.

(d) If $S \subseteq [I]^2$ and $k \in \mathbb{N}$, then either there is a $K \in [A]^k$ such that $[K]^2 \cap (S \cap [A]^2) = \emptyset$ (in which case $K \in [I]^k$ and $[K]^2 \cap S = \emptyset$) or there is a $B \in \mathcal{F}\restriction A$ such that $[B]^2 \subseteq S \cap [A]^2$ (in which case $B \in \mathcal{F}$ and $[B]^2 \subseteq S$).

6C Proof of Theorem 1Mc Every weakly Ramsey ultrafilter is an arrow ultrafilter.

PROOF. **(a)(i)** To begin with, I will suppose that $\mathcal{F}$ is a uniform κ-complete weakly Ramsey ultrafilter on an infinite cardinal κ. Note that it follows at once that κ is regular, since if $A \in [\kappa]^{<\kappa}$ then $\bigcap_{\xi\in A} \kappa \setminus \xi$ belongs to $\mathcal{F}$ and cannot be empty. I aim to show, by induction on k, that if $S \subseteq [\kappa]^2$ is such that $[K]^2$ meets S for every $K \in [\kappa]^k$ then there is an $A \in \mathcal{F}$ such that $[A]^2 \subseteq S$.

(ii) The cases $k = 0$ and $k = 1$ are vacuous, since there is a $K \in [\kappa]^k$ and $[K]^2 = \emptyset$; and the case $k = 2$ is trivial, since if $[K]^2 \cap S \neq \emptyset$ for every $K \in [\kappa]^2$ then $S = [\kappa]^2$. For the inductive step to $k \geq 3$, take $S \subseteq [\kappa]^2$ such that $[K]^2$ meets S for every $K \in [\kappa]^k$. For each $\xi < \kappa$ set $A_\xi = \{\eta : \eta < \kappa,\ \{\xi, \eta\} \in S\} \cup \{\xi\}$. If there is any ξ such that $A_\xi \notin \mathcal{F}$, then there is an $A \in \mathcal{F}$ such that $[A]^2 \subseteq S$. **P** Set $B = \kappa \setminus A_\xi$, $S' = S \cup ([\kappa]^2 \setminus [B]^2)$; because $\mathcal{F}$ is an ultrafilter, $B \in \mathcal{F}$. If $K \in [\kappa]^{k-1}$ and $K \not\subseteq B$, then certainly $[K]^2$ meets S'. If $K \in [B]^{k-1}$, then $K' = K \cup \{\xi\}$ belongs to $[\kappa]^k$ and there is an $L \in S$ such that $L \subseteq K'$; now $\xi \notin L$ so $L \subseteq K$ and $[K]^2$ meets S'. By the inductive hypothesis, there is a $C \in \mathcal{F}$ such that $[C]^2 \subseteq S'$; now $A = B \cap C$ belongs to $\mathcal{F}$ and $[A]^2 \subseteq S$. **Q**

(iii) So we may suppose that $A_\xi \in \mathcal{F}$ for every $\xi < \kappa$. By 6Ba, there is an $A \in \mathcal{F}$ such that $\#(A \setminus A_\xi) < \kappa$ for every $\xi < \kappa$. Because κ is regular, we can define inductively a strictly increasing family $\langle \xi_\alpha \rangle_{\alpha<\kappa}$ in κ by saying that ξ_α is to be the least ordinal such that $\xi_\alpha > \xi_\beta$ for every $\beta < \alpha$ and $A \setminus A_\eta \subseteq \xi_\alpha$ whenever $\eta < \sup_{\beta<\alpha} \xi_\beta$. Note that $\xi_\alpha = \sup_{\beta<\alpha} \xi_\beta$ whenever $\alpha < \kappa$ is a limit ordinal. Set $D_\alpha = \xi_{\alpha+1} \setminus \xi_\alpha$ for each α; then $\langle D_\alpha \rangle_{\alpha\in\kappa}$ is a partition of κ. Write Q for $\bigcup_{\alpha<\kappa} [D_\alpha]^2$.

One of $\bigcup_{\alpha<\kappa \text{ is even}} D_\alpha$, $\bigcup_{\alpha<\kappa \text{ is odd}} D_\alpha$ belongs to $\mathcal{F}$. Call this B. Then $[A \cap B]^2 \subseteq S \cup Q$. **P** If ξ, $\eta \in A \cap B$ and $\xi < \eta$, let $\alpha \leq \beta$ be such that $\xi \in D_\alpha$ and $\eta \in D_\beta$. If $\alpha = \beta$ then $\{\xi, \eta\} \in [D_\alpha]^2 \subseteq Q$. Otherwise, $\xi < \alpha + 1 < \beta$ and $A \setminus A_\xi \subseteq \xi_{\alpha+2}$ does not meet D_β, so $\eta \in A_\xi$ and $\{\xi, \eta\} \in S$. **Q**

(iv) There is a $C \in \mathcal{F}$ such that $Q \cap [C]^2 \subseteq S$. **P** By 6Bb, there is a $C_0 \in \mathcal{F}$ such that $Q \cap [C_0]^2$ is either included in S or disjoint from S. In the former case, we can take $C = C_0$ and stop. In the latter case, $\#(C_0 \cap D_\alpha) < k$ for every $\alpha < \kappa$, so there is a $C \in \mathcal{F}$ such that $\#(C \cap D_\alpha) \leq 1$ for every $\alpha < \kappa$. But in this case $Q \cap [C]^2$ is empty so is included in S. **Q**

(v) Now $A \cap B \cap C \in \mathcal{F}$ and

$$[A \cap B \cap C]^2 = [A \cap B]^2 \cap [C]^2 \subseteq S \cup (Q \cap [C]^2) = S.$$

So we have a suitable member of $\mathcal{F}$. Thus the induction continues, and it is the case for every $k \in \mathbb{N}$ that if $S \subseteq [\kappa]^2$ is such that $S \cap [K]^2 \neq \emptyset$ for every $K \in [\kappa]^k$, then there is an $A \in \mathcal{F}$ such that $[A]^2 \subseteq S$; that is, $\mathcal{F}$ is an arrow ultrafilter.

(b) Accordingly the theorem is proved in the case of a uniform κ-complete ultrafilter on an infinite cardinal κ. Now suppose that $\mathcal{F}$ is a uniform weakly Ramsey ultrafilter on an infinite cardinal κ, and is not κ-complete. Then there are a cardinal $\lambda < \kappa$ and a family $\langle A_\lambda \rangle_{\alpha<\lambda}$ in $\mathcal{F}$ such that $\bigcap_{\alpha<\lambda} A_\lambda$ is empty. For $\xi < \kappa$, set $h(\xi) = \min\{\alpha : \alpha < \lambda,\ \xi \notin A_\alpha\}$; for $\alpha < \lambda$, set $D_\alpha = h^{-1}[\{\alpha\}]$, so that $\langle D_\alpha \rangle_{\alpha<\lambda}$ is a partition of κ into sets not belonging to $\mathcal{F}$. Set $Q = \bigcup_{\alpha<\lambda} [D_\alpha]^2$.

Let $S \subseteq [\kappa]^2$ and $k \in \mathbb{N}$ be such that $[K]^2 \cap S \neq \emptyset$ whenever $K \in [\kappa]^k$. Set $S_0 = [\kappa]^2 \setminus (S \cup Q)$ and $S_1 = S \setminus Q$. Then there is a $B \in \mathcal{F}$ such that $[B]^2$ is disjoint from at least one of Q, S_0 and S_1. Since $\#(B) = \kappa > \lambda$, there is some $\alpha < \lambda$ such that $\#(B \cap D_\alpha) \geq 2$, and $[B]^2$ meets Q. Since B meets infinitely many D_α, there is a $K \in [B]^k$ such that $\#(K \cap D_\alpha) \leq 1$ for every α; now $[B]^2 \cap S_1 \supseteq [K]^2 \cap S$ is non-empty. So $[B]^2 \cap S_0 = \emptyset$, that is, $[B]^2 \subseteq Q \cup S$.

By 6Bb again, there is a $C \in \mathcal{F}$ such that $Q \cap [C]^2$ is either included in S or disjoint from S. Since $\#(C) = \kappa > \lambda$, there must be an $\alpha < \lambda$ such that $\#(C \cap D_\alpha) \geq k$, in which case $[C \cap D_\alpha]^2$ meets S; so we must have $Q \cap [C]^2 \subseteq S$. So if we set $A = B \cap C$, A belongs to $\mathcal{F}$ and $[A]^2 \subseteq S \cup (Q \cap [C]^2) = S$. As S and k are arbitrary, $\mathcal{F}$ is an arrow ultrafilter.

(c) Putting (a) and (b) together, we see that any uniform weakly Ramsey ultrafilter on an infinite set is an arrow ultrafilter. Of course any principal ultrafilter is an arrow ultrafilter, so all uniform weakly Ramsey ultrafilters are arrow ultrafilters. Finally, if $\mathcal{F}$ is any weakly Ramsey ultrafilter on any set I, let $A \in \mathcal{F}$ be a set of minimal size; then $\mathcal{F}\lceil A$ is uniform; by 6Bc, $\mathcal{F}\lceil A$ is weakly Ramsey, therefore an arrow ultrafilter; by 6Bd, $\mathcal{F}$ is an arrow ultrafilter. So the proof is complete.

6D The following lemma is very well known in essence, though it is usually expressed in less quantitative forms.

Lemma Let $\mathfrak{A}$ be a Boolean algebra, and $\nu : \mathfrak{A} \to [0,1]$ an additive functional such that $\nu 1 = 1$. Suppose that ϵ, $\delta \in [0,1]$ are such that $\delta < \epsilon^2$, and that $\langle a_i \rangle_{i \in I}$ is a family in $\mathfrak{A}$ such that $\nu a_i \geq \epsilon$ and $\nu(a_i \cap a_j) \leq \delta$ for all distinct i, $j \in I$. Then $\#(I) \leq \frac{1}{\epsilon^2 - \delta}$.

PROOF. It is enough to deal with the case in which I is finite and $\mathfrak{A}$ is generated by $\{a_i : i \in I\}$, so that $\mathfrak{A}$ is finite and can be identified with a power set $\mathcal{P}X$. Consider $u = \sum_{i \in I} \chi a_i \in \mathbb{R}^X$. By Cauchy's ineqality,

$$\begin{aligned}(\sum_{x \in X} u(x)\nu\{x\})^2 &\leq \sum_{x \in X} \nu\{x\} \cdot \sum_{x \in X} u(x)^2 \nu\{x\} \\ &= \nu X \cdot \sum_{x \in X} u(x)^2 \nu\{x\} = \sum_{x \in X} u(x)^2 \nu\{x\}.\end{aligned}$$

Now, setting $m = \#(I)$,

$$\begin{aligned}\sum_{x \in X} u(x)\nu\{x\} &= \sum_{x \in X} \sum_{i \in I} \chi a_i(x)\nu\{x\} \\ &= \sum_{i \in I} \sum_{x \in X} \chi a_i(x)\nu\{x\} = \sum_{i \in I} \nu a_i \geq \epsilon m,\end{aligned}$$

$$\begin{aligned}\sum_{x \in X} u(x)^2 \nu\{x\} &= \sum_{x \in X, i, j \in I} \chi a_i(x)\chi a_j(x)\nu\{x\} = \sum_{i,j \in I} \nu(a_i \cap a_j) \\ &= \sum_{i \in I} \nu a_i + \sum_{i \neq j} \nu(a_i \cap a_j) \leq m + \delta m(m-1).\end{aligned}$$

So we get

$$\epsilon^2 m^2 \leq m + \delta m^2, \quad m \leq \tfrac{1}{\epsilon^2 - \delta},$$

as claimed.

6E Proof of Theorem 1Md (i) An arrow ultrafilter is strongly measure-linking.

(ii) An arrow ultrafilter on $\mathbb{N}$ is nowhere dense.

PROOF. **(i)** Let $\mathcal{F}$ be an arrow ultrafilter on a set I, $\mathfrak{A}$ a Boolean algebra, $\nu : \mathfrak{A} \to [0,1]$ an additive functional such that $\nu(1_{\mathfrak{A}}) = 1$, and $\langle a_i \rangle_{i \in I}$ a family in $\mathfrak{A}$ such that $\inf_{i \in I} \nu a_i = \epsilon > 0$. Take any $\delta \in \left]0, \epsilon^2\right[$ and set $S = \{\{i, j\} : i,\, j \in I$ are distinct, $\nu(a_i \cap a_j) \geq \delta\}$. Let $k \in \mathbb{N}$ be such that $k > \frac{1}{\epsilon^2 - \delta}$. If $K \in [I]^k$ then by Lemma 6D there are distinct $i,\, j \in K$ such that $\nu(a_i \cap a_j) \geq \delta$, that is, $[K]^2$ meets S. Because $\mathcal{F}$ is an arrow ultrafilter, there is a $J \in \mathcal{F}$ such that $[J]^2 \subseteq S$, that is, $\nu(a_i \cap a_j) \geq \delta$ for all $i,\, j \in J$. As $\mathfrak{A}$, ν and $\langle a_i \rangle_{i \in I}$ are arbitrary, $\mathcal{F}$ is strongly measure-linking.

(ii)(α) The key to the proof is the following: if $D \subseteq \left]0, 1\right]$ is a countable set, there is a set $S \subseteq [D]^2$ such that $[K]^2 \cap S \neq \emptyset$ for every $K \in [D]^3$ and A is nowhere dense whenever $A \subseteq D$, $[A]^2 \subseteq S$ and $0 \in \overline{A}$. **P** Let $\preccurlyeq$ be a well-ordering of D in order type at most ω. Let $\langle \epsilon_t \rangle_{t \in D}$ be a family of strictly positive real numbers with sum at most 1. For $m,\, k \in \mathbb{N}$ set $H_{mk} = \left]2^{-m}k, 2^{-m}(k+1)\right[$; for $t \in D$ let $m_t \in \mathbb{N}$ be such that $2^{-m_t - 1} < t \leq 2^{-m_t}$. Define G_t inductively, for $t \in D$, following the well-ordering $\preccurlyeq$, in such a way that, writing m for m_t,

$G_t \subseteq \left]2^{-m}, 1\right[$,
$s \notin G_t$ if $s \prec t$,
$G_t \cap G_s = \emptyset$ if $s \prec t$ and $m_s > m$,
for $1 \leq k < 2^m$, $G_t \cap H_{mk}$ is a non-empty open interval of length at most $2^{-m}\epsilon_t$.

To see that this is possible, note that when we come to choose G_t the forbidden points in $\bigcup_{1 \leq k < 2^m} H_{mk}$ consist of some of the the finitely many $s \prec t$, together with $\bigcup_{s \prec t, m_s > m} G_s$; and the latter meets each H_{mk} in a finite union of intervals of total length at most $\sum_{s \prec t} 2^{-m}\epsilon_s < 2^{-m}$, so there must be a gap remaining.

Note that $m_s < m_t$ whenever $s \in G_t$. On completing the inductive construction, set

$$S = \{\{s, t\} : s,\, t \in D \text{ are distinct, } s \notin G_t \text{ and } t \notin G_s\}.$$

If $K \in [D]^3$ then either there are distinct $s,\, t \in K$ such that $m_s = m_t$, in which case $\{s, t\} \in S \cap [K]^2$, or $K = \{s, t, u\}$ where $m_s < m_t < m_u$. If $\{t, u\} \notin S$, then $t \in G_u$ so $u \prec t$ and $G_t \cap G_u$ is empty. But this means that one of $\{s, t\}$, $\{s, u\}$ belongs to S. So in all cases we have $[K]^2 \cap S \neq \emptyset$.

Now suppose that $A \subseteq D$ is such that $0 \in \overline{A}$ and $[A]^2 \subseteq S$. For each $m \in \mathbb{N}$ and $k \geq 1$ there is a $t \in A$ such that $m_t > m$. In this case, A cannot meet G_t, so $\overline{A}$ cannot include H_{mk}. As m and k are arbitrary, A is nowhere dense. **Q**

(β) Let $\mathcal{F}$ be an arrow ultrafilter on $\mathbb{N}$, and $f:\mathbb{N}\to\mathbb{R}$ a function. Set $g(n)=\arctan f(n)$ for $n\in\mathbb{N}$; then g is bounded; set $z=\lim_{n\to\mathcal{F}}g(n)$. Set $h(n)=\frac{1}{\pi}|g(n)-z|$ for $n\in\mathbb{N}$; then h takes values in $[0,1]$ and $\lim_{n\to\mathcal{F}}h(n)=0$.

If $A_0=\{i:h(i)=0\}$ belongs to $\mathcal{F}$, then $f[A_0]$ is a singleton, and is certainly nowhere dense. Otherwise, set $D=h[\mathbb{N}]\setminus\{0\}$, and let $S\subseteq[D]^2$ be as in (α). Set

$$\begin{aligned}S_1=\{\{i,j\}:i,\,j\in\mathbb{N},\ &\text{either } i\in A_0 \text{ or } j\in A_0\\ &\text{or } h(i)=h(j) \text{ or } \{h(i),h(j)\}\in S\}.\end{aligned}$$

If $K\subseteq\mathbb{N}$ and $[K]^2\cap S_1=\emptyset$, then $h{\restriction}K$ is injective, $h[K]\subseteq D$ and $[h[K]]^2\cap S=\emptyset$, so $\#(K)\le 2$. Because $\mathcal{F}$ is an arrow ultrafilter, there is an $A\in\mathcal{F}$ such that $[A]^2\subseteq S_1$; we can suppose that $A\cap A_0$ is empty, so that $h[A]\subseteq D$ and $[h[A]]^2\subseteq S$. But now recall that $\lim_{n\to\mathcal{F}}h(n)=0$, so $0\in\overline{h[A]}$ and $h[A]$ is nowhere dense, by the choice of S. In this case, $B=h[A]\cup(-h[A])$ is nowhere dense, $g[A]\subseteq z+\pi B$ is nowhere dense, and $f[A]=\tan[g[A]]$ is nowhere dense. As f is arbitrary, $\mathcal{F}$ is a nowhere dense filter.

6F The next lemma is a version of one which I learnt from Michel Talagrand when visiting him in 1987. He claims to have no recollection of it.

Lemma (M.Talagrand) (a) For finite I, $J\subseteq\mathbb{N}$, say that $I\preccurlyeq J$ if $\#(I)=\#(J)$ and $\#(I\setminus k)\le\#(J\setminus k)$ for every $k\in\mathbb{N}$.

(i) $\preccurlyeq$ is a transitive relation.

(ii) If $I\preccurlyeq J$, there are $I_0,\dots,I_r$ such that $I=I_0\preccurlyeq I_1\preccurlyeq\ldots\preccurlyeq I_r=J$ and $\#(I_k\triangle I_{k+1})=2$ for every $k<r$.

(iii) Suppose that $m\ge 1$. For $n\in\mathbb{N}$, set

$$\mathcal{I}_n=\{I:I\subseteq n,\ \#(I\cap J)\ge m \text{ whenever } J\subseteq n \text{ and } J\succcurlyeq I\}.$$

Then $\#(\mathcal{I}_n)\le\dfrac{2^n\sqrt{n}}{m}$.

(b) If $1\le m\le n\in\mathbb{N}$, $\mathcal{I}\subseteq\mathcal{P}n$ and $\#(I\cap J)\ge m$ for all I, $J\in\mathcal{I}$, then $\#(\mathcal{I})\le\dfrac{2^n\sqrt{n}}{m}$.

PROOF. **(a)(i)** is trivial.

(ii) Induce on $r=\#(I\setminus J)$. If $r=0$ then $I=J$ and we can stop. For the inductive step to $r+1$, set $i_0=\max(I\setminus J)$, $j_0=\max(J\setminus I)$. Then $i_0<j_0$. **P**

$$\#(I\setminus j_0)\le\#(J\setminus j_0)=\#(I\cap J\setminus j_0)$$

so $I\setminus j_0=I\cap J\setminus j_0$ does not contain i_0, and $i_0\le j_0$; but of course $i_0\ne j_0$. **Q**

Set $J'=J\triangle\{i_0,j_0\}$. Then

$$\begin{aligned}\#(I\setminus k)&\le\#(J\setminus k)=\#(J'\setminus k) \text{ if } k\le i_0 \text{ or } j_0<k,\\ &=\#(I\cap J\setminus k)\le\#((J\setminus\{j_0\})\setminus k)=\#(J'\setminus k) \text{ if } i_0<k\le j_0,\end{aligned}$$

so $I\preccurlyeq J'\preccurlyeq J$. Also $\#(I\setminus J')=r$ and $\#(J'\triangle J)=2$. By the inductive hypothesis, there are $I_0,\dots,I_r$ such that $I=I_0\preccurlyeq I_1\preccurlyeq\ldots\preccurlyeq I_r=J'$ and $\#(I_k\triangle I_{k+1})=2$

for every $k < r$; setting $I_{r+1} = J$, we have an appropriate chain to complete the inductive step.

(iii) For $I \subseteq \mathbb{N}$ and $n \in \mathbb{N}$, set $h_n(I) = \min\{\#(I \cap J) : I \cap n \preccurlyeq J \subseteq n\}$. Then we find that

$$\begin{aligned} h_{n+1}(I) &= h_n(I) + 1 \text{ if } n \in I, \\ &= \max(0, h_n(I) - 1) \text{ otherwise.} \end{aligned}$$

P (α) If $n \in I$, then for $J \subseteq n+1$ we have $I \cap (n+1) \preccurlyeq J$ iff $n \in J$ and $I \cap n \preccurlyeq J \cap n$; so $h_{n+1}(I) = h_n(I) + 1$. (β) If $n \notin I$, let J be such that $I \cap n \preccurlyeq J \subseteq n$ and $\#(I \cap J) = h_n(I)$. If $h_n(I) = 0$ then J witnesses that $h_{n+1}(I) = 0$. Otherwise, take any $i_0 \in I \cap J$, and consider $J' = J \triangle \{i_0, n\}$; we shall have

$$I \cap (n+1) = I \cap n \preccurlyeq J \preccurlyeq J' \subseteq n+1,$$

so $h_{n+1}(I) \le \#(I \cap J') = h_n(I) - 1$. Thus $h_{n+1}(I) \le \max(0, h_n(I) - 1)$. ($\gamma$) Again supposing that $n \notin I$, take $J \subseteq n+1$ such that $I \cap (n+1) \preccurlyeq J$ and $\#(I \cap J) = h_{n+1}(I)$. If $n \notin J$, then

$$I \cap n = I \cap (n+1) \preccurlyeq J \subseteq n$$

so $h_n(I) \le \#(I \cap J) = h_{n+1}(I)$. If $n \subseteq J$, then $I \cap n \preccurlyeq J \cap n$ and again $h_n(I) \le h_{n+1}(I)$. If $n \in J$ but $n \not\subseteq J$, set $j_0 = \max(n \setminus J)$ and $J' = J \triangle \{j_0, n\}$. For $k \in \mathbb{N}$,

$$\begin{aligned} \#(I \cap n \setminus k) &\le \#(J \setminus k) = \#(J' \setminus k) \text{ if } k \le j_0, \\ &\le n - k = \#(J' \setminus k) \text{ if } j_0 \le k \le n, \\ &= 0 = \#(J' \setminus k) \text{ if } k \ge n. \end{aligned}$$

So $I \cap n \preccurlyeq J'$ and

$$h_n(I) \le \#(I \cap J') \le \#(I \cap J) + 1 = h_{n+1}(I) + 1.$$

Thus we must have $h_{n+1}(I) \ge h_n(I) - 1$, and of course $h_{n+1}(I) \ge 0$, so $h_{n+1}(I) \ge \max(0, h_n(I) - 1)$. **Q**

We have $h_{n+1} = h_n + g_n$ where

$$\begin{aligned} g_n(I) &= 1 \text{ if } n \in I, \\ &= -1 \text{ if } h_n(I) > 0 \text{ and } n \notin I, \\ &= 0 \text{ if } h_n(I) = 0 \text{ and } n \notin I. \end{aligned}$$

Giving $\mathcal{P}\mathbb{N}$ its usual probability measure, matching $\nu_{\mathbb{N}}$ on $\{0,1\}^{\mathbb{N}}$, and writing $\mathbb{E}$ for expectation with respect to this measure, we have

$$\mathbb{E}(h_{n+1}^2) = \mathbb{E}(h_n^2) + 2\mathbb{E}(h_n \times g_n) + \mathbb{E}(g_n^2).$$

Now $\mathbb{E}(h_n \times g_n) = 0$, because given that $h_n(I) \ne 0$ then $g_n(I)$ is equally likely to be ± 1; while $\mathbb{E}(g_n^2) \le 1$. Since $h_0(I) = 0$ for every I, we see by induction that

$\mathbb{E}(h_n^2) \le n$ for every $n \in \mathbb{N}$. Consequently $\mathbb{E}(h_n) \le \sqrt{n}$. Now $\mathcal{I}_n = \{I \cap n : h_n(I) \ge m\}$, so

$$\#(\mathcal{I}_n) = 2^n \Pr(h_n \ge m) \le \frac{2^n}{m}\mathbb{E}(h_n) \le \frac{2^n\sqrt{n}}{m}.$$

(b) Induce on $w(\mathcal{I}) = \sum_{I\in\mathcal{I}}\sum_{i\in I}(n-i)$. If $w(\mathcal{I}) = 0$ then $\mathcal{I}$ must be empty and the result is trivial. For the inductive step to $w(\mathcal{I}) = k+1$, then if $\mathcal{I} \subseteq \mathcal{I}_n$, as defined in (a-iii), we can stop. Otherwise, there must be an $I_0 \in \mathcal{I}\setminus\mathcal{I}_n$ and $J_0 \subseteq n$ such that $J_0 \succcurlyeq I_0$ and $\#(I_0 \cap J_0) < m$, so that $J_0 \notin \mathcal{I}$.

By (a-ii), $\mathcal{I}_0$ and $\mathcal{J}_0$ are linked by a $\preccurlyeq$-chain of sets each differing from the preceding one at just two points; so there must be $I_1 \in \mathcal{I}$, $J_1 \in \mathcal{P}n\setminus\mathcal{I}$ such that $I_1 \preccurlyeq J_1$ and $\#(I_1\triangle J_1) = 2$. Let i_1 be the member of $I_1 \setminus J_1$ and j_1 the member of $J_1 \setminus I_1$; of course $i_1 < j_1$. Define $\phi : \mathcal{I} \to \mathcal{P}n$ by saying that

$$\begin{aligned}\phi(I) &= I\triangle\{i_1, j_1\} \text{ if } i_1 \in I,\ j_1 \notin I \text{ and } I\triangle\{i_1, j_1\} \notin \mathcal{I},\\ &= I \text{ otherwise.}\end{aligned}$$

Then ϕ is injective. Set $\mathcal{J} = \phi[\mathcal{I}]$, so that $\#(\mathcal{I}) = \#(\mathcal{J})$, while $w(\mathcal{J}) < w(\mathcal{I})$. Now $\#(J \cap J') \ge m$ for all J, $J' \in \mathcal{J}$. **P** Set $I = \phi^{-1}(J)$, $I' = \phi^{-1}(J')$. Then

— if $I = J$ and $I' = J'$, surely $\#(J \cap J') = \#(I \cap I') \ge m$;

— if $I \ne J$ and $I' \ne J'$, then

$$\#(J \cap J') = \#((I\triangle\{i_1, j_1\}) \cap (I'\triangle\{i_1, j_1\}) = \#(I \cap I') \ge m$$

because $i_1 \in I \cap I'$ and $j_1 \notin I \cup I'$;

— if $I = J$ and $I' \ne J'$ and $j_1 \in I$, then $j_1 \notin I'$, $J' = I'\triangle\{i_1, j_1\}$ so

$$\#(J \cap J') = \#(I \cap J') \ge \#(I \cap I') \ge m;$$

— if $I = J$ and $I' \ne J'$ and $i_1 \notin I$, then $i_1 \in I'$, $J' = I'\triangle\{i_1, j_1\}$ so

$$\#(J \cap J') = \#(I \cap J') \ge \#(I \cap I') \ge m;$$

— if $I = J$ and $I' \ne J'$ and $i_1 \in I$ and $j_1 \notin I$, then $I'' = I\triangle\{i_1, j_1\} \in \mathcal{I}$ so

$$\#(J \cap J') = \#((I''\triangle\{i_1, j_1\}) \cap (I'\triangle\{i_1, j_1\})) = \#(I'' \cap I') \ge m$$

because $i_1 \in I' \setminus I''$ and $j_1 \in I'' \setminus I'$.

Similarly, $\#(J\cap J') \ge m$ if $I' = J'$ and $I \ne J$. Thus in all cases we have $\#(J\cap J') \ge m$. **Q**

By the inductive hypothesis,

$$\frac{2^n\sqrt{n}}{m} \ge \#(\mathcal{J}) = \#(\mathcal{I})$$

and the induction continues.

6G Proposition (V.Bergelson-M.Talagrand) There are a probability algebra $(\mathfrak{A}, \bar{\mu})$ and a sequence $\langle a_n\rangle_{n\in\mathbb{N}}$ in $\mathfrak{A}$ such that $\bar{\mu}a_n \ge \frac{1}{2}$ for every n and I has zero asymptotic density whenever $I \subseteq \mathbb{N}$ and $\inf_{m,n\in I}\bar{\mu}(a_m \cap a_n) > 0$.

Proof. For $r \in \mathbb{N}$, set $K_r = \{n : 2^r - 1 \le n \le 2^{r+1} - 2\}$ and

$$\mathcal{K}_r = \{I : I \subseteq r+1,\ \#(I) > \tfrac{r+1}{2}\} \cup \{I : 0 \in I \subseteq r+1,\ \#(I) = \tfrac{r+1}{2}\},$$

so that $\#(K_r) = \#(\mathcal{K}_r) = 2^r$. For each $r \in \mathbb{N}$ let $h_r : K_r \to \mathcal{K}_r$ be a bijection. Set $X = \prod_{r\in\mathbb{N}}(r+1)$, with the product μ of the uniform probabilities on the factors; let $(\mathfrak{A}, \bar{\mu})$ be the measure algebra of μ. Define $A_n \subseteq X$, for $n \in \mathbb{N}$, by saying that $A_n = \{x : x \in X,\ x(r) \in h_r(n)\}$ for that r such that $n \in K_r$; as $\#(h_r(n)) \geq \frac{r+1}{2}$, $\mu A_n \geq \frac{1}{2}$. Set $a_n = A_n^\bullet \in \mathfrak{A}$, so that $\bar{\mu} a_n \geq \frac{1}{2}$ for every n.

Let $I \subseteq \mathbb{N}$ and $\epsilon > 0$ be such that $\bar{\mu}(a_m \cap a_n) \geq \epsilon$ for all $m, n \in I$. For $r \in \mathbb{N}$, set $\mathcal{J}_r = h_r[I \cap K_r]$. If $J, J' \in \mathcal{J}_r$, then $\#(J \cap J') \geq \epsilon(r+1)$. So, setting $m_r = \lceil \epsilon(r+1) \rceil$,

$$\#(I \cap K_r) = \#(\mathcal{J}_r) \leq \frac{2^{r+1}\sqrt{r+1}}{m_r}$$

for each r, by 6Fb, and

$$2^{-r}\#(I \cap K_r) \leq \frac{2\sqrt{r+1}}{\epsilon(r+1)} \to 0$$

as $r \to \infty$. So I has zero asymptotic density, as claimed.

6H Proof of Theorem 1M, parts (e)-(g)

(e) A strongly measure-linking ultrafilter on $\mathbb{N}$ contains a set of zero asymptotic density.

(f) A closed Lebesgue null ultrafilter on $\mathbb{N}$ contains a set of zero asymptotic density.

(g) If $\operatorname{cov}\mathcal{N}_{\text{Leb}} = \mathfrak{c}$, there is a measure-centering ultrafilter on $\mathbb{N}$ which is neither strongly measure-linking nor closed Lebesgue null.

PROOF. For $I \subseteq \mathbb{N}$ write $d^*(I)$ for its upper asymptotic density; write $\mathcal{Z}$ for the asymptotic density ideal $\{I : d^*(I) = 0\}$.

(e) Let $(\mathfrak{A}, \bar{\mu})$ and $\langle a_n \rangle_{n\in\mathbb{N}}$ be as in 6G. If $\mathcal{F}$ is a strongly measure-linking ultrafilter on $\mathbb{N}$, there is an $I \in \mathcal{F}$ such that $\inf_{m,n\in I} \bar{\mu}(a_m \cap a_n) > 0$, so $I \in \mathcal{Z}$.

(f) Let $\langle t_n \rangle_{n\in\mathbb{N}}$ be a sequence in $[0,1]$ which is equidistributed for Lebesgue measure, and $\mathcal{F}$ a closed Lebesgue null ultrafilter on $\mathbb{N}$. Then there is an $I \in \mathcal{F}$ such that $F = \overline{\{t_n : n \in I\}}$ is negligible. So $d^*(I) = 0$ ([13], 491B).

(g) By Theorem 1Bc, we have a measure-centering ultrafilter on $\mathbb{N}$ which contains no set of zero asymptotic density, so can be neither strongly measure-linking nor closed Lebesgue null.

6I I turn now to the proofs of 1Mh-1Mi. While these are expected results, they seem to depend on some non-trivial combinatorial probability theory. For the rest of this section, I will write $\mathcal{N}\mathsf{wd}$ for the ideal of nowhere dense subsets of $\mathbb{R}$ and $\mathcal{I}$ for the ideal $\{A : A \subseteq \mathbb{R},\ A \cap [0,\epsilon] \in \mathcal{N}\mathsf{wd} \text{ for some } \epsilon > 0\}$.

Lemma Suppose that $D \subseteq \mathbb{R}$.

(a)(i) If $D \notin \mathcal{N}\mathsf{wd}$ there is a $D' \subseteq D$, with no isolated points, such that $D' \notin \mathcal{N}\mathsf{wd}$.

(ii) If $D \notin \mathcal{I}$ there is a $D' \subseteq D$, with no isolated points, such that $D' \notin \mathcal{I}$.

(b) Suppose that $S \subseteq [D]^2$ is such that $\{t : \{s,t\} \notin S\} \in \mathcal{N}\mathsf{wd}$ for every $s \in D$.

(i) If $D \notin \mathcal{N}\mathsf{wd}$ there is a $D' \subseteq D$ such that $[D']^2 \subseteq S$ and $D' \notin \mathcal{N}\mathsf{wd}$.

(ii) If $D \notin \mathcal{I}$ there is a $D' \subseteq D$ such that $[D']^2 \subseteq S$ and $D' \notin \mathcal{I}$.

PROOF. **(a)** Set $G = \operatorname{int}\overline{D}$, $D' = D \cap G$; then D' has no isolated points and $D \setminus D' \subseteq \overline{D} \setminus G$ is nowhere dense. So $D' \notin \mathcal{N}\mathsf{wd}$ if $D \notin \mathcal{N}\mathsf{wd}$ and $D' \notin \mathcal{I}$ if $D \notin \mathcal{I}$.

(b) By (a), we can suppose that D has no isolated points. Let $\langle V_n\rangle_{n\in\mathbb{N}}$ enumerate a base for the topology of $\mathbb{R}$ and choose $\langle s_n\rangle_{n\in\mathbb{N}}$ inductively so that

$$s_n \in D, \quad \{s_i, s_n\} \in S \text{ for } i < n, \quad \text{if } V_n \cap D \neq \emptyset \text{ then } s_n \in V_n$$

for each n. Set $D' = \{s_n : n \in \mathbb{N}\}$; then $[D']^2 \subseteq S$ and D' is dense in D. So $D' \notin \mathcal{N}\mathsf{wd}$ if $D \notin \mathcal{N}\mathsf{wd}$ and $D' \notin \mathcal{I}$ if $D \notin \mathcal{I}$.

6J Lemma Suppose that $D \subseteq \mathbb{R}$, and that $S \subseteq [D]^2$ and $k \in \mathbb{N}$ are such that $[K]^2 \cap S \neq \emptyset$ for every $K \in [D]^k$.

(a) If $D \notin \mathcal{N}\mathsf{wd}$, then there is a $D' \subseteq D$ such that $D' \notin \mathcal{N}\mathsf{wd}$ and $[D']^2 \subseteq S$.

(b) If $D \notin \mathcal{I}$, there is a $D' \subseteq D$ such that $D' \notin \mathcal{I}$ and $\{t : t \in D', \{s,t\} \notin S\} \in \mathcal{I}$ for every $s \in D'$.

PROOF. For $t \in D$ set $S'_t = \{s : s \in D \setminus \{t\}, \{s,t\} \notin S\}$.

(a) Induce on k. If $k \leq 1$ the result is vacuous and for $k = 2$ it is trivial. For the inductive step to $k \geq 3$, if $S'_t \in \mathcal{N}\mathsf{wd}$ for every $t \in D$, then 6I(b-i) tells us that there is a $D' \subseteq D$ such that $[D']^2 \subseteq S$ and $D' \notin \mathcal{N}\mathsf{wd}$. Otherwise, take t such that $S'_t \notin \mathcal{N}\mathsf{wd}$. If $K \in [S'_t]^{k-1}$ then $K \cup \{t\} \in [D]^k$ so there is a $J \in [K \cup \{t\}]^2 \cap S$; since $K \subseteq S'_t$, $t \notin J$ and $J \subseteq K$. By the inductive hypothesis, there is a $D' \subseteq S'_t$ such that $[D']^2 \subseteq S$ and $D \notin \mathcal{N}\mathsf{wd}$.

(b) Again induce on k; as before, the case $k \leq 2$ is trivial. For the inductive step to $k \geq 3$, if there is a $t \in D$ such that S'_t does not belong to $\mathcal{I}$, then we can apply the inductive hypothesis to S'_t.

6K Lemma Suppose that $D \subseteq \mathbb{R}$, $(\mathfrak{A}, \bar{\mu})$ is a probability algebra and that $\langle a_t\rangle_{t\in D}$ is a family in $\mathfrak{A}$ such that $\epsilon = \inf_{t\in D} \bar{\mu}a_t$ is greater than 0.

(a) Take any $\delta < \epsilon^2$, and set $S = \{\{s,t\} : s, t \in D \text{ are distinct}, \bar{\mu}(a_s \cap a_t) \geq \delta\}$. If $k > \frac{1}{\epsilon^2 - \delta}$, then $[K]^2 \cap S \neq \emptyset$ for every $K \in [D]^k$.

(b) If $D \notin \mathcal{N}\mathsf{wd}$ and δ, $\eta > 0$ then there are an $e \in \mathfrak{A}$ and a $D' \subseteq D$ such that $D' \notin \mathcal{N}\mathsf{wd}$, $\bar{\mu}(a_s \setminus e) \leq \eta$ for every $s \in D'$, and whenever $d \subseteq e$, $\bar{\mu}d \geq \delta$ then $\{s : s \in D', \bar{\mu}(a_s \cap d) \leq \eta\bar{\mu}d\} \in \mathcal{N}\mathsf{wd}$.

(c) If $D \notin \mathcal{I}$ there are a $D' \subseteq D$ and an $\eta > 0$ such that $D' \notin \mathcal{I}$ and $\{s : s \in D', s > t, \bar{\mu}(a_s \cap a_t) \leq \eta\} \in \mathcal{N}\mathsf{wd}$ for every $t \in D'$.

(d) If $D \notin \mathcal{I}$ there are a $D' \subseteq D$ and an $\eta > 0$ such that $D' \notin \mathcal{I}$ and $\{t : t \in D', t < s, \bar{\mu}(a_s \cap a_t) \leq \eta\} \in \mathcal{N}\mathsf{wd}$ for every $s \in D'$.

(e) If $D \notin \mathcal{I}$ there are a $D' \subseteq D$ and an $\eta > 0$ such that $D' \notin \mathcal{I}$ and $\bar{\mu}(a_s \cap a_t) \geq \eta$ for all s, $t \in D'$.

PROOF. **(a)** This is immediate from Lemma 6D.

(b) Set

$$\gamma = \inf\{\bar{\mu}d : d \in \mathfrak{A}, \{s : s \in D, \bar{\mu}(a_s \setminus d) \leq \eta(1 - \bar{\mu}d)\} \notin \mathcal{N}\mathsf{wd}\}.$$

Let $e \in \mathfrak{A}$ be such that $D' = \{s : s \in D, \bar{\mu}(a_s \setminus e) \leq \eta(1 - \bar{\mu}e)\}$ is not in $\mathcal{N}\mathsf{wd}$ and $\bar{\mu}e < \gamma + \delta$. If $d \subseteq e$ and $\bar{\mu}d \geq \delta$, set $C = \{s : s \in D', \bar{\mu}(a_s \cap d) \leq \eta\bar{\mu}d\}$. For any $s \in C$,

$$\bar{\mu}(a_s \setminus (e \setminus d)) = \bar{\mu}(a_s \setminus e) + \bar{\mu}(a_s \cap d) \leq \eta(1 - \bar{\mu}e) + \eta\bar{\mu}d = \eta(1 - \bar{\mu}(e \setminus d)).$$

But also $\bar{\mu}(e \setminus d) < \gamma$; by the definition of γ, $C \in \mathcal{N}\mathsf{wd}$, as required.

(c) The proof proceeds by induction on the least $n > \frac{1}{\epsilon}$. The case $n \leq 2$ is trivial, because if $\epsilon > \frac{1}{2}$ then we can take $D' = D$ and $\eta = \epsilon - \frac{1}{2}$. For the inductive step to $n \geq 3$, take $\mathfrak{A}$, $\bar{\mu}$ and $\langle a_t \rangle_{t \in D}$ as described. Let $\delta > 0$ be such that $\delta < \epsilon - \frac{1}{n}$; set $\eta_0 = \delta^2$.

case 1 Suppose that whenever A, $B \subseteq D$, $A \notin \mathcal{N}\mathsf{wd}$ and $B \notin \mathcal{I}$ there are $A' \subseteq A$, $B' \subseteq B$ such that $A' \notin \mathcal{N}\mathsf{wd}$, $B' \notin \mathcal{I}$ and $\{s : s \in A', \bar{\mu}(a_s \cap a_t) \leq \eta_0\} \in \mathcal{N}\mathsf{wd}$ for every $t \in B'$. In this case, choose $\langle A_n \rangle_{n \in \mathbb{N}}$, $\langle B_n \rangle_{n \in \mathbb{N}}$ inductively, as follows. Start with $B_0 = D \cap [0, \infty[$. Given that $B_n \notin \mathcal{I}$, let $\alpha \in]0, 2^{-n}]$ be such that $B_n \setminus [0, \alpha] \notin \mathcal{N}\mathsf{wd}$. Then there are $A_n \subseteq B_n \setminus [0, \alpha]$ and $B_{n+1} \subseteq B_n \cap [0, \alpha]$ such that $A_n \notin \mathcal{N}\mathsf{wd}$, $B_{n+1} \notin \mathcal{I}$ and $\{s : s \in A_n, \bar{\mu}(a_s \cap a_t) \leq \eta_0\} \in \mathcal{N}\mathsf{wd}$ for every $t \in B_n$. Continue.

At the end of the induction, choose for each $n \in \mathbb{N}$ an $A_n^* \subseteq A_n$ such that $A_n^* \notin \mathcal{N}\mathsf{wd}$ and $\bar{\mu}(a_s \cap a_t) \geq \eta_0$ for every s, $t \in A_n^*$; this is possible by (a) above and Lemma 6Ja. Set $D' = \bigcup_{n \in \mathbb{N}} A_n^*$, $\eta = \eta_0$; this works.

case 2 Otherwise, take A, $B \subseteq D$ such that $A \notin \mathcal{N}\mathsf{wd}$, $B \notin \mathcal{I}$ and whenever $A' \subseteq A$, $B' \subseteq B$, $A' \notin \mathcal{N}\mathsf{wd}$ and $B' \notin \mathcal{I}$, then $\{s : s \in A', \bar{\mu}(a_s \cap a_t) \leq \eta_0\} \notin \mathcal{N}\mathsf{wd}$ for some $t \in B'$. By (b), there are $e \in \mathfrak{A}$, $A' \subseteq A$ such that $A' \notin \mathcal{N}\mathsf{wd}$, $\bar{\mu}(a_s \setminus e) \leq \delta$ for every $s \in A'$, and whenever $d \subseteq e$ and $\bar{\mu}d \geq \eta_0$ then $\{s : s \in A', \bar{\mu}(a_s \cap d) \leq \delta\bar{\mu}d\} \in \mathcal{N}\mathsf{wd}$.

Set $B' = \{t : t \in B, \bar{\mu}(a_t \cap e) \geq \delta\}$. **?** If $B' \notin \mathcal{I}$, then there is a $t \in B'$ such that $C = \{s : s \in A', \bar{\mu}(a_s \cap a_t) \leq \eta_0\} \notin \mathcal{N}\mathsf{wd}$. But $C \subseteq \{s : s \in A', \bar{\mu}(a_s \cap (a_t \cap e)) \leq \delta\bar{\mu}(a_t \cap e)\}$, which is nowhere dense, by the choice of A' and e. **X**

Accordingly $B' \in \mathcal{I}$; set $D_1 = B \setminus B' \notin \mathcal{I}$. Since there are $s \in A'$ and $t \in D_1$, we have

$$\epsilon - \delta \leq \bar{\mu}a_s - \delta \leq \bar{\mu}e = \bar{\mu}(e \cup a_t) + \bar{\mu}(e \cap a_t) - \bar{\mu}a_t \leq 1 + \delta - \epsilon < 1,$$

and $1 \setminus e \neq 0$. Consider the principal ideal $\mathfrak{A}_{1 \setminus e}$ with the normalized measure $\bar{\nu}$ where $\bar{\nu}a = \frac{\bar{\mu}a}{1 - \bar{\mu}e}$ for $a \in \mathfrak{A}_{1 \setminus e}$; set $a'_t = a_t \setminus e$ for $t \in D_1$. Then

$$\bar{\nu}a'_t \geq \frac{\epsilon - \delta}{1 - \epsilon + \delta} > \frac{1}{n - 1}$$

(by the choice of δ) for every $t \in D_1$. So the inductive hypothesis tells us that there are a $D' \subseteq D_1$ and an $\eta > 0$ such that $D' \notin \mathcal{I}$ and $\{s : s \in D', s > t, \bar{\mu}(a_s \cap a_t) \leq \eta\} \in \mathcal{N}\mathsf{wd}$ for every $t \in D'$.

(d) Set $\delta = \frac{1}{3}\epsilon^2$, $\eta_0 = \frac{1}{2}\delta^2$. Putting (a) here and Lemma 6Jb together, there is a $D_1 \subseteq D$ such that $D_1 \notin \mathcal{I}$ and $\{t : t \in D_1, \bar{\mu}(a_s \cap a_t) \leq \delta\} \in \mathcal{I}$ for every $s \in D_1$. Let $\langle \alpha_n \rangle_{n \in \mathbb{N}}$ be a strictly decreasing sequence in $]0, 1]$ such that $\alpha_{n+1} \leq \frac{1}{2}\alpha_n$ and $A_n = D_1 \cap]\alpha_{n+1}, \alpha_n[\notin \mathcal{N}\mathsf{wd}$ for every $n \in \mathbb{N}$. By (b), we can find for each $n \in \mathbb{N}$ an $A'_n \subseteq A_n$ and an $e_n \in \mathfrak{A}$ such that $A'_n \notin \mathcal{N}\mathsf{wd}$, $\bar{\mu}(a_s \setminus e_n) \leq \frac{1}{2}\delta$ for every $s \in A'_n$ and whenever $d \subseteq e_n$ and $\bar{\mu}d \geq \eta_0$ then $\{s : s \in A'_n, \bar{\mu}(a_s \cap d) \leq \frac{1}{2}\delta\bar{\mu}d\} \in \mathcal{N}\mathsf{wd}$; moreover, by Lemma 6I(a-i), we can suppose that A'_n has no isolated points. Of course $\bar{\mu}e_n \geq \epsilon - \delta$ for every n; since $\delta < (\epsilon - \delta)^2$, there is an infinite $J \subseteq \mathbb{N}$ such that $\bar{\mu}(e_m \cap e_n) \geq \delta$ whenever m, $n \in J$ and $m < n$. (Apply Ramsey's theorem to $\{\{m, n\} : m < n, \bar{\mu}(e_m \cap e_n) < \delta\}$.)

If $m \in J$ and $J' \subseteq J$ is infinite, there are an infinite $J'' \subseteq J'$ and a set $A''_m \subseteq A'_m$ such that A''_m is dense in A'_m and $\bar\mu(a_s \cap e_n) \ge \eta_0$ whenever $s \in A''_m$ and $n \in J''$. **P** Let $\langle V_n\rangle_{n\in\mathbb{N}}$ enumerate a base for the topology of $\mathbb{R}$. Choose $\langle s_n\rangle_{n\in\mathbb{N}}$, $\langle j_n\rangle_{n\in\mathbb{N}}$ inductively, as follows. j_0 is to be any member of J' greater than m. Given that $j_i > m$ for $i \le n$,

$$\{s : s \in A'_m,\ \bar\mu(a_s \cap e_{j_i}) \le \eta_0\} \subseteq \{s : s \in A'_m,\ \bar\mu(a_s \cap e_{j_i}) \le \frac{1}{2}\delta\bar\mu(e_m \cap e_{j_i})\}$$
$$\in \mathcal{N}\mathsf{wd}$$

for every $i \le n$, so there is an $s_n \in A_m$ such that $\bar\mu(a_{s_n} \cap e_{j_i}) \ge \eta_0$ for every $i \le n$, and if $A_m \cap V_n$ is not empty then we can take $s_n \in V_n$. Given s_i for $i < n$, where $n \ge 1$, then there is an $r \in \mathbb{N}$ such that $\{t : t \in D_1 \cap [0, \alpha_r],\ \bar\mu(a_{s_i} \cap a_t) \le \delta\}$ is nowhere dense for each $i < n$; so if we take $j_n \in J'$ such that $j_n > \max(j_{n-1}, r)$, there is a $t \in A'_{j_n}$ such that $\bar\mu(a_{s_i} \cap a_t) \ge \delta$ for every $i < n$, in which case $\bar\mu(a_{s_i} \cap e_{j_n}) \ge \delta - \frac{1}{2}\delta \ge \eta_0$ for every $i < n$. Continue; at the end, set $A''_m = \{s_i : i \in \mathbb{N}\}$ and $J'' = \{j_i : i \in \mathbb{N}\}$. **Q**

We can therefore find an infinite $I \subseteq J$ and a family $\langle A''_m\rangle_{m\in I}$ of sets such that A''_m is dense in A'_m and $\bar\mu(a_s \cap e_n) \ge \eta_0$ whenever $m < n$ in I and $s \in A''_m$. But this will mean that $\{t : t \in A''_n,\ \bar\mu(a_s \cap a_t) \le \frac{1}{2}\delta\eta_0\} \in \mathcal{N}\mathsf{wd}$ whenever $m < n$ in I and $s \in A''_m$.

Finally, by Lemma 6Ja, there is for each $m \in I$ an $A^*_m \subseteq A''_m$ such that $A^*_m \notin \mathcal{N}\mathsf{wd}$ and $\{t : t \in A^*_m,\ \bar\mu(a_s \cap a_t) \le \delta\} \in \mathcal{N}\mathsf{wd}$ for every $s \in A^*_m$. So setting $D' = \bigcup_{m\in I} A^*_m$ we shall have a suitable set, with $\eta = \frac{1}{2}\delta\eta_0$.

(e) Putting (c) and (d) together, we see that there are an $\eta > 0$ and a $D_1 \subseteq D$ such that $D_1 \notin \mathcal{I}$ and $\{t : t \in D_1,\ \bar\mu(a_s \cap a_t) \le \eta\} \in \mathcal{N}\mathsf{wd}$ for every $s \in D_1$. By Lemma 6I(b-ii) there is a $D' \subseteq D_1$ such that $D' \notin \mathcal{I}$ and $\bar\mu(a_s \cap a_t) \ge \eta$ for all s, $t \in D'$.

6L Proof of Theorem 1Mh If $\mathfrak{c} = \omega_1$, there is a strongly measure-linking ultrafilter on $\mathbb{N}$ which is not nowhere dense, therefore not measure-centering nor an arrow ultrafilter.

PROOF. **(a)** Set $D = \mathbb{Q} \cap [0,1]$. Let $\langle\langle a_{\xi t}\rangle_{t\in D}\rangle_{\xi<\omega_1}$ run over all families $\langle a_t\rangle_{t\in D}$ in $\mathfrak{B}_\omega$ such that $\inf_{t\in D} \bar\nu_\omega a_t > 0$. Choose a family $\langle D_\xi\rangle_{\xi<\omega_1}$ of subsets of D inductively, as follows. The inductive hypothesis will be that $\bigcap_{\eta\in J} D_\eta \notin \mathcal{I}$ for every finite $J \subseteq \xi$. Start with $D_0 = D$. Given $\langle D_\eta\rangle_{\eta<\xi}$, where $1 \le \xi < \omega_1$, let $\langle\theta(\xi,n)\rangle_{n\in\mathbb{N}}$ run over ξ, and take a strictly decreasing sequence $\langle\alpha_n\rangle_{n\in\mathbb{N}}$, with infimum 0, such that $C_n =]\alpha_{n+1}, \alpha_n[\cap \bigcap_{i\le n} D_{\theta(\xi,i)} \notin \mathcal{N}\mathsf{wd}$ for every $n \in \mathbb{N}$. Then $C = \bigcup_{n\in\mathbb{N}} C_n \notin \mathcal{I}$. By Lemma 6Ke, there is a $D_\xi \subseteq C$ such that $D_\xi \notin \mathcal{I}$ and $\inf_{s,t\in D_\xi} \bar\nu_\omega(a_{\xi s} \cap a_{\xi t}) > 0$. Since $D_\xi \cap [0, \alpha_n[\subseteq \bigcap_{i\le n} D_{\theta(\xi,i)}$ for every n, $D_\xi \cap \bigcap_{\eta\in J} D_\eta \notin \mathcal{I}$ for every finite $J \subseteq \xi$, and $\bigcap_{\eta\in J} D_\eta \notin \mathcal{I}$ for every finite $J \subseteq \xi+1$. Continue.

(b) At the end of the induction, let $\mathcal{F}$ be an ultrafilter on D containing every D_ξ and no nowhere dense set; then $\mathcal{F}$ is not nowhere dense. But $\mathcal{F}$ is strongly measure-linking. **P** If $\mathfrak{A}$ is a Boolean algebra, $\nu : \mathfrak{A} \to [0,1]$ an additive functional such that $\nu 1 = 1$, and $\langle a_t\rangle_{t\in D}$ is a family in $\mathfrak{A}$ such that $\inf_{t\in D} \nu a_t > 0$, then (as in (iv)⇒(i) of the proof in 3C) there is a Radon probability measure μ on the Stone space Z of $\mathfrak{A}$, identified with the algebra of open-and-closed subsets of Z, which extends ν. Taking T to be the σ-algebra of subsets of Z generated by $\{a_t : t \in D\}$, $(\mathfrak{B}, \bar{\mu})$ to be the measure algebra of $(Z, \mathrm{T}, \mu{\restriction}\mathrm{T})$ and $a_t^\bullet \in \mathfrak{B}$ the equivalence class of $a_t \in \mathrm{T}$ for each $t \in D$, we see that $\inf_{t\in D} \bar{\mu} a_t^\bullet > 0$. As in part (ii)$\Rightarrow$(iv) of the proof in 3C, there is a measure-preserving Boolean homomorphism $\pi : \mathfrak{B} \to \mathfrak{B}_\omega$, and now there must be a $\xi < \omega_1$ such that $\pi a_t^\bullet = a_{\xi t}$ for every $t \in D$. In this case, $D_\xi \in \mathcal{F}$ and

$$\inf_{s,t\in D_\xi} \nu(a_s \cap a_t) = \inf_{s,t\in D_\xi} \bar{\nu}_\omega(a_{\xi s} \cap a_{\xi t}) > 0.$$

As $\mathfrak{A}$, ν and $\langle a_t\rangle_{t\in D}$ are arbitrary, $\mathcal{F}$ is strongly measure-linking. **Q**

(c) Thus we have a strongly measure-linking ultrafilter on the countably infinite set D which is not nowhere dense. Of course it follows at once that there is such an ultrafilter on $\mathbb{N}$. By parts (b) and (d-ii) of Theorem 1M it cannot be either measure-centering or an arrow ultrafilter.

6M To convert the last result into a proof that there can be a Hausdorff p-point ultrafilter which is not measure-centering, I use the language of 'game strategies'. Let $G^{\mathrm{H|m\text{-}c}}$ be the game for two players, Empty and Non-empty, in which

Empty chooses $m \geq 1$,

Non-empty chooses $k \in \mathbb{N}$,

Empty chooses $n \geq 1$, a set B with mn members, and a set $L_0 \subseteq [B]^n$,

given $i < k$ and L_i, Non-empty chooses f_i, $g_i : L_i \to \mathbb{N}$ with $f_i(a) \neq g_i(a)$ for every $a \in L_i$,

given $i < k$ and L_i, f_i and g_i, Empty chooses $L_{i+1} \subseteq L_i$ such that $f_i[L_{i+1}] \cap g_i[L_{i+1}] = \emptyset$.

A run of the game ends when Empty has chosen L_k; Empty wins if $\bigcap L_k = \emptyset$; otherwise Non-empty wins.

Note that the game is determined, that is, one of the players has a winning strategy. (Since the game always terminates after finitely many moves, it is an 'open' game in the usual terminology of infinite games.)

6N Lemma Empty has a winning strategy in the game $G^{\mathrm{H|m\text{-}c}}$.

PROOF. **(a)** To begin with, suppose that $\mathfrak{c} = \omega_1$.

(i) By Theorem 1Mh, there is a strongly measure-linking ultrafilter $\mathcal{F}$ on $\mathbb{N}$ which is not measure-centering. By Theorem 1Ma, $\mathcal{F}$ is a Hausdorff ultrafilter. Let $\langle a_j\rangle_{j\in\mathbb{N}}$ be a sequence in $\mathfrak{B}_\omega$ such that $\epsilon = \inf_{j\in\mathbb{N}} \bar{\nu}_\omega a_i$ is greater than 0, but there is no $A \in \mathcal{F}$ such that $\{a_j : j \in A\}$ is centered

(ii) Let $\mathcal{L}_0$ be the family of finite subsets L of $\mathbb{N}$ such that $\inf_{j\in L} a_j = 0$ in $\mathfrak{B}_\omega$, and for $i \in \mathbb{N}$ set

$$\mathcal{L}_{i+1} = \{L : L \in \mathcal{L}_i \text{ and for every pair } f,\ g \text{ of nowhere equal functions defined on } L \text{ there is an } L' \in \mathcal{L}_i \text{ such that } L' \subseteq L \text{ and } f[L'] \cap g[L'] = \emptyset\}.$$

By the choice of $\langle a_j\rangle_{j\in\mathbb{N}}$, every member of $\mathcal{F}$ has a finite subset belonging to $\mathcal{L}_0$. In fact, if $A \in \mathcal{F}$ and $i \in \mathbb{N}$, A has a finite subset belonging to $\mathcal{L}_i$. **P** Induce on i. For the inductive step to $i + 1$, **?** suppose, if possible, that $A \in \mathcal{F}$ has no finite subset belonging to $\mathcal{L}_{i+1}$. For each $r \in \mathbb{N}$, $A \cap r \notin \mathcal{L}_{i+1}$, so there are nowhere equal functions f_r, g_r defined on $A \cap r$ such that if $L \subseteq A_r$ is such that $f_r[L] \cap g_r[L]$ is empty, then $L \notin \mathcal{L}_i$. Adjusting f_r, g_r if necessary, we can suppose that $f_r(j) \leq 2j$

and $g_r(j) \le 2j{+}1$ for every $j \in A{\cap}r$. In this case, there will be functions $f : A \to \mathbb{N}$, $g : A \to \mathbb{N}$ such that for every finite $L \subseteq A$ there is an $r \in \mathbb{N}$ such that $L \subseteq r$, $f{\restriction}L = f_r{\restriction}L$ and $g{\restriction}L = g_r{\restriction}L$; of course f and g are nowhere equal. Because $\mathcal{F}$ is a Hausdorff ultrafilter, there is an $A' \subseteq A$ such that $A' \in \mathcal{F}$ and $f[A'] \cap g[A'] = \emptyset$. Now A' has a finite subset $L \in \mathcal{L}_i$, by the inductive hypothesis. Take r such that $f_r{\restriction}L = f{\restriction}L$ and $g_r{\restriction}L = g{\restriction}L$; since $f_r[L] \cap g_r[L] = \emptyset$, $L \notin \mathcal{L}_i$, which is absurd. **XQ**

(iii) I am now in a position to describe a winning strategy for Empty. His first move should be $m \ge 1$ such that $\frac{1}{m} < \epsilon$. Suppose that Non-empty responds with $k \in \mathbb{N}$. By (ii), $\mathcal{L}_k$ is not empty; take $L_0 \in \mathcal{L}_k$. Let $\mathfrak{B}$ be the finite subalgebra of $\mathfrak{B}_\omega$ generated by $\{a_j : j \in L_0\}$; let r be the number of atoms of $\mathfrak{B}$. Let $n \ge r$ be such that $\dfrac{r}{mn} \le \epsilon - \dfrac{1}{m}$ and every atom of $\mathfrak{B}$ has measure at least $\dfrac{1}{mn}$, and let C be a partition of unity in $\mathfrak{B}_\omega$ such that every member of C has measure $\frac{1}{mn}$ and for every atom b of $\mathfrak{B}$ the number of members of C included in b is the maximum possible value $\lfloor mn\bar{\mu}(b) \rfloor$. For $j \in L_0$, set $K_j = \{c : c \in C,\ c \subseteq a_j\}$; then $\#(K_j) \ge n$ (because the number of members of C not included in any atom of $\mathfrak{B}$ is at most $r \le mn\epsilon - n$, so $\#(K_j) \ge mn\bar{\mu}(a_j) - (mn\epsilon - n) \ge n$). Take $K'_j \subseteq K_j$ to be a set of size n for each $j \in L_0$; because $n \ge r$, we can suppose that whenever b is an atom of $\mathfrak{B}$ included in a_j, there is a $c \in K'_j$ included in b. Consequently $j \mapsto K'_j : L_0 \to [C]^n$ is injective. Finally, Empty plays (n, C, L'_0) for his second move, where $L'_0 = \{K'_j : j \in L_0\}$.

For subsequent moves, given that $i < k$, Empty has played $L'_i \subseteq L'_0$ and Non-empty has played nowhere-equal functions f'_i, g'_i, the rule for Empty is as follows. The inductive hypothesis will be that $L_i = \{j : K'_j \in L'_i\}$ belongs to $\mathcal{L}_{k-i}$. Define f_i, g_i on L_i by saying that $f_i(j) = f'_i(K'_j)$, $g_i(j) = g'_i(K'_j)$ for $j \in L_i$; because $L_i \in \mathcal{L}_{k-i}$, there is an $L_{i+1} \in \mathcal{L}_{k-i-1}$ such that $L_{i+1} \subseteq L_i$ and $f_i[L_{i+1}] \cap g_i[L_{i+1}]$ is empty. Now Empty plays $L'_{i+1} = \{K'_j : j \in L_{i+1}\}$, and the run continues.

At the end of the run, we get $L_k \in \mathcal{L}_0$. But this means that $\inf_{j \in L_k} a_j = 0$, so $\bigcap_{j \in L_k} K'_j$ must be empty, and Empty has won the run. Thus we have a winning strategy for Empty.

(b) This proves the result on the assumption that $\mathfrak{c} = \omega_1$. But now look at the logical nature of the statement 'Empty has a winning strategy in $G^{\text{H|m-c}}$'. It makes no difference if Empty is required to choose a member of $\mathbb{N}$ for the set B in his second move, following which all Non-empty's moves will have to belong to the countable set $\mathbb{N}^{[[\mathbb{N}]^{<\omega}]^{<\omega}}$, all Empty's moves will be in the countable set $[[\mathbb{N}]^{<\omega}]^{<\omega}$, and the deciding move L_k wins iff it too belongs to a specific countable set (the family of finite subsets of $[\mathbb{N}]^{<\omega}$ with empty intersection). We therefore have in fact a Borel code for the set of winning strategies for Empty. By Shoenfield's theorem ([18], Theorem 98, or [19], 25.20), the assertion that it is non-empty is absolute for inner models of ZFC. Consequently it is absolute for forcing, that is, if $\mathbb{P}$ is any forcing notion, then

$$\begin{aligned}&\text{Empty has a winning strategy in } G^{\text{H|m-c}}\\ &\qquad\qquad \text{iff } \Vdash_{\mathbb{P}} \text{Empty has a winning strategy in } G^{\text{H|m-c}}.\end{aligned}$$

Now take any forcing notion $\mathbb{P}$ such that

$$\Vdash_{\mathbb{P}} \mathfrak{c} = \omega_1;$$

for instance, take $\mathbb{P}$ to be the partially ordered set of functions from countable ordinals to $\mathbb{R}$, active upwards. Then we shall have

$$\Vdash_{\mathbb{P}} \mathfrak{c} = \omega_1, \text{ so Empty has a winning strategy in } G^{\text{H|m-c}},$$

and it follows that Empty has a winning strategy in $G^{\text{H|m-c}}$ in the ordinary universe.

Remark No doubt there is a more illuminating proof of this lemma which does not employ considerations of absoluteness.

6O Proof of Theorem 1Mi If $\mathfrak{p} = \mathfrak{c}$, there is a Hausdorff p-point ultrafilter on $\mathbb{N}$ which is not measure-centering.

PROOF. **(a)** By Lemma 6N, Empty has a winning strategy in $G^{\text{H|m-c}}$; let $m \geq 1$ be such that Empty has a winning strategy with first move m. For each $n \in \mathbb{N}$, fix a set B_n of size mn. For $n \in \mathbb{N}$, set $I_n = [B_n]^n$; set $I = \bigcup_{n\in\mathbb{N}} I_n$,

$$\mathcal{L}_0 = \bigcup_{n\geq 1}\{L : L \subseteq I_n, \bigcap L = \emptyset\},$$

and for $i \in \mathbb{N}$ set

$$\mathcal{L}_{i+1} = \{L : L \in \mathcal{L}_i \text{ and whenever } f,\ g \text{ are nowhere equal functions defined on } L, \text{ there is an } L' \in \mathcal{L}_i \text{ such that } L' \subseteq L \text{ and } f[L'] \cap g[L'] = \emptyset\}.$$

An easy induction shows that if i, $n \in \mathbb{N}$, $L \subseteq L' \subseteq I_n$ and $L \in \mathcal{L}_i$, then $L' \in \mathcal{L}_i$.

For $L \in \mathcal{L}_0$, say that the **depth** $\text{depth}(L)$ of L is the greatest k such that $L \in \mathcal{L}_k$; observe that $\text{depth}(L) \leq \#(L)$ for every L. (The point is that every member of $\mathcal{L}_0$ has at least two members, so that for every $L \in \mathcal{L}_0$ there are nowhere equal functions f and g defined on L such that if $L' \subseteq L$ and $f[L']$ is disjoint from $g[L']$, L' is a proper subset of L.) Say that $A \subseteq I$ is **deep** if $\sup_{n\geq 1, A\cap I_n \in \mathcal{L}_0} \text{depth}(A\cap I_n)$ is infinite. Note that if $A \subseteq B \subseteq I$, then $\text{depth}(B \cap I_n) \geq \text{depth}(A \cap I_n)$ whenever the latter is defined, so B will be deep if A is.

Now let σ be a winning strategy for Empty in $G^{\text{H|m-c}}$ with first move m; we can suppose that σ is such that whenever Empty plays $(n, B, L_0) = \sigma(k)$ for his second move, he actually chooses $B = B_n$.

(b) Suppose that $(m, k, (n, B_n, L_0), (f_0, g_0), L_1, (f_1, g_1), \ldots, L_i)$, where $i \leq k$, is any partial run of the game in which Empty follows the strategy σ. Then $L_i \in \mathcal{L}_{k-i}$. **P** Induce downwards on i, starting with $i = k$. At the end of the run, Empty must win, so we certainly have L_k a subset of $L_0 \subseteq I_n$ with empty intersection, in which case $L_k \in \mathcal{L}_0$. For the inductive step down to $i < k$, given that Empty has just played L_i, let f, g be any nowhere-equal functions defined on L_i. Then Empty will reply with $L_{i+1} = \sigma(k, (f_0, g_0), \ldots, (f, g)) \subseteq L_i$ such that $f[L_{i+1}] \cap g[L_{i+1}] = \emptyset$; by the inductive hypothesis, $L_{i+1} \in \mathcal{L}_{k-i-1}$; as f and g are arbitrary, $L_i \in \mathcal{L}_{k-i}$, as required. **Q**

It follows that I is deep. **P** For every $k \in \mathbb{N}$ there is a partial run, following Empty's strategy, of the form $(m, k, (n, B_n, L_0))$, where $(n, B_n, L_0) = \sigma(k)$, so that $L_0 \in \mathcal{L}_k$ and the depth of I_n is at least k. **Q**

(c) If $A \subseteq I$ is deep and $A' \subseteq A$, then at least one of A', $A \setminus A'$ is deep. **P** Define f, $g : I \to \{0, 1\}$ by setting

$$f(a) = 1 \text{ if } a \in A',$$
$$= 0 \text{ otherwise},$$
$$g(a) = 1 - f(a) \text{ for every } a \in I.$$

For every k, there is an $n_k \in \mathbb{N}$ such that $A \cap I_{n_k} \in \mathcal{L}_{k+1}$. Since f and g are nowhere equal on $A \cap I_{n_k}$, there is an $L_k \subseteq A \cap I_{n_k}$ such that $L_k \in \mathcal{L}_k$ and $f[L_k] \cap g[L_k] = \emptyset$, that is, L_k is either included in A' or disjoint from A'. If $L_k \subseteq A'$ for infinitely many k, then A' is deep; otherwise, $L_k \subseteq A \setminus A'$ for infinitely many k and $A \setminus A'$ is deep. **Q**

(d) If $A \subseteq I$ is deep and f, g are nowhere equal functions defined on I, then there is a deep $A' \subseteq A$ such that $f[A']$ and $g[A']$ are disjoint. **P** (i) If there is some j such that $A \cap f^{-1}[\{j\}]$ is deep, we can take this for A'; similarly, if there is some j such that $A \cap g^{-1}[\{j\}]$ is deep, this will serve for A'. By (c), we can therefore restrict our attention to the case in which $A \setminus (f^{-1}[M] \cup g^{-1}[M])$ is deep for every finite set M. (ii) Choose $\langle n_k \rangle_{k \in \mathbb{N}}$, $\langle M_k \rangle_{k \in \mathbb{N}}$ inductively in such a way that, for each $k \in \mathbb{N}$,

$$M_k = \bigcup_{i<k} f[I_{n_i}] \cup g[I_{n_i}],$$
$$n_k > n_i \text{ whenever } i < k,$$
$$\operatorname{depth}(A \cap I_{n_k} \setminus (f^{-1}[M_k] \cup g^{-1}[M_k])) \geq k + 1.$$

Then we have for each k an $L_k \subseteq A \cap I_{n_k} \setminus (f^{-1}[M_k] \cup g^{-1}[M_k])$ such that $\operatorname{depth}(L_k) \geq k$ and $f[L_k]$ is disjoint from $g[L_k]$. Setting $A' = \bigcup_{k \in \mathbb{N}} L_k$, we see that $A' \subseteq A$ is deep and $f[A'] \cap g[A'] = \emptyset$. **Q**

(e) If $\xi < \mathfrak{p}$ and $\langle A_\eta \rangle_{\eta < \xi}$ is a family of deep subsets of I such that $A_\eta \setminus A_\zeta$ is finite whenever $\zeta \leq \eta < \xi$, then there is a deep set $A \subseteq I$ such that $A \setminus A_\eta$ is finite for every $\eta < \xi$. **P** Let P be the set of pairs (J, D) where $J \subseteq I$, $D \subseteq \xi$ are finite, ordered by saying that $(J, D) \leq (J', D')$ if $J \subseteq J'$, $D \subseteq D'$ and $J' \setminus J \subseteq A_\eta$ for every $\eta \in D$. Then P is a partially ordered set, σ-centered upwards. For every $k \in \mathbb{N}$, $Q_k = \{(J, D) : (J, D) \in P, \text{ there is an } n \in \mathbb{N} \text{ such that } \operatorname{depth}(J \cap I_n) \geq k\}$ is cofinal with P, because if $D \subseteq \xi$ is finite then $I \cap \bigcap_{\eta \in D} A_\eta$ is deep. For every $\eta < \xi$, $Q'_\eta = \{(J, D) : (J, D) \in P, \eta \in D\}$ is cofinal with P. By Bell's theorem ([9], 14C) there is an upwards-directed $R \subseteq P$ meeting every Q_k and every Q'_η; set $A = \bigcup \{J : (J, D) \in R\}$. **Q**

(f) Let $\langle (f_\xi, g_\xi) \rangle_{\xi < \mathfrak{c}}$ enumerate the set of pairs (f, g) of nowhere equal functions from I to $\mathbb{N}$. Because $\mathfrak{p} = \mathfrak{c}$, we can use (d) and (e) to find a family $\langle A_\xi \rangle_{\xi < \mathfrak{c}}$ of deep subsets of I such that

$$A_\xi \setminus A_\eta \text{ is finite whenever } \eta \leq \xi < \mathfrak{c},$$
$$f_\xi[A_{\xi+1}] \text{ is disjoint from } g_\xi[A_{\xi+1}] \text{ for every } \xi < \mathfrak{c}.$$

Let $\mathcal{F}$ be the filter on I generated by $\{A_\xi \setminus J : \xi < \mathfrak{c}, J \in [I]^{<\omega}\}$. If f, g are any two nowhere-equal functions on I, then (because $f[I] \cup g[I]$ must be countable) there is a $\xi < \mathfrak{c}$ such that for all a, $b \in I$, $f(a) = g(b)$ iff $f_\xi(a) = g_\xi(b)$; so $f[A_{\xi+1}] \cap g[A_{\xi+1}] = \emptyset$. It follows that $\mathcal{F}$ is a Hausdorff ultrafilter. (It is an ultrafilter by the argument in (c) above.) By construction, $\mathcal{F}$ is a p-point ultrafilter.

(g) $\mathcal{F}$ is not measure-centering. **P** Let $(\mathfrak{A}, \bar{\mu})$ be any atomless probability algebra. For each $n \in \mathbb{N}$ let $\pi_n : \mathcal{P}B_n \to \mathfrak{A}$ be a Boolean homomorphism which is measure-preserving for the uniform probability measure on $\mathcal{P}B_n$. For $n \in \mathbb{N}$ and

$a \in I_n$ set $d_a = \pi_n a$; then $\bar{\mu} d_a = \frac{\#(a)}{\#(B_n)} = \frac{1}{m}$. If $A \in \mathcal{F}$, there is a $\xi < \mathfrak{c}$ such that $A_\xi \setminus A$ is finite, so A is deep; in particular, there is some $n \in \mathbb{N}$ such that $A \cap I_n \in \mathcal{L}_0$, that is, $A \cap I_n$ has empty intersection. But now $\{d_a : a \in A\} \supseteq \pi_n[A \cap I_n]$ is not centered. So $\langle d_a \rangle_{a \in I}$ witnesses that $\mathcal{F}$ is not measure-centering. **Q**

Transferring $\mathcal{F}$ from the countably infinite set I to $\mathbb{N}$, we have the required example.

7 Problems

7A If Martin's axiom is true, the constructions used in the proof of Theorem 1B, given in 3F-3G, and in the proof of existence of Ramsey ultrafilters in [14], provide three essentially different classes of free measure-centering ultrafilters on $\mathbb{N}$, and further measure-centering ultrafilters on κ for $\omega < \kappa < \mathfrak{c}$; moreover, since there are $2^{\mathfrak{c}}$ Ramsey ultrafilters ([9], 26Ed), we get further measure-centering ultrafilters on $\mathbb{N}$ from Theorem 1J. I know of no construction for a uniform measure-centering ultrafilter on $\mathfrak{c}$. Indeed, Martin's axiom ensures that there is no such ultrafilter, for the following reason. A cardinal κ is a **measure-precaliber of probability algebras** if whenever $(\mathfrak{A}, \bar{\mu})$ is a probability algebra and $\langle a_\xi \rangle_{\xi < \kappa}$ is a family in $\mathfrak{A}$ such that $\inf_{\xi < \kappa} \bar{\mu} a_\xi > 0$, there is a set $J \subseteq \kappa$, of cardinal κ, such that $\{a_\xi : \xi \in J\}$ is centered. If $\mathcal{F}$ is a uniform measure-centering ultrafilter on κ, there will always be such a set belonging to $\mathcal{F}$, so κ will certainly be a measure-precaliber of probability algebras. Now Martin's axiom (or much less) ensures that $\mathfrak{c}$ is not a measure-precaliber of probability algebras (see [14], 525D and 525O); so it is certainly consistent to suppose that there is no uniform measure-centering ultrafilter on $\mathfrak{c}$. On the other hand, it is also consistent to suppose that $\mathfrak{c}$ is a measure-precaliber of probability algebras (this happens, for instance, if $\mathfrak{c} = \omega_2$ and there is a subset of $\mathbb{R}$ of cardinal ω_1 which is not Lebesgue negligible; see [14], 525L). So my question is: is it consistent with ZFC to suppose that there is a uniform measure-centering ultrafilter on $\mathfrak{c}$? Could it even be consistent to suppose that for every infinite cardinal κ there is a uniform measure-centering ultrafilter on κ?

7B I have been unable to answer the following question from Andreas Blass: is a weakly Ramsey ultrafilter on $\mathbb{N}$ necessarily measure-centering?

7C Under what circumstances is the product $\mathcal{F} \ltimes \mathcal{G}$ of two measure-centering ultrafilters measure-centering? For instance, when the product is Hausdorff, will it always be measure-centering?

7D Is every Hausdorff ultrafilter measure-linking?

Acknowledgments I am greatly indebted to Dr Di Nasso and the organizers of the conference for inviting me to speak there. In addition I should like to mention conversations with M.Benedikt, V.Bergelson, A.Blass, I.Farah, R.Jin, D.Raghavan and M.Talagrand, and correspondence with W.Comfort. Some of the ideas of §6 were worked out while I was a guest of the Fields Institute at the University of Toronto.The referee made several helpful suggestions.

References

[1] Baumgartner J.E. 'Ultrafilters on ω', J. Symbolic Logic 60 (1995) 624-639.

[2] Baumgartner J.E. & Taylor A.D. 'Partition theorems and ultrafilters', Trans. Amer. Math. Soc. 241 (1978) 283-309.

[3] Benedikt M. 'Ultrafilters which extend measures', J. Symbolic Logic 63 (1998) 638-662.

[4] Benedikt M. 'Hierarchies of measure-theoretic ultrafilters', Ann. Pure Appl. Logic 97 (1999) 203-219.

[5] Blass A. 'Ultrafilter mappings and their Dedekind cuts', Trans. Amer. Math. Soc. 188 (1974) 327-340.

[6] Blass A. 'Selective ultrafilters and homogeneity', Ann. Pure Appl. Logic 38 (1988) 252-285.

[7] Comfort W.W. & Negrepontis S. *The Theory of Ultrafilters.* Springer, 1974.

[8] Daguenet-Tessier M. 'Ultrafiltres à la façon de Ramsey', Trans. Amer. Math. Soc. 250 (1979) 91-120.

[9] Fremlin D.H. *Consequences of Martin's Axiom.* Cambridge U.P., 1984.

[10] Fremlin D.H. 'Real-valued-measurable cardinals', pp. 151-304 in [20].

[11] Fremlin D.H. *Measure Theory, Vol. 2: Broad Foundations.* Torres Fremlin, 2001.

[12] Fremlin D.H. *Measure Theory, Vol. 3: Measure Algebras.* Torres Fremlin, 2002.

[13] Fremlin D.H. *Measure Theory, Vol. 4: Topological Measure Spaces.* Torres Fremlin, 2003.

[14] Fremlin D.H. *Measure Theory, Vol. 5: Set-theoretic Measure Theory.* Torres Fremlin, 2008.

[15] Fremlin D.H. 'Dependently selective filters', note of 30.1.09 (`http://www.essex.ac.uk/maths/staff/fremlin/preprints.htm`).

[16] Grigorieff S. 'Combinatorics on ideals and forcing', Ann. Math. Logic 3 (1971) 363-394.

[17] Henson C.W. & Wattenberg B. 'Egoroff's theorem and the distribution of standard points in a non-standard model', Proc. Amer. Math. Soc. 81 (1981) 455-461.

[18] Jech T. *Set Theory.* Academic, 1978 (ISBN 0-120-381950-4).

[19] Jech T. *Set Theory, Millennium Edition.* Springer, 2003 (ISBN 3540440852).

[20] Judah H. (ed.) *Proceedings of the Bar-Ilan Conference on Set Theory and the Reals, 1991.* Amer. Math. Soc. (Israel Mathematical Conference Proceedings 6), 1993.

[21] Loeb P.A. 'Conversion from nonstandard to standard measure spaces and applications in probability theory', Trans. Amer. Math. Soc. 211 (1975) 113-122.

[22] Makowsky J.A. & Ravve E.V. (eds.) *Logic colloquium '95.* Springer, 1998 (Lecture Notes in Logic 11).

[23] Sazonov V.V. 'On perfect measures', Amer. Math. Soc. Transl. (Ser. 2) 48 (1966) 229-254.

[24] Shelah S. 'There may be no nowhere dense ultrafilter', pp. 305-324 in [22].

[25] Sierpiński W. 'Sur une suite infinie de fonctions de classe 1 dont toute fonction d'accumulation est non mesurable', Fund. Math. 33 (1945) 104-105.

E-mail address: `fremdh@essex.ac.uk`

UNIVERSITY OF ESSEX, COLCHESTER, ENGLAND

Contemporary Mathematics
Volume **530**, 2010

Algebra in the Space of Ultrafilters and Ramsey Theory

Neil Hindman and Dona Strauss

ABSTRACT. We survey developments in the algebraic theory of the Stone-Čech compactification βS of a discrete semigroup S and its applications to Ramsey Theory that have occurred since the publication of [**46**].

1. Introduction

If S is a discrete topological space, we view its Stone-Čech compactification, βS, as the set of all ultrafilters on S, the points of S being identified with the principal ultrafilters. Given $A \subseteq S$, $\overline{A} = c\ell_{\beta S} A = \{p \in \beta S : A \in p\}$. The set $\{\overline{A} : A \subseteq S\}$ is a basis for the open sets of βS as well as a basis for the closed sets of βS.

If $(S, \cdot)$ is a discrete semigroup, then the operation extends uniquely to βS making $(\beta S, \cdot)$ a right topological semigroup with S contained in its topological center. To say that (βS) is right topological is to say that for each $p \in \beta S$, the function $\rho_p : \beta S \to \beta S$ defined by $\rho_p(q) = q \cdot p$ is continuous. The topological center of βS is $\{q \in \beta S : \lambda_q \text{ is continuous}\}$, where $\lambda_q : \beta S \to \beta S$ is defined by $\lambda_q(p) = q \cdot p$. The fact that this extension can be done was implicitly established by M. Day [**22**] using a multiplication on the second conjugate of a Banach algebra, in this case $l_1(S)$, first introduced by R. Arens [**5**] for arbitrary Banach algebras. P. Civin and B. Yood [**19**, Theorem 3.4] explicitly stated that if S is a discrete group, then the above operation produced an operation on the Stone-Čech compactification of S, viewed as a subspace of that second dual. R. Ellis [**27**] carried out the extension in βS viewed as a space of ultrafilters, again assuming that S is a group. He also proved the important fact that any compact Hausdorff right topological semigroup has an idempotent [**27**, Corollary 2.10].

The fact that $(\beta\mathbb{N}, +)$ has an idempotent provided the first application of the algebra of βS to Ramsey Theory, namely a very simple proof due to F. Galvin and S. Glazer of the Finite Sums Theorem. (The original, elementary but complicated, proof is in [**38**].) Given a set X, $\mathcal{P}_f(X)$ is the set of finite nonempty subsets of X.

2010 *Mathematics Subject Classification.* Primary 05D10, 22A15, 54D35.

The first author acknowledges support received from the National Science Foundation via Grants DMS-0554803 and DMS-0852512.

THEOREM 1.1 (Finite Sums Theorem). *Let $r \in \mathbb{N}$ and let $\mathbb{N} = \bigcup_{i=1}^{r} C_i$. There exist $i \in \{1, 2, \ldots, r\}$ and a sequence $\langle x_n \rangle_{n=1}^{\infty}$ in $\mathbb{N}$ such that for each $F \in \mathcal{P}_f(\mathbb{N})$, $\sum_{n \in F} x_n \in C_i$.*

This proof, while never published by either of the originators, is widely available. For their original proof and a newer one see [**45**, Theorem 5.8]. For a detailed discussion of the history of the discovery of this proof see [**40**].

Any compact right topological semigroup T has a smallest two sided ideal, $K(T)$, which is the union of all of the minimal left ideals of T and is also the union of all of the minimal right ideals of T. The intersection of any minimal left ideal and any minimal right ideal is a group, and any two such groups are isomorphic. If S is a discrete semigroup, then it may happen that $K(\beta S)$ has only trivial structure. For example, if S is a right zero semigroup, (i.e., $st = t$ for all s and t in S), then so is βS. In this case the minimal left ideals are precisely the singletons. But it often happens that the structure of $K(\beta S)$ is very rich. For example, in $(\beta\mathbb{N}, +)$ the groups in the smallest ideal all contain a copy of the free group on $2^{\mathfrak{c}}$ generaters.

We shall be concerned in this paper with results about the algebraic structure of βS and with continued application of that structure to the branch of combinatorics known as Ramsey Theory. We will concentrate on results obtained since the preparation of the last survey [**46**] which we wrote on this subject.

Section 2 will be concerned with results establishing that there are many objects (such as minimal left or right ideals in βS) or that certain such objects can be very large or very small.

Section 3 will deal with more general results about the algebra of βS for either arbitrary semigroups S, or particular semigroups other than $(\mathbb{N}, +)$ or $(\mathbb{N}, \cdot)$ while Section 4 will deal with results about $\beta\mathbb{N}$.

Section 5 will deal with *central* sets as well as other notions of size such as *syndetic* or *piecewise syndetic* that had their origins in topological dynamics.

Section 6 will consist of new applications of the algebra of βS to Ramsey Theory.

Of course we will necessarily not mention most of the results in the papers we are surveying. We will choose sample results which are reasonably easy to state without introducing too much notation, and the reader should decide whether she is interested in consulting the original papers.

2. Number and cardinality of substructures of βS

We remark that the following theorem was known previously for countable cancellative semigroups [**45**, Corollary 6.41]. However, this result was completely new for uncountable groups.

THEOREM 2.1 (Zelenyuk). *Let G be an infinite discrete abelian group and let $\kappa = |G|$. Then βG contains $2^{2^{\kappa}}$ minimal right ideals.*

PROOF. [**79**, Theorem 1]. □

THEOREM 2.2 (Zelenyuk and Zelenyuk). *Let G be an infinite group and let $\kappa = |G|$. If G can be embedded into a direct sum of countable groups, in particular if G is abelian, then every maximal group in $K(\beta G)$ contains a free group on $2^{2^{\kappa}}$ generators.*

PROOF. [**82**, Theorem 1.1 and Corollary 1.2]. □

The "in particular" part of Theorem 2.2 was obtained independently in [**28**].

DEFINITION 2.3. A *digital representation* of a semigroup $(S, \cdot)$ is a family $\langle F_t \rangle_{t \in I}$, where I is a linearly ordered set, each F_t is a finite non-empty subset of S and every element of S is uniquely representable in the form $\prod_{t \in H} x_t$ where H is a finite subset of I, each $x_t \in F_t$, and products are taken in increasing order of indices.

A *strong digital representation* of S is a digital representation with the additional property that, for each $t \in I$, $F_t = \{x_t, x_t^2, \cdots, x_t^{m_t - 1}\}$ for some $x_t \in S$ and some $m_t > 1$ in $\mathbb{N}$, where $m_t = 2$ if x_t has infinite order and m_t is a prime, with the order of x_t being a power of m_t, if x_t has finite order.

THEOREM 2.4 (Ferri, Hindman and Strauss). *Every Abelian group has a strong digital representation.*

PROOF. [**28**, Theorem 3.9]. □

THEOREM 2.5 (Ferri, Hindman, and Strauss). *Let S be a left cancellative semigroup with identity of cardinality κ which has a digital representation $\langle F_t \rangle_{t \in \kappa}$ such that whenever $s < t < \kappa$, $x \in F_s$, and $y \in F_t$, one has $xy = yx$. Then every maximal group in the smallest ideal of βS contains a free group on 2^{2^κ} generators. In particular if S is an abelian group, then every maximal group in $K(\beta S)$ contains a free group on 2^{2^κ} generators.*

PROOF. [**28**, Theorem 3.9 and Corollary 4.16]. □

Recall that a set theoretic (or Boolean algebra) *ideal* is the dual of a filter. That is an ideal $\mathcal{I}$ on a set X is a nonempty set of subsets of X which does not include X and is closed under finite unions and subsets. Thus $\mathcal{I}$ is an ideal on X if and only if $\{X \setminus A : A \in \mathcal{I}\}$ is a filter on X. Interestingly enough, a special kind of ideal in a discrete group G has consequences for algebraic ideals in βG.

DEFINITION 2.6. Let G be an infinite group. A *Boolean group ideal* is an ideal $\mathcal{I}$ on G such that

(a) every finite subset of G is a member of $\mathcal{I}$;
(b) if $A \in \mathcal{I}$, then $A^{-1} \in \mathcal{I}$; and
(c) if $A, B \in \mathcal{I}$, then $AB \in \mathcal{I}$.

Recall that, given a filter $\mathcal{F}$ on a set X, $\widehat{\mathcal{F}} = \{p \in \beta X : \mathcal{F} \subseteq p\}$.

THEOREM 2.7 (Protasov and Protasova). *Let G be an infinite group, let $\mathcal{I}$ be a Boolean group ideal on G, and let $\mathcal{F} = \{G \setminus A : A \in \mathcal{I}\}$. Then $\widehat{\mathcal{F}}$ is a closed two sided ideal of the semigroup βG.*

PROOF. [**68**, Theorem 1]. □

COROLLARY 2.8. *Let G be an infinite abelian group with $|G| = \kappa$. There are 2^{2^κ} distinct closed two sided ideals in βG.*

PROOF. If $\mathcal{F}$ and $\mathcal{G}$ are distinct filters on G, then $\widehat{\mathcal{F}} \neq \widehat{\mathcal{G}}$. (If say $A \in \mathcal{F} \setminus \mathcal{G}$, then $\mathcal{G} \cup \{G \setminus A\}$ has the finite intersection property so there is an ultrafilter p on G such that $\mathcal{G} \cup \{G \setminus A\} \subseteq p$. Then $p \in \widehat{\mathcal{G}} \setminus \widehat{\mathcal{F}}$.) By [**67**, Theorem 3] any infinite abelian group of cardinality κ has 2^{2^κ} distinct Boolean group ideals. □

The following theorem drops the commutativity requirement from Corollary 2.8, but applies only when $\kappa = \omega$.

THEOREM 2.9 (Filali, Lutsenko, and Protasov). *Let G be a countably infinite group. There are 2^{2^ω} distinct Boolean group ideals on G. Consequently, there are 2^{2^ω} distinct closed left ideals of βG.*

PROOF. Theorem 2.7 and [**31**, Theorem 1.1]. □

DEFINITION 2.10. Let S be an infinite semigroup with cardinality κ.

(a) A subset A of S is a *left solution set of* S if and only if there exist $w, z \in S$ such that $A = \{x \in S : w = zx\}$.
(b) S is *very weakly left cancellative* if the union of fewer than κ left solution sets of S must have cardinality less than κ.

Note that if κ is regular, S is very weakly left cancellative if and only if every left solution set of S has cardinality less than κ. If κ is singular, S is very weakly left cancellative if and only if there is a cardinal less than κ which is an upper bound for the cardinalities of all left solution sets of S.

THEOREM 2.11 (Carlson, Hindman, McLeod, and Strauss). *Let S be an infinite very weakly left cancellative semigroup with cardinality κ. There is a collection of 2^{2^κ} pairwise disjoint left ideals of βS. In particular, βS has 2^{2^κ} minimal idempotents.*

PROOF. [**18**, Theorem 1.7]. □

The subsemigroup $\mathbb{H}$ of $(\beta\mathbb{N}, +)$ contains much of the known algebraic structure of $\beta\mathbb{N}$. (See [**45**, Section 6.1].)

DEFINITION 2.12. $\mathbb{H} = \bigcap_{n=1}^{\infty} c\ell_{\beta\mathbb{N}}(\mathbb{N}2^n)$.

DEFINITION 2.13. Let κ be an infinite cardinal.

(a) $G_\kappa = \bigoplus_{\sigma<\kappa} \mathbb{Z}_2$.
(b) For $x \in G_\kappa \setminus \{0\}$, $\operatorname{supp}(x) = \{\sigma < \kappa : x_\sigma \neq 0\}$.
(c) $\mathbb{H}_\kappa = \bigcap_{\sigma<\kappa} c\ell_{G_\kappa}(\{x \in G_\kappa \setminus \{0\} : \min \operatorname{supp}(x) > \sigma\}$.

It is an easy fact that if $f : G_\omega \to \mathbb{N}$ is defined by $f(x) = \sum_{n\in\operatorname{supp}(x)} 2^n$ and $\widetilde{f} : \beta G_\omega \to \beta\mathbb{N}$ is its continuous extension, then the restriction of $\widetilde{f}$ to $\mathbb{H}_\omega$ is both an isomorphism and a homeomorphism onto $\mathbb{H}$. More generally, we have the following.

THEOREM 2.14 (Hindman, Strauss, and Zelenyuk). *Let S be an infinite cancellative discrete semigroup with $|S| = \kappa$. Then S^* contains a topological and algebraic copy of $\mathbb{H}_\kappa$.*

PROOF. [**55**, Theorem 2.7]. □

We remark that the following theorem follows immediately from Zelenyuk's Theorem and from a previously known theorem due to I. Protasov, in the case in which $\kappa = \omega$ [**66**], but that it is completely new if $\kappa > \omega$.

THEOREM 2.15 (Zelenyuk). *Let κ be an infinite cardinal. Then $\mathbb{H}_\kappa$ contains no nontrivial finite group.*

PROOF. [**78**, Theorem 1.2]. □

However, we see that $\mathbb{H}_\kappa$ has plenty of large semigroups.

THEOREM 2.16 (Hindman, Strauss, and Zelenyuk). *Let κ be an infinite cardinal. Let L be a left zero semigroup and let R be a right zero semigroup with $|L| = |R| = 2^{2^\kappa}$. There is an algebraic copy of $L \times R$ contained in $\mathbb{H}_\kappa$. If $\kappa = \omega$, there is an algebraic copy of $L \times R$ contained in $K(\mathbb{H})$. In particular, there is an algebraic copy of $L \times R$ contained in $K(\beta\mathbb{N})$.*

PROOF. [**55**, Corollaries 3.10 and 3.11]. □

DEFINITION 2.17. Let G be a group with topology τ, let e be the identity of G, and let G_d be G with the discrete topology. Then $\mathrm{Ult}_\tau(G) = \{p \in \beta G_d : p$ converges to e with respect to $\tau\}$.

By a *left topological group* we mean a group with a topology making it a left topological semigroup. If (G, τ) is a left topological group, then $\mathrm{Ult}_\tau(G)$ is a compact subsemigroup of $G^* = \beta G_d \setminus G$ [**45**, Exercise 9.2.3].

THEOREM 2.18 (Zelenyuk). *Let $(G, +, \tau)$ be a countable abelian nondiscrete topological group and let $B = \{x \in G : x + x = 0\}$. If B is not open, then $\mathrm{Ult}_\tau(G)$ contains $2^{\mathfrak{c}}$ minimal right ideals. In particular, $\big|K\big(\mathrm{Ult}_\tau(G)\big)\big| = 2^{\mathfrak{c}}$.*

PROOF. [**75**, The Theorem and Corollary]. □

On the other hand, it is possible for $\mathrm{Ult}_\tau(G)$ to be small. A *chain of idempotents* is a set of idempotents linearly ordered by the relation $p \leq q$ if and only if $p = pq = qp$.

THEOREM 2.19 (Zelenyuk). *Let G be a countable group and let $n \in \mathbb{N}$. There is a topology τ on G such that (G, τ) is a regular Hausdorff left topological group without isolated points and $\mathrm{Ult}_\tau(G)$ is an n-element chain of idempotents.*

PROOF. [**76**, Theorem 6.1]. □

Recall that the *Bohr compactification* of $\mathbb{Z}$, also known as $\mathcal{SAP}(\mathbb{Z}) = \mathcal{AP}(\mathbb{Z})$, is the largest compact topological group containing an algebraic copy of $\mathbb{Z}$ as a dense subgroup. Recall also that a P-point in a topological space is a point with the property that any countable intersection of neighborhoods is again a neighborhood. While the existence of P-points in $\mathbb{N}^*$ follows from Martin's axiom, it is consistent that there are none [**71**, VI, §].

THEOREM 2.20 (Zelenyuk). *Assume that there are no P-points in $\mathbb{N}^*$ and let (G, τ) be a countable nondiscrete topological group. Then $\mathrm{Ult}_\tau(G)$ can be partitioned into closed right ideals, each of which admits a continuous homomorphism onto the Bohr compactification of $\mathbb{Z}$.*

PROOF. [**74**, The Theorem]. □

THEOREM 2.21 (Zelenyuk). *Let (G, τ) be a countable regular Hausdorff left topological group and assume that G contains a discrete subset with exactly one accumulation point. There exists a continuous surjective homomorphism from $\mathrm{Ult}_\tau(G)$ onto $\beta\mathbb{N}$.*

PROOF. [**77**, Lemma 1.1 and Corollary 2.3]. □

DEFINITION 2.22. Let S be an infinite semigroup and let p be an idempotent in βS. Then $H(p)$ is the largest subgroup of βS with p as its identity.

If S is cancellative and $|S| = \kappa$, then by [**45**, Corollary 7.39] there exists an idempotent $p \in \beta S$ such that $H(p)$ contains a copy of the free group on 2^{2^κ} generators. We see here that it is consistent that maximal groups in such semigroups are as small as possible.

THEOREM 2.23 (Legette). *Let S and G be respectively the free semigroup and the free group on a countably infinite set of generators. For an idempotent $p \in \beta S$, let $H_S(p)$ and $H_G(p)$ be the maximal groups associated with p in βS and βG respectively. Assume Martin's Axiom. Then there is an idempotent $p \in \beta S$ such that $H_S(p) = H_G(p) = \{p\}$.*

PROOF. [**58**, Theorem 4.3 and Corollary 4.4]. □

The ultrafilters which Legette produces for the proof of Theorem 2.23 are essentially equivalent to *ordered union ultrafilters* introduced in [**12**], and the existence of ordered union ultrafilters is known to be independent of ZFC.

3. General algebra and topology of βS

Recall that for any ultrafilter p, $||p|| = \min\{|A| : A \in p\}$.

DEFINITION 3.1. Let S be an infinite semigroup and let $\omega \leq \kappa \leq |S|$. Then $P_\kappa(S) = \{p \in \beta S : ||p|| = \kappa\}$ and $U_\kappa(S) = \{p \in \beta S : ||p|| \geq \kappa\}$.

If $\kappa = |S|$, then $P_\kappa(S) = U_\kappa(S)$, the space of κ-uniform ultrafilters on S. $P_\kappa(S)$ need not be a subsemigroup of βS, but if S is weakly left cancellative it is [**30**, Proposition 2.3].

THEOREM 3.2 (Filali). *Let S be an infinite discrete semigroup and let $\omega \leq \kappa \leq |S|$. If either S is right cancellative and weakly left cancellative and $\kappa = |S|$ or S is a subset of a group, then $\{p \in P_\kappa(S) : p$ is right cancelable in $\beta S\}$ has dense interior in $P_\kappa(S)$.*

PROOF. [**30**, Theorems 3.2 and 3.3]. □

THEOREM 3.3 (Hindman and Strauss). *Let S be a discrete semigroup. The following statements are equivalent:*

(a) *βS is simple.*
(b) *S is a simple semigroup with a minimal left ideal containing an idempotent. Furthermore, the structure group of S is finite and S has only a finite number of minimal left ideals or only a finite number of minimal right ideals.*
(c) *S contains a finite group G, a left zero semigroup X and a right zero semigroup Y such that S is isomorphic to the semigroup $X \times G \times Y$ with the semigroup operation defined by $(x, g, y)(x', g', y') = (x, gyx'g', y')$ for every $x, x' \in X, g, g' \in G, y, y' \in Y$. Furthermore, either X or Y is finite.*

PROOF. [**49**, Theorem 4]. □

In [**49**, Theorems 5 and 6] several equivalent conditions to βS being right or left cancellative are obtained. As a consequence, one has the following.

THEOREM 3.4 (Hindman and Strauss). *Let S be a discrete semigroup. If $K(\beta S)$ contains an element left cancelable in βS and an element right cancelable in βS, then S is a finite group.*

PROOF. [**49**, Corollary 7]. □

DEFINITION 3.5. Let S be a semigroup.

(a) Let $f : S \to \mathbb{C}$, let $\epsilon > 0$, and let $F \in \mathcal{P}_f(S)$. Then
$S(f,\epsilon,F) = \{t \in S : \mathrm{diam} f[Ft \cup \{t\}] < \epsilon\}$.

(b) Let $f : S \to \mathbb{C}$. Then
$\mathrm{so}(f) = \{A \subseteq S : (\exists \epsilon > 0)\big(\exists F \in \mathcal{P}_f(S)\big)\big(S(f,\epsilon,F) \subseteq A\big)\}$.

Notice that $\mathrm{so}(f)$ is a filter. The following theorem characterizes closed left ideals of βS for certain countable semigroups. (Notice that these include any right cancellative semigroup which has a left identity.)

THEOREM 3.6 (Alaste and Filali). *Let S be a countable semigroup, assume that there exists $s \in S$ such that λ_s has no fixed point in S, and let L be a closed left ideal of βS. There exists a family $\mathcal{F}$ of bounded functions from S to $\mathbb{C}$ such that $L = \bigcap_{f \in \mathcal{F}} \{p \in \beta S : \mathrm{so}(f) \subseteq p\}$.*

PROOF. [**4**, Corollary 2.1]. □

THEOREM 3.7 (Protasov). *Let G be a countable discrete group, let $E = \{(p,q) \in G^* \times G^* : (\exists g \in G)(gp = q)\}$, let $\check{E} = \bigcap\{R : R$ is an equivalence relation on G^*, R is closed in $G^* \times G^*$, and $E \subseteq R\}$, and for $p \in G^*$, let $\check{p} = \{q \in G^* : (p,q) \in \check{E}\}$. For any $p \in G^*$, $\beta G p \subseteq \check{p}$. If p is a P-point in G^*, then $\beta G p = \check{p}$.*

PROOF. [**64**, Theorem 4.1]. □

Protasov then asked [**64**, Question 4.2] whether one can prove in ZFC that there exist a countable group G and $p \in G^*$ such that $\beta G p = \check{p}$.

Recall that ultrafilters p and q on a discrete space X are *of the same type* if and only if there exists $f : X \xrightarrow[\text{onto}]{\text{1-1}} X$ such that $\widetilde{f}(p) = q$, where $\widetilde{f} : \beta X \to \beta X$ is the continuous extension of f.

THEOREM 3.8 (Protasov). *Let G be a countable discrete group and for $p \in G^*$, let $G(p)$ be the finest topology on G with respect to which G is a left topological group and p converges to the identity of G, and let p and q be right cancelable members of G^*. Then $G(p)$ is homeomorphic to $G(q)$ if and only if p and q are of the same type.*

PROOF. [**65**, Theorem 2]. □

DEFINITION 3.9. Let G be an infinite group with identity e and let p be an idempotent in G^*. Then $\mathcal{T}_p$ is the topology on G such that $\{A \cup \{e\} : A \in p\}$ is the filter of neighborhoods of e and $(G, \mathcal{T}_p)$ is a left topological group.

DEFINITION 3.10. Let S be a semigroup and let p be an idempotent in S^*. Then p is *strongly right maximal in S^** if and only if $\{q \in S^* : qp = p\} = \{p\}$.

The existence of strongly right maximal idempotents in $\mathbb{N}^*$ is a ZFC theorem [**45**, Theorem 9.10] due to Protasov. It has been an open question as to whether there exists a uniform strongly right maximal idempotent on an uncountable semigroup.

Similarly, it has been an open question as to whether there exists an uncountable homogeneous regular space with topology which is (1) maximal among topologies with no isolated points and (2) has uncountable dispersion character. (Recall that the *dispersion character* of a space is the minimum cardinality of a nonempty open subset.)

THEOREM 3.11 (Protasov). *Let G be an infinite group and let p be an idempotent in G^*. Then $\mathcal{T}_p$ is Hausdorff. Also, $\mathcal{T}_p$ is regular if and only if p is strongly right maximal in G^*.*

PROOF. [**45**, Theorem 9.15]. □

THEOREM 3.12 (Zelenyuk). *Let S be an infinite cancellative semigroup. There is a uniform strongly right maximal idempotent in S^*. Consequently, if G is an infinite group with cardinality κ, then $(G, \mathcal{T}_p)$ is a homogeneous regular space with dispersion character κ such that $\mathcal{T}_p$ is maximal among topologies with no isolated points.*

PROOF. [**80**, Corollary 1.4]. □

Given an infinite set J, let $\mathcal{I} = \mathcal{P}_f(J)$ and consider the semigroup $(\mathcal{I}, \cup)$. Grainger characterized the closure of the smallest ideal of $\mathcal{I}$.

THEOREM 3.13 (Grainger). *Let J be an infinite set and let $\mathcal{I} = \mathcal{P}_f(J)$. Then $c\ell K(\beta\mathcal{I}) = \{p \in \beta\mathcal{I} : (\forall F \in \mathcal{I})(\{G \in \mathcal{I} : F \subseteq G\} \in p)\}$.*

PROOF. [**36**, Theorem 7.5]. □

Koppelberg showed that in fact the smallest ideal of $\beta\mathcal{I}$ is closed, and extended the result of Theorem 3.13 to *semilattices*.

DEFINITION 3.14. A *upper semilattice* is a commutative semigroup $(S, +)$, every member of which is an idempotent. If S is an upper semilattice, and $x, y \in S$, then $x \leq y$ if and only if $x + y = y$.

THEOREM 3.15 (Koppelberg). *Let S be an upper semilattice. Then $K(\beta S) = \{p \in \beta S : (\forall x \in S)(\{y \in S : x \leq y\} \in p)\}$.*

PROOF. [**57**, Theorem 2.4]. □

She also showed that two distinct notions of size are equivalent for an upper semilattice.

THEOREM 3.16 (Koppelberg). *Let S be an upper semilattice and let $A \subseteq S$. Then A is central if and only if A is piecewise syndetic.*

PROOF. [**57**, Corollary 2.5]. □

Notice that, if J is an infinite set and $\mathcal{I} = \mathcal{P}_f(J)$, then $\{p \in \beta\mathcal{I} : (\forall F \in \mathcal{I})(\{G \in \mathcal{I} : F \subseteq G\} \in p)\} = \{p \in \beta\mathcal{I} : (\forall j \in J)(\{G \in \mathcal{I} : j \in G\} \in p)\}$.

DEFINITION 3.17. Let J be an infinite set, let $\mathcal{I} = \mathcal{P}_f(J)$, and let $A \subseteq J$. Then $\beta_A(\mathcal{I}) = \{p \in \mathcal{I} : (\forall j \in A)(\{G \in \mathcal{I} : j \in G\} \in p)$ and $(\forall j \in J \setminus A)(\{G \in \mathcal{I} : j \notin G\} \in p)\}$.

Notice that by Theorem 3.15, $K(\beta\mathcal{I}) = \beta_J(\mathcal{I})$.

THEOREM 3.18 (Grainger). *Let J be an infinite set, let $\mathcal{I} = \mathcal{P}_f(J)$, and let $A \subseteq J$. Let $V = \bigcup\{\beta_B(\mathcal{I}) : B \subseteq A\}$. If A and $J \setminus A$ are infinite, then $\beta_A(\mathcal{I})$ and V are subsemigroups of $\beta(\mathcal{I})$, $c\ell K(V)$ is a proper subset of $\beta_A(\mathcal{I})$, and $K(V) = K(\beta_A(\mathcal{I}))$.*

PROOF. [**36**, Proposition 5.4 and Theorem 5.14] and [**37**, Remark 4.3 and Proposition 4.8]. □

It has been known for some years that, if S is a countable cancellative semigroup, there are elements in the closure of the set of minimal idempotents in βS which are not in S^*S^* [**45**, Theorem 8.22]. The proof in [**45**] depends very essentially on countability. Y. Zelenyuk has obtained an analogous result for a large class of uncountable semigroups.

THEOREM 3.19 (Zelenyuk). *Let S be an infinite discrete semigroup which can be embedded algebraically in a compact topological group. Then $K(\beta S)$ and $E(K(\beta S))$ are not closed.*

PROOF. [**81**, Theorem 1]. □

The topological properties of minimal left ideals and minimal right ideals in compact right topological semigroups can be very different. Every minimal left ideal is compact; but this statement is far from being true in general for minimal right ideals. E. Glasner has recently shown that the minimal right ideals of $\beta\mathbb{Z}$ are not Borel measurable [**34**].

I. Protasov and J. Pym have obtained results about the discontinuity of the mappings λ_x on the remainder spaces $G^* = G^{LUC} \setminus G$, where G denotes a locally compact non-compact group and G^{LUC} denotes its largest semigroup compactification. In the case in which G is discrete, $G^{LUC} = \beta G$. They have shown that, for every $q \in G^*$, λ_q is discontinuous at some point $p \in G^*$. If G is σ-compact, there is one element p which will serve for every q [**69**, Theorem 1].

If S is a discrete semigroup, the dual of $C(\beta S)$ is $M(\beta S)$, the Banach space of complex valued regular Borel measures defined on βS. Since $C(\beta S)$ can be identified with $l_\infty(S)$, the dual of $l_1(S)$, $M(\beta S)$ is the second dual of a Banach algebra and, as such, it is a compact right topological semigroup for the Arens product $\Box$. This is a semigroup in which βS is embedded as a compact subsemigroup. H. G. Dales, A. T.-M. Lau and D. Strauss have shown that, for a class of semigroups S significantly larger than the class of cancellative semigroups, there are two points $p, q \in S^*$ such that, for every non-zero $\mu \in M(S^*)$, λ_μ cannot be continuous at both p and q [**20**]. This result was obtained independently by M. Neufang for cancellative semigroups. Note that this claim cannot be made in general for fewer than two points, since, if S is commutative, λ_μ is continuous at μ for every $\mu \in M(\beta S)$.

There is another interesting compact right topological semigroup in which βS can be embedded as a compact subsemigroup. If X is a topological space, an inclusion hyperspace of X is a family $\mathcal{F}$ of closed non-empty subsets of X which is closed in the Vietoris topology and has the property that, for any two non-empty closed subsets A and B of X, $A \in \mathcal{F}$ and $A \subseteq B$, implies that $B \in \mathcal{F}$. $G(X)$ denotes the space of all inclusion hyperspaces of X endowed with the topology defined by choosing the sets of the form U^+ and U^- as a subbase, where U is an open subset of X, $U^+ = \{\mathcal{A} \in G(X) : \exists B \in \mathcal{A} \text{ with } B \subseteq U\}$ and $U^- = \{\mathcal{A} \in G(X) : \forall B \in \mathcal{A}, B \cap U \neq \emptyset\}$. V. Gavrilkiv has shown that, if S is a discrete semigroup, the semigroup operation of S can be extended to $G(S)$ so that $G(S)$ becomes a compact right topological semigroup [**33**]. $G(S)$ then contains βS, as well as many other interesting spaces, as compact subsemigroups.

4. Algebra of $\beta\mathbb{N}$

One of the most challenging open questions about the algebra of $(\beta\mathbb{N},+)$ is whether there are any nontrivial continuous homomorphisms from $\beta\mathbb{N}$ into $\mathbb{N}^* = \beta\mathbb{N}\setminus\mathbb{N}$. It is known that if ϕ is such a homomorphism, then $\phi[\beta\mathbb{N}]$ is finite, and $|\phi[\mathbb{N}^*]| = 1$ [**45**, Theorem 18.10]. There are also two simple equivalent conditions to the existence of such a homomorphism.

THEOREM 4.1. *The following statements are equivalent.*

(a) *There is a nontrivial continuous homomorphism from* $\beta\mathbb{N}$ *into* $\mathbb{N}^*$.
(b) *There exist* $p \neq q$ *in* $\mathbb{N}^*$ *such that* $p+p=p+q=q+p=q+q=q$.
(c) *There is a finite subsemigroup of* $\mathbb{N}^*$ *whose elements are not all idempotents.*

PROOF. [**45**, Corollary 10.20]. □

DEFINITION 4.2. For $k\in\mathbb{N}$, define $\phi_k:\mathbb{N}\to\mathbb{N}$ by $\phi_k(n)=k\cdot n$, let $\widetilde{\phi_k}:\beta\mathbb{N}\to\beta\mathbb{N}$ be its continuous extension, and let ϕ_k^* be the restriction of $\widetilde{\phi_k}$ to $\mathbb{N}^*$.

Each ϕ_k^* is a continuous homomorphism from $\mathbb{N}^*$ to $\mathbb{N}^*$. It is not known whether these are the only nontrivial continuous homomorphisms from $\mathbb{N}^*$ to $\mathbb{N}^*$.

THEOREM 4.3 (Adams and Strauss). *Let* $\phi:\mathbb{N}^*\to\mathbb{N}^*$ *be a continuous homomorphism which is not equal to* ϕ_k^* *for any* $k\in\mathbb{N}$ *and let* $C=\phi[\mathbb{N}^*]$. *Then*

(1) $|K(C)|=1$;
(2) C *has a unique idempotent* q *and* $C+C=\{q\}$; *and*
(3) $|\phi^2[\mathbb{N}^*]|=1$.

PROOF. [**3**, Theorems 3.10 and 3.11 and Corollary 3.12] □

The list from Theorem 4.1 can be extended.

THEOREM 4.4 (Adams and Strauss). *The following statements are equivalent.*

(a) *There is a nontrivial continuous homomorphism from* $\mathbb{N}^*$ *to* $\mathbb{N}^*$ *which is not equal to* ϕ_k^* *for any* $k\in\mathbb{N}$.
(b) *There is a nontrivial continuous homomorphism from* $\beta\mathbb{N}$ *into* $\mathbb{N}^*$.
(c) *There exist* $p\neq q$ *in* $\mathbb{N}^*$ *such that* $p+p=p+q=q+p=q+q=q$.
(d) *There is a finite subsemigroup of* $\mathbb{N}^*$ *whose elements are not all idempotents.*

PROOF. [**3**, Theorem 3.14]. □

This investigation was extended in [**29**] to consideration of continuous homomorphisms from countable semigroups to countable groups.

THEOREM 4.5 (Ferri and Strauss). *Let* S *be a countably infinite commutative semigroup and let* T *be a countably infinite group. If* $\varphi:\beta S\to T^*$ *is a continuous homomorphism, then* $K(\varphi[\beta S])$ *is a finite group.*

PROOF. [**29**, Corollary 1.13]. □

Given any idempotent $p=p+p$ in $\beta\mathbb{N}$, $\{n+p : n\in\mathbb{Z}\}$ is an algebraic copy of $\mathbb{Z}$. However it cannot be discrete – indeed it has no isolated points.

THEOREM 4.6 (Hindman and Strauss). *Let H be a maximal group in the smallest ideal of $\beta\mathbb{N}$. There are $2^{\mathfrak{c}}$ discrete copies of $\mathbb{Z}$ contained in H, any two of which intersect only at the identity. There are $2^{\mathfrak{c}}$ discrete copies of the free group on 2 generators contained in H, any two of which intersect only at the identity.*

PROOF. [**48**, Corollaries 2.2 and 3.4]. □

The identity function $\iota : \mathbb{N}\times\mathbb{N} \to \mathbb{N}\times\mathbb{N} \subseteq \beta\mathbb{N}\times\beta\mathbb{N}$ has a continuous extension $\widetilde{\iota} : \beta(\mathbb{N}\times\mathbb{N}) \to \beta\mathbb{N}\times\beta\mathbb{N}$, and $\widetilde{\iota}\big[K\big(\beta(\mathbb{N}\times\mathbb{N})\big)\big] = K(\beta\mathbb{N})\times K(\beta\mathbb{N}) = K(\beta\mathbb{N}\times\beta\mathbb{N})$. It has been known since the early 1970's that there are points $(p,q) \in \beta\mathbb{N}\times\beta\mathbb{N}$ such that $|\widetilde{\iota}^{-1}[\{(p,q)\}]| = 2^{\mathfrak{c}}$ and that it follows from the Continuum Hypothesis that there are points $(p,q) \in \beta\mathbb{N}\times\beta\mathbb{N}$ such that $|\widetilde{\iota}^{-1}[\{(p,q)\}]| = 2$.

THEOREM 4.7 (Moche). *Let $(p,q) \in K(\beta\mathbb{N})\times K(\beta\mathbb{N})$. Then $\{r \in K\big(\beta(\mathbb{N}\times\mathbb{N})\big) : \widetilde{\iota}(r) = (p,q)\}$ is infinite.*

PROOF. [**60**, Corollary 3.10]. □

It has been known for some time that the smallest ideals of $(\beta\mathbb{N},+)$ and $(\beta\mathbb{N},\cdot)$ are disjoint while $K(\beta\mathbb{N},\cdot) \cap c\ell K(\beta\mathbb{N},+) \neq \emptyset$. (See [**45**, Corollaries 13.15 and 16.25].)

THEOREM 4.8 (Strauss). *$c\ell K(\beta\mathbb{N},\cdot)\cap(\mathbb{N}^*+\mathbb{N}^*) = \emptyset$. In particular, $c\ell K(\beta\mathbb{N},\cdot)\cap K(\beta\mathbb{N},+) = \emptyset$.*

PROOF. [**73**, Corollary 2.3]. □

Recall that, given an finite or infinite sequence $\vec{x}$ in $\mathbb{N}$, by $FS(\vec{x})$ we mean the set of sums of finitely many distinct terms of $\vec{x}$.

DEFINITION 4.9. For $r \in \mathbb{N}$,

$$\Sigma_r = \{p \in \beta\mathbb{N} : (\forall A \in p)(\exists \langle x_t\rangle_{t=1}^{r})(FS(\langle x_t\rangle_{t=1}^{r}) \subseteq A)\}\,.$$

Also $\Sigma = \bigcap_{r=1}^{\infty} \Sigma_r$.

It is easy to see that Σ is a compact subsemigroup of $(\beta\mathbb{N},+)$ and an ideal of $(\beta\mathbb{N},\cdot)$ and all of the idempotents of $(\beta\mathbb{N},+)$ are in Σ. Bergelson suggested in a personal communication that perhaps Σ is the smallest such object.

THEOREM 4.10 (Hindman and Strauss). *There is a compact subsemigroup of $(\beta\mathbb{N},+)$ which contains the idempotents of $(\beta\mathbb{N},+)$, is a two sided ideal of $(\beta\mathbb{N},\cdot)$, and is properly contained in Σ.*

PROOF. [**50**, Theorem 3.1]. □

THEOREM 4.11 (Maleki). *Let $p \in \mathbb{N}^*$.*

(1) *If $a,b,c,d \in \mathbb{N}$ and $a\cdot p + b\cdot p = c\cdot p + d\cdot p$, then $a = c$ and $b = d$.*
(2) *If $n,m,a_1,a_2,\ldots,a_n,b_1,b_2,\ldots,b_m \in \mathbb{N}$, $a_1\cdot p + \ldots a_n\cdot p = b_1\cdot p + \ldots b_m\cdot p$ and p is right cancelable in $\beta\mathbb{N}$, then $m = n$ and for each $i \in \{1,2,\ldots,n\}$, $a_i = b_i$.*

PROOF. [**59**, Theorems E and H]. □

Given $f : \mathbb{N} \to \mathbb{N}$, let $\widetilde{f} : \beta\mathbb{N} \to \beta\mathbb{N}$ be its continuous extension.

DEFINITION 4.12. An *ultrafilter semiring* is a triple $(A,\oplus,\odot)$ such that

(1) $\emptyset \neq A \subseteq \beta\mathbb{N}$;

(2) $(\forall\mathcal{U}\in A)(\forall f:\mathbb{N}\to\mathbb{N})(\widetilde{f}(\mathcal{U})\in A)$;
(3) $\oplus$ is associative and commutative;
(4) $\odot$ is associative and distributive with respect to $\oplus$;
(5) $(\forall\mathcal{U},\mathcal{V}\in A)(\exists!\,\mathcal{W}\in A)(\mathcal{U}\oplus\mathcal{W}=\mathcal{V}$ or $\mathcal{V}\oplus\mathcal{W}=\mathcal{U})$;
(6) $(\forall\mathcal{U},\mathcal{V},\mathcal{W}\in A)(\mathcal{U}\oplus\mathcal{W}=\mathcal{V}$ and $\mathcal{V}\oplus\mathcal{W}=\mathcal{U}\Rightarrow\mathcal{U}=\mathcal{V})$; and
(7) $(\forall\mathcal{U}\in A)(\forall f,g:\mathbb{N}\to\mathbb{N})(\widetilde{f}(\mathcal{U})\oplus\widetilde{g}(\mathcal{U})=\widetilde{f+g}(\mathcal{U})$ and $\widetilde{f}(\mathcal{U})\odot\widetilde{g}(\mathcal{U})=\widetilde{f\cdot g}(\mathcal{U}))$.

DEFINITION 4.13. A *Hausdorff* ultrafilter is an ultrafilter $\mathcal{U}$ on $\mathbb{N}$ such that $(\forall f,g\colon\mathbb{N}\to\mathbb{N})(\widetilde{f}(\mathcal{U})=\widetilde{g}(\mathcal{U})\Rightarrow\{x\in\mathbb{N}:f(x)=g(x)\}\in\mathcal{U})$.

THEOREM 4.14 (DiNasso and Forti). *Let $\emptyset\neq A\subseteq\beta\mathbb{N}$. Then A is an ultrafilter semiring if and only if*
(a) $(\forall\mathcal{U}\in A)(\forall f\colon\mathbb{N}\to\mathbb{N})(\widetilde{f}(\mathcal{U})\in A)$;
(b) $(\forall\mathcal{U},\mathcal{V}\in A)(\exists f,g:\mathbb{N}\to\mathbb{N})(\exists\mathcal{W}\in A)(\widetilde{f}(\mathcal{W})=\mathcal{U}$and $g^*(\mathcal{W})=\mathcal{V})$; *and*
(c) $(\forall\mathcal{U}\in A)(\mathcal{U}$ *is Hausdorff*).

PROOF. [**26**, Theorem 1.6]. □

Since $(\mathbb{N},\cdot)$ is a semigroup, $(\beta\mathbb{N},\cdot)$ is a compact right topological semigroup. Further, the fact that $(\mathbb{N},+,\cdot)$ is distributive leads to some strong combinatorial applications of the algebra of $(\beta\mathbb{N},+)$ and $(\beta\mathbb{N},\cdot)$. (See [**45**, Sections 5.3 and 17.1].) However, there is very limited interaction in $\mathbb{N}^*$. For example, by [**45**, Corollary 13.27], $\{p\in\mathbb{N}^*:(\forall q,r\in\mathbb{N}^*)(q\cdot p+r\cdot p\neq(q+r)\cdot p$ and $r\cdot(q+p)\neq r\cdot q+r\cdot p)\}$ has dense interior in $\mathbb{N}^*$. There is another way to define a multiplication on $\beta\mathbb{N}$ which is in some respects better behaved. On the other hand, that multiplication is not associative.

THEOREM 4.15 (Hindman, Pym, and Strauss). *For $q\in\beta\mathbb{N}$, define $r_q:\mathbb{N}\to\beta\mathbb{N}$ by $r_q(n)=q+q+\ldots+q$, n times and let $\widetilde{r_q}:\beta\mathbb{N}\to\beta\mathbb{N}$ be its continuous extension. Define for $p,q\in\beta\mathbb{N}$, $p*q=\widetilde{r_q}(p)$. Then*
(1) *1 is a two sided identity for $(\beta\mathbb{N},*)$;*
(2) *for $p,q\in\beta\mathbb{N}$ and $x\in\mathbb{N}$, $(p+q)*x=p*x+q*x$;*
(3) *for $p\in\beta\mathbb{N}$ and $x,y\in\mathbb{N}$, $p*(x+y)=p*x+p*y$;*
(4) *the topological and algebraic centers of $(\beta\mathbb{N},*)$ are both equal to $\{1\}$;*
(5) *$*$ is not associative on $\beta\mathbb{N}$; and*
(6) *neither the left nor right distributive laws hold in $(\beta\mathbb{N},+,*)$.*

PROOF. [**44**, Theorems 2.1, 6.1, 6.2, and 6.4]. □

5. Central sets and other large subsets of S

In [**32**] Furstenberg defined a *central* subset of the set $\mathbb{N}$ of positive integers in terms of some notions from topological dynamics. He showed that if $\mathbb{N}$ is partitioned into finitely many classes, one of these classes contains a central set. Then he proved the following theorem.

THEOREM 5.1 (Furstenberg). *Let $l\in\mathbb{N}$ and for each $i\in\{1,2,\ldots,l\}$, let $f_i:\mathbb{N}\to\mathbb{Z}$. Let C be a central subset of $\mathbb{N}$. Then there exist sequences $\langle a_n\rangle_{n=1}^\infty$ in $\mathbb{N}$ and $\langle H_n\rangle_{n=1}^\infty$ in $\mathcal{P}_f(\mathbb{N})$ such that*
(1) *for all n, $\max H_n<\min H_{n+1}$ and*
(2) *for all $F\in\mathcal{P}_f(\mathbb{N})$ and all $i\in\{1,2,\ldots,l\}$, $\sum_{n\in F}\big(a_n+\sum_{t\in H_n}f_i(t)\big)\in C$.*

PROOF. [**32**, Proposition 8.21]. □

As an immediate consequence of this theorem, one sees that any central subset of $\mathbb{N}$ contains $FS(\langle x_n\rangle_{n=1}^{\infty})$ for some sequence $\langle x_n\rangle_{n=1}^{\infty}$, and similarly, must contain arbitrarily long arithmetic progressions. As shown in [**32**], this theorem also implies Rado's Theorem [**70**]. In particular, Theorem 5.1 was strong enough to show that central subsets of $\mathbb{N}$ contain solutions to any partition regular system of homogeneous linear equations.

Subsequently Vitaly Bergelson had the idea that one might be able to derive the conclusion of the Central Sets Theorem for a set $C \subseteq \mathbb{N}$ which is a member of an idempotent in $K(\beta\mathbb{N})$. This suggested the following definition of central, which makes sense in any semigroup. It turned out that in fact this definition agrees with Furstenberg's original definition, as was shown in [**72**].

DEFINITION 5.2. Let S be a discrete semigroup and let C be a subset of S.

(a) The set C is *central* if and only there exists an idempotent $p \in K(\beta S)\cap\overline{C}$.
(b) The set C is *quasi-central* if and only there exists an idempotent $p \in c\ell K(\beta S)\cap\overline{C}$.

Notice that a virtue of Definition 5.2 is that it makes obvious the facts that central sets are partition regular and preserved under passage to supersets.

In [**14**, Theorem 3.4] a characterization of quasi-central sets in terms of dynamical notions similar to Furstenberg's original definition of central sets was obtained.

Theorem 5.1 dealt with finitely many sequences at a time and sums from one of those sequences at a time. In [**45**, Theorems 14.11 and 14.15] versions for commutative and noncommutative semigroups respectively were proved that dealt with countably many sequences at a time and allowed the sequence whose sums were taken to change as n changed. The following is the currently strongest version of the Central Sets Theorem for commutative semigroups. (For the general version for arbitrary semigroups see [**24**, Corollary 3.10].)

THEOREM 5.3 (De, Hindman, and Strauss). *Let $(S,+)$ be a commutative semigroup and let $\mathcal{T} = {}^{\mathbb{N}}S$, the set of sequences in S. Let C be a central subset of S. There exist functions $\alpha : \mathcal{P}_f(\mathcal{T}) \to S$ and $H : \mathcal{P}_f(\mathcal{T}) \to \mathcal{P}_f(\mathbb{N})$ such that*

(1) *if $F, G \in \mathcal{P}_f(\mathcal{T})$ and $F \subsetneq G$, then $\max H(F) < \min H(G)$ and*
(2) *whenever $m \in \mathbb{N}$, $G_1, G_2, \ldots, G_m \in \mathcal{P}_f(\mathcal{T})$, $G_1 \subsetneq G_2 \subsetneq \ldots \subsetneq G_m$, and for each $i \in \{1, 2, \ldots, m\}$, $\langle y_{i,n}\rangle_{n=1}^{\infty} \in G_i$, one has $\sum_{i=1}^{m}(\alpha(G_i) + \sum_{t\in H(G_i)} y_{i,t}) \in C$.*

PROOF. [**24**, Theorem 2.2]. □

The following version is superficially stronger than Theorem 5.3, replacing $\mathbb{N}$ by an arbitrary directed set with no largest element. However, as is shown in [**25**, Theorem 3.6], any set C satisfying the conclusion of Theorem 5.3 also satisfies the conclusion of Theorem 5.4.

THEOREM 5.4 (De, Hindman, and Strauss). *Let $(S,+)$ be a commutative semigroup, let $(D, \leq)$ be a directed set with no largest element, and let $\mathcal{T} = {}^{D}S$. Let C be a central subset of S. There exist functions $\alpha : \mathcal{P}_f(\mathcal{T}) \to S$ and $H : \mathcal{P}_f(\mathcal{T}) \to \{K \in \mathcal{P}_f(D) : K$ is linearly ordered$\}$ such that*

(1) *if $F, G \in \mathcal{P}_f(\mathcal{T})$ and $F \subsetneq G$, then $\max H(F) < \min H(G)$ and*

(2) *whenever* $m \in \mathbb{N}$, $G_1, G_2, \ldots, G_m \in \mathcal{P}_f(\mathcal{T})$, $G_1 \subsetneq G_2 \subsetneq \ldots \subsetneq G_m$, *and for each* $i \in \{1, 2, \ldots, m\}$, $\langle y_{i,n} \rangle_{n=1}^{\infty} \in G_i$, *one has* $\sum_{i=1}^{m} (\alpha(G_i) + \sum_{t \in H(G_i)} y_{i,t}) \in C$.

PROOF. [**25**, Corollary 3.5]. □

From the point of view of combinatorial applications, what one cares most about central sets is that they satisfy the conclusion of Theorem 5.3. We call such sets C-sets. We give now the definition of such sets for commutative semigroups. We observe that J-sets and C-sets are defined in arbitrary semigroups. We only state the definitions for commutative semigroups here because the general definitions are significantly more complicated. They can be found in [**51**].

DEFINITION 5.5. Let $(S, +)$ be a commutative semigroup and let $\mathcal{T} = {}^{\mathbb{N}}S$.

(a) Let $A \subseteq S$. Then A is a C-set if and only if there exist functions $\alpha : \mathcal{P}_f(\mathcal{T}) \to S$ and $H : \mathcal{P}_f(\mathcal{T}) \to \mathcal{P}_f(\mathbb{N})$ such that
   (1) if $F, G \in \mathcal{P}_f(\mathcal{T})$ and $F \subsetneq G$, then $\max H(F) < \min H(G)$ and
   (2) whenever $m \in \mathbb{N}$, $G_1, G_2, \ldots, G_m \in \mathcal{P}_f(\mathcal{T})$, $G_1 \subsetneq G_2 \subsetneq \ldots \subsetneq G_m$, and for each $i \in \{1, 2, \ldots, m\}$, $\langle y_{i,n} \rangle_{n=1}^{\infty} \in G_i$, one has $\sum_{i=1}^{m} (\alpha(G_i) + \sum_{t \in H(G_i)} y_{i,t}) \in A$.

(b) Let $A \subseteq S$. Then A is a J-set if and only if whenever $F \in \mathcal{P}_f(\mathcal{T})$ there exist $a \in S$ and $H \in \mathcal{P}_f(S)$ such that for each $f \in F$, $a + \sum_{t \in H} f(t) \in A$.

(c) $J(S) = \{p \in \beta S : (\forall A \in p)(A \text{ is a } J\text{-set})\}$.

The following four theorems are in fact valid for arbitrary semigroups (with the more general definitions of C-sets and J-sets).

THEOREM 5.6 (De, Hindman and Strauss). *Let S be a commutative semigroup. Then $J(S)$ is an ideal of βS.*

PROOF. [**24**, Theorem 3.5]. □

THEOREM 5.7 (Hindman and Strauss). *Let $(S, +)$ be a commutative semigroup and let $A \subseteq S$. Then A is a C-set if and only if there is an idempotent in $J(S) \cap \overline{A}$.*

PROOF. [**53**, Theorem 1.13]. □

THEOREM 5.8 (Hindman and Strauss). *Let S be a commutative semigroup and let $A \subseteq S$ be a J-set in S. If $A = A_1 \cup A_2$ then A_1 is a J-set in S or A_2 is a J-set in S.*

PROOF. [**51**, Theorem 2.14]. □

THEOREM 5.9 (Hindman and Strauss). *Let S and T be commutative semigroups, let $A \subseteq S$ and let $B \subseteq T$. If A and B are central, so is $A \times B$. If A and B are C-sets, so is $A \times B$. If A and B are J-sets, so is $A \times B$.*

PROOF. [**51**, Corollary 2.2 and Theorems 2.11 and 2.16]. □

In [**51**] a characterization of when the arbitrary product of central sets is central was also obtained.

While the following notion of density was introduced by Polya in [**62**], it is commonly called "Banach density".

DEFINITION 5.10. Let $A \subseteq \mathbb{N}$. Then

$$\begin{aligned} d^*(A) &= \sup\{\alpha \in [0,1] : (\forall k \in \mathbb{N})(\exists n \geq k)(\exists a \in \mathbb{N}) \\ &\qquad (|A \cap \{a, a+1, \ldots, a+n-1\}| \geq \alpha \cdot n)\} \text{ and} \\ \Delta^* &= \{p \in \beta\mathbb{N} : (\forall A \subseteq \mathbb{N})(p \in c\ell(A) \Rightarrow d^*(A) > 0)\}. \end{aligned}$$

Since Δ^* is a two sided ideal of $(\beta\mathbb{N}, +)$, one has that if C is a central subset of $\mathbb{N}$, then $d^*(C) > 0$. And the following result establishes that a set need not be central in order to satisfy the conclusion of the original Central Sets Theorem. (It is a consequence of Theorems 5.18 and 5.19, that there are idempotents in $\Delta^* \setminus c\ell(K(\beta\mathbb{N}))$.)

THEOREM 5.11 (Beiglböck, Bergelson, Downarowicz, and Fish). *Let $C \subseteq \mathbb{N}$ and assume that there is an idempotent in $\Delta^* \cap c\ell(C)$. Let $l \in \mathbb{N}$ and for each $i \in \{1, 2, \ldots, l\}$, let f_i be a sequence in $\mathbb{Z}$. Then there exist sequences $\langle a_n \rangle_{n=1}^{\infty}$ in $\mathbb{N}$ and $\langle H_n \rangle_{n=1}^{\infty}$ in $\mathcal{P}_f(\mathbb{N})$ such that*

(1) *for all n, $\max H_n < \min H_{n+1}$ and*
(2) *for all $F \in \mathcal{P}_f(\mathbb{N})$ and all $i \in \{1, 2, \ldots, l\}$, $\sum_{n \in F} \big(a_n + \sum_{t \in H_n} f_i(t)\big) \in C$.*

PROOF. [**8**, Theorem 11]. □

The question then naturally arose as to whether any subset C of $\mathbb{N}$ which satisfies the conclusion of Theorem 5.11 must satisfy $d^*(C) > 0$. This question was answered in the negative in [**42**, Theorem 2.1], where it was shown that there is a C-set $C \subseteq \mathbb{N}$ with $d^*(C) = 0$.

Banach density extends naturally to any semigroup satisfying the *Strong Følner condition*.

DEFINITION 5.12. Let $(S, \cdot)$ be a semigroup.

(a) S satisfies the *Strong Følner Condition* (SFC) if and only if $\big(\forall H \in \mathcal{P}_f(S)\big)\big(\forall \epsilon > 0\big)\big(\exists K \in \mathcal{P}_f(S)\big)\big(\forall s \in H\big)\big(|K \triangle sK| < \epsilon \cdot |K|\big)$.
(b) For $A \subseteq S$, $d(A) = \sup\{\alpha \in [0,1] : \big(\forall H \in \mathcal{P}_f(S)\big)(\forall \epsilon > 0)\big(\exists K \in \mathcal{P}_f(S)\big)$ $\big((\forall s \in H)(|K \setminus sK| < \epsilon \cdot |K|) \text{ and } |A \cap K| \geq \alpha \cdot |K|\big)\}$.
(c) $\Delta^*(S) = \{p \in \beta S : (\forall A \subseteq S)(p \in c\ell(A) \Rightarrow d(A) > 0)\}$.

Sets satisfying SFC include all commutative semigroups [**6**, Theorem 4]. By [**53**, Theorem 1.9] for $A \subseteq \mathbb{N}$, the densities as defined in Definitions 5.10 and 5.12 are the same, and so $\Delta^*(\mathbb{N}) = \Delta^*$.

THEOREM 5.13 (Hindman and Strauss). *Let S and T be left cancellative semigroups satisfying SFC, let $A \subseteq S$, and let $B \subseteq T$. Then $S \times T$ satisfies SFC and $d(A \times B) = d(A) \cdot d(B)$.*

PROOF. [**54**, Lemma 3.1 and Theorems 2.12 and 3.4]. □

A stronger version of Theorem 5.11 holds.

THEOREM 5.14 (Hindman and Strauss). *Let $(S, +)$ be a commutative semigroup and let $A \subseteq S$. If A is central, then $d(A) > 0$. If $d(A) > 0$, then A is a J-set in S. If there is an idempotent in $\Delta^*(S) \cap \overline{A}$, then A is a C-set in S.*

PROOF. [**53**, Theorem 5.12]. □

In view of Theorem 5.14 and the fact already mentioned that there are C-sets contained in $\mathbb{N}$ with density zero, one asks how common are semigroups that contain C-sets with density zero.

THEOREM 5.15 (Hindman and Strauss). *If S is a subsemigroup of $(\mathbb{R},+)$ such that $\mathbb{Z} \subseteq S$ or S is the direct sum of countably many finite abelian groups, then there is a C-set $C \subseteq S$ such that $d(C) = 0$.*

PROOF. [**53**, Theorems 3.5 and 4.3]. □

Elementary characterizations of central sets have been known for some time. (See [**45**, Section 14.5].) These characterizations have, however, limited utility because they all involve showing that some collection of sets is *collectionwise piecewise syndetic*, a notion with a very complicated definition. The corresponding characterization of C-sets, only recently discovered, involves the much simpler notion of a J-set.

THEOREM 5.16 (Hindman and Strauss). *Let S be an infinite semigroup and let $A \subseteq S$. Statements (a) and (b) are equivalent and are implied by statement (c). If S is countable, then all three statements are equivalent.*

(a) *A is a C-set.*
(b) *There is a downward directed family $\langle C_F\rangle_{F\in I}$ of subsets of A such that*
  (i) *for all $F \in I$ and all $x \in C_F$, there exists $G \in I$ such that $C_G \subseteq x^{-1}C_F$ and*
  (ii) *for each $F \in I$, C_F is a J-set.*
(c) *There is a decreasing sequence $\langle C_n\rangle_{n=1}^{\infty}$ of subsets of A such that*
  (i) *for all $n \in \mathbb{N}$ and all $x \in C_n$, there exists $m \in \mathbb{N}$ such that $C_m \subseteq x^{-1}C_n$ and*
  (ii) *for all $n \in \mathbb{N}$, C_n is a J-set.*

PROOF. [**52**, Theorem 2.6]. □

The notions of *syndetic* and *piecewise syndetic* have reasonably simple combinatorial characterizations. (See [**45**, Definition 4.38].) We take the following even simpler algebraic characterizations as the definitions here.

DEFINITION 5.17. Let S be a semigroup and let $A \subseteq S$.

(a) A is *syndetic* if and only if for every left ideal L of βS, $\overline{A} \cap L \neq \emptyset$.
(b) A is *piecewise syndetic* if and only if $\overline{A} \cap K(\beta S) \neq \emptyset$.

If $\langle x_n\rangle_{n=1}^{\infty}$ is a sequence in $\mathbb{N}$ such that for each $n \in \mathbb{N}$, $x_{n+1} > \sum_{t=1}^{n} x_t$, then $\bigcap_{m=1}^{\infty} c\ell_{\beta S} FS(\langle x_n\rangle_{n=m}^{\infty})$ contains much of the known algebraic structure of $K(\beta\mathbb{N})$. The following result says that several notions of size are equivalent for such nicely behaved sequences in $\mathbb{N}$.

THEOREM 5.18 (Adams, Hindman, and Strauss). *Let $\langle x_n\rangle_{n=1}^{\infty}$ be a sequence in $\mathbb{N}$ such that for each $n \in \mathbb{N}$, $x_{n+1} > \sum_{t=1}^{n} x_t$. The following statements are equivalent:*

(a) *For all $m \in \mathbb{N}$, $FS(\langle x_n\rangle_{n=m}^{\infty})$ is central.*
(b) *$FS(\langle x_n\rangle_{n=1}^{\infty})$ is central.*
(c) *For all $m \in \mathbb{N}$, $FS(\langle x_n\rangle_{n=m}^{\infty})$ is piecewise syndetic.*
(d) *$FS(\langle x_n\rangle_{n=1}^{\infty})$ is piecewise syndetic.*
(e) *$\{x_{n+1} - \sum_{t=1}^{n} x_t : n \in \mathbb{N}\}$ is bounded.*
(f) *$FS(\langle x_n\rangle_{n=1}^{\infty})$ is syndetic.*
(g) *For all $m \in \mathbb{N}$, $FS(\langle x_n\rangle_{n=m}^{\infty})$ is syndetic.*
(h) *$\bigcap_{m=1}^{\infty} c\ell_{\beta S} FS(\langle x_n\rangle_{n=m}^{\infty}) \cap K(\beta\mathbb{N}) \neq \emptyset$.*

PROOF. [**2**, Theorem 2.8 and Corollary 4.2] □

As a consequence of the following theorem, one has much of the algebraic structure of $K(\beta\mathbb{N})$, specifically all of the structure of $K(\mathbb{H})$, close to, but disjoint from, $K(\beta\mathbb{N})$.

THEOREM 5.19 (Adams). *Let $\epsilon > 0$. There exists a sequence $\langle x_n\rangle_{n=1}^\infty$ in $\mathbb{N}$ such that for each $n \in \mathbb{N}$, $x_{n+1} > \sum_{t=1}^n x_t$, $\{x_{n+1} - \sum_{t=1}^n x_t : n \in \mathbb{N}\}$ is unbounded, and the density $d\big(FS(\langle x_n\rangle_{n=1}^\infty)\big) > 1 - \epsilon$.*

PROOF. [**1**, Theorem 2.15] □

Of course if A is a subset of $\mathbb{N}$ which is piecewise syndetic, then (since $K(\beta\mathbb{N}) \subseteq \Delta^*$) one has that $d^*(A) > 0$. The following theorem was established using non-standard analysis.

THEOREM 5.20 (Jin). *Let $A, B \subseteq \mathbb{N}$ such that $d^*(A) > 0$ and $d^*(B) > 0$. Then $A + B$ is piecewise syndetic.*

PROOF. [**56**, Corollary 3] □

Recall that a set $\mathcal{A}$ of subsets of a set X is said to be a *set of almost disjoint subsets of* X if and only if for each $A \in \mathcal{A}$, $|A| = |X|$ and if A and B are distinct members of $\mathcal{A}$, then $|A \cap B| < |X|$. As is well known, there is a set $\mathcal{A}$ of $\mathfrak{c}$ almost disjoint subsets of $\mathbb{N}$. If $|S| = \kappa > \omega$, there may not exist *any* set of 2^κ almost disjoint subsets of S. (Baumgartner proved [**7**, Theorem 2.8] that there is always a family of κ^+ almost disjoint subsets of S, and also showed that it is consistent with ZFC that if $\kappa = \omega_1$, there is no family of 2^κ almost disjoint subsets of S.)

THEOREM 5.21 (Carlson, Hindman, McLeod, and Strauss). *Let S be an infinite very weakly left cancellative semigroup with cardinality κ. Assume that κ has a set of μ almost disjont sets. Then every central set in S has a set of μ almost disjoint central subsets. Furthermore, every central set in S contains κ pairwise disjoint central subsets.*

PROOF. [**18**, Theorem 3.4]. □

6. Combinatorial applications

Given a number k we write $\overline{k}$ for a vector with all terms equal to k.

DEFINITION 6.1. Let $u, v \in \mathbb{N}$ and let A be a $u \times v$ matrix with entries from $\mathbb{Q}$. Let S be a subsemigroup of $(\mathbb{R}, +)$.

(a) A is *kernel partition regular over* S (KPR/S) if and only if, whenever $S \setminus \{0\}$ is finitely colored, there must exist monochromatic $\vec{x} \in S^v$ such that $A\vec{x} = \overline{0}$.

(b) A is *image partition regular over* S (IPR/S) if and only if, whenever $S\setminus\{0\}$ is finitely colored, there must exist $\vec{x} \in S^v$ such that the entries of $A\vec{x}$ are monochromatic.

A survey of results on image and kernel partition regular matrices can be found in [**41**].

A particularly simple class of image partition regular matrices is the class of *first entries* matrices.

DEFINITION 6.2. Let $u, v \in \mathbb{N}$ and let A be a $u \times v$ matrix. Then A is a *first entries matrix* if and only if

(1) the entries of A are from $\mathbb{Q}$;
(2) no row of A is $\overline{0}$;
(3) the first (leftmost) nonzero entry of each row is positive; and
(4) the first nonzero entries of any two rows are equal if they occur in the same column.

The first non-zero entry in any row of A is called a *first entry* of A.

The following theorem, which extends Theorem 15.5 in [**45**], illustrates the combinatorial richness of C-sets.

THEOREM 6.3 (Hindman and Strauss). *Let $u, v \in \mathbb{N}$ and let A be a $u \times v$ first entries matrix with entries from ω. Let $(S, +)$ be a commutative semigroup with an identity 0. For $n \in \mathbb{N}$ and $s \in S$, let ns denote the sum $s + s + \cdots + s$ with n terms and let $0s = 0$. Let C be a C-set in S and let p be an idempotent in $J(S) \cap \overline{C}$ such that $nS \in p$ for every first entry n of A. Then there exist sequences $\langle x_{1,n}\rangle_{n=1}^{\infty}, \langle x_{2,n}\rangle_{n=1}^{\infty}, \ldots, \langle x_{v,n}\rangle_{n=1}^{\infty}$ such that for every $F \in \mathcal{P}_f(\mathbb{N})$, $\vec{x}_F \in (S \setminus \{0\})^v$ and $A\vec{x}_F \subseteq C^u$, where $\vec{x}_F = \sum_{n\in F} \begin{pmatrix} x_{1,n} \\ x_{2,n} \\ \vdots \\ x_{v,n} \end{pmatrix}$. In the case in which S is a commutative group, this statement holds if A is a first entries matrix with entries from $\mathbb{Z}$.*

PROOF. [**53**, Theorem 2.8]. □

In his 1933 paper [**70**] Rado characterized the kernel partition regularity of matrices, which of course correspond to linear transformations. In that same paper he also characterized the kernel partition regularity of affine transformations. These characterizations are not as well known as his linear characterizations, probably because, with the exception of Theorem 6.4(b)(ii), the answer is that the affine transformation is kernel partition regular if and only if it is trivially so, that is it has a constant solution.

THEOREM 6.4 (Rado). *Let $u, v \in \mathbb{N}$, let A be a $u \times v$ matrix with entries from $\mathbb{Q}$, and let $\vec{b} \in \mathbb{Q}^u \setminus \{\overline{0}\}$.*

(a) *Whenever $\mathbb{Z}$ is finitely colored, there exists a monochromatic $\vec{x} \in \mathbb{Z}^v$ such that $A\vec{x} + \vec{b} = \overline{0}$ if and only if there exists $k \in \mathbb{Z}$ such that $A\overline{k} + \vec{b} = \overline{0}$.*
(b) *Whenever $\mathbb{N}$ is finitely colored, there exists a monochromatic $\vec{x} \in \mathbb{N}^v$ such that $A\vec{x} + \vec{b} = \vec{0}$ if and only if either*
   (i) *there exists $k \in \mathbb{N}$ such that $A\overline{k} + \vec{b} = \overline{0}$ or*
   (ii) *there exists $k \in \mathbb{Z}$ such that $A\overline{k} + \vec{b} = \overline{0}$ and the linear mapping $\vec{x} \mapsto A\vec{x}$ is kernel partition regular.*

PROOF. (a) [**70**, Satz VIII].
(b) [**70**, Satz V]. □

The following characterization of image partition regularity of an affine transformation over $\mathbb{Z}$ is nearly identical to Rado's characterization of kernel partition regularity of affine transformations.

THEOREM 6.5 (Moshesh). *Let $u, v \in \mathbb{N}$, let A be a $u \times v$ matrix with entries from $\mathbb{Q}$, and let $\vec{b} \in \mathbb{Q}^u \setminus \{\overline{0}\}$. Whenever $\mathbb{Z}$ is finitely colored, there exists $\vec{x} \in \mathbb{Z}^v$ such that the entries of $A\vec{x} + \vec{b}$ are monochromatic if and only if there exist $\vec{x} \in \mathbb{Z}^v$ and $k \in \mathbb{Z}$ such that $A\vec{x} + \vec{b} = \overline{k}$.*

PROOF. [**61**, Theorem 4.8]. □

The characterization in the following is significantly more interesting. (Note in particular the appearance of central sets.)

THEOREM 6.6 (Hindman and Moshesh). *Let $u, v \in \mathbb{N}$, let A be a $u \times v$ matrix with entries from $\mathbb{Q}$, and let $\vec{b} \in \mathbb{Q}^u \setminus \{\overline{0}\}$. Whenever $\mathbb{N}$ is finitely colored there exists $\vec{x} \in \mathbb{Z}^v$ such that the entries of $A\vec{x} + \vec{b}$ are monochromatic if and only if either*

(i) *there exists $k \in \mathbb{N}$ and $\vec{x} \in \mathbb{Z}^v$ such that $A\vec{x} + \vec{b} = \overline{k}$ or,*
(ii) *there exists $k \in \mathbb{Z}$ and $\vec{x} \in \mathbb{Z}^v$ such that $A\vec{x} + \vec{b} = \overline{k}$ and for every central set C in $\mathbb{N}$, there exists $\vec{x} \in \mathbb{Z}^v$ such that $A\vec{x} \in C^u$.*

PROOF. [**43**, Theorem 4.5]. □

DEFINITION 6.7. Let S be a subsemigroup of $(\mathbb{R}, +)$ with $0 \in c\ell S$, let $u, v \in \mathbb{N}$, and let A be a $u \times v$ matrix with entries from $\mathbb{Q}$. Then A is *image partition regular over S near zero* (abbreviated IPR/S_0) if and only if, whenever $S \setminus \{0\}$ is finitely colored and $\delta > 0$, there exists $\vec{x} \in S^v$ such that the entries of $A\vec{x}$ are monochromatic and lie in the interval $(-\delta, \delta)$.

$\mathbb{D}$ denotes the set of dyadic rational numbers.

THEOREM 6.8. *Let $u, v \in \mathbb{N}$ and let A be a $u \times v$ matrix with entries from $\mathbb{Q}$. The seven statements in (I) below are equivalent and are strictly stronger than the seven equivalent statements in (II).*

(I)
  (a) *A is IPR/$\mathbb{N}$.*
  (b) *A is IPR/$\mathbb{D}^+$.*
  (c) *A is IPR/$\mathbb{Q}^+$.*
  (d) *A is IPR/$\mathbb{R}^+$.*
  (e) *A is IPR/$\mathbb{D}^+_0$.*
  (f) *A is IPR/$\mathbb{Q}^+_0$.*
  (g) *A is IPR/$\mathbb{R}^+_0$.*

(II)
  (a) *A is IPR/$\mathbb{Z}$.*
  (b) *A is IPR/$\mathbb{D}$.*
  (c) *A is IPR/$\mathbb{Q}$.*
  (d) *A is IPR/$\mathbb{R}$.*
  (e) *A is IPR/$\mathbb{D}_0$.*
  (f) *A is IPR/$\mathbb{Q}_0$.*
  (g) *A is IPR/$\mathbb{R}_0$.*

PROOF. [**23**, Theorem 2.6]. □

In [**23**] it is also demonstrated that there are many distinct notions of image partion regularity near zero for infinite matrices.

Let A denote a nonempty finite set (the *alphabet*). We choose a set $V = \{v_n : n \in \omega\}$ (of *variables*) such that $A \cap V = \emptyset$ and define W to be the semigroup of words over the alphabet $A \cup V$ (including the empty word), with concatenation as the semigroup operation. (Formally a *word* w is a function with domain $k \in \omega$ to the alphabet and the length $\ell(w)$ of w is k. We shall need to resort to this formal meaning, so that if $i \in \{0, 1, \ldots, \ell(w) - 1\}$, then $w(i)$ denotes the $(i + 1)^{\text{st}}$ letter of w.)

DEFINITION 6.9. Let $n \in \omega$ and let $k \in \{0, 1, \ldots, n\}$. Then $[A]\binom{n}{k}$ is the set of all words w over the alphabet $A \cup \{v_0, v_1, \ldots, v_{k-1}\}$ of length n such that

(1) for each $i \in \{0, 1, \ldots, k - 1\}$, if any, v_i occurs in w and
(2) for each $i \in \{0, 1, \ldots, k-2\}$, if any, the first occurrence of v_i in w precedes the first occurrence of v_{i+1}.

Let $k \in \mathbb{N}$. Then the set of *k-variable words* is $S_k = \bigcup_{n=k}^{\infty} [A]\binom{n}{k}$. Also S_0 is the semigroup of words over A. Given $w \in S_n$ and $u \in W$ with $\ell(u) = n$, we define $w\langle u\rangle$ to be the word with length $\ell(w)$ such that for $i \in \{0, 1, \ldots, \ell(w) - 1\}$

$$w\langle u\rangle(i) = \begin{cases} w(i) & \text{if } w(i) \in A \\ u(j) & \text{if } w(i) = v_j \,. \end{cases}$$

That is, $w\langle u\rangle$ is the result of substituting $u(j)$ for each occurrence of v_j in w. For example, let $A = \{a, b\}$, let $w = av_0bv_1v_0abv_2bv_0$, and let $u = bv_0v_1$. Then $w\langle u\rangle = abbv_0babv_1bb$.

The following theorem is commonly known as the Graham-Rothschild Parameter Sets Theorem. The original theorem [**35**] (or see [**63**]) is stated in a significantly stronger fashion. However this stronger version is derivable from the version stated here in a reasonably straightforward manner. (See [**17**, Theorem 5.1].)

THEOREM 6.10 (Graham-Rothschild). *Let $m, n \in \omega$ with $m < n$, and let S_m be finitely colored. There exists $w \in S_n$ such that $\{w\langle u\rangle : u \in [A]\binom{n}{m}\}$ is monochromatic.*

Section 9 of [**35**] contains 13 corollaries. Included among these are four results that were known at the time (namely the Hales-Jewett Theorem, van der Waerden's Theorem, Ramsey's Theorem, and the finite version of the Finite Sums Theorem). We believe that the other nine were new at the time. These include the finite version of the Finite Unions Theorem. While the infinite version of the Finite Unions Theorem is obviously derivable from the infinite version of the Finite Sums Theorem, the finite version of the Finite Unions Theorem is not obviously derivable from the finite version of the Finite Sums Theorem. (It is derivable therefrom – we just don't think that the derivation is obvious. In particular, we don't believe that anyone had derived it at that time.)

DEFINITION 6.11. For $r, n \in \mathbb{N}$ with $r > n$ and $u \in [A]\binom{r}{n}$ define $h_u : S_r \to S_n$ by, for $w \in S_r$, $h_u(w) = w\langle u\rangle$, and let $\widetilde{h_u} : \beta S_r \to \beta S_n$ be the continuous extension of h_u.

The following algebraic result was used in [**17**] to derive an infinitary extension of Theorem 6.10.

THEOREM 6.12 (Carlson, Hindman, and Strauss). *Let A be a nonempty alphabet. Let p be a minimal idempotent in βS_0. There is a sequence $\langle p_n\rangle_{n=0}^{\infty}$ such that*

(1) $p_0 = p$;

(2) *for each* $n \in \mathbb{N}$, p_n *is a minimal idempotent in* βS_n;
(3) *for each* $n \in \mathbb{N}$, $p_n \leq p_{n-1}$; *and*
(4) *for each* $n \in \mathbb{N}$ *and each* $u \in [A]\binom{n}{n-1}$, $\widetilde{h_u}(p_n) = p_{n-1}$.

Further, p_1 *can be any minimal idempotent of* βS_1 *such that* $p_1 \leq p_0$.

PROOF. [**17**, Theorem 2.12]. □

THEOREM 6.13 (Carlson, Hindman, and Strauss). *Let* A *be a nonempty alphabet, let* $n \in \mathbb{N}$, *and let*

$$T_n = \{p \in \beta S_n : (\forall r > n)(\exists q \in \beta S_r)(\forall u \in [A]\tbinom{r}{n})(\widetilde{h_u}(q) = p)\}\,.$$

Let $\kappa = \max\{|A|, \omega\}$. *Then* T_n *is a subsemigroup of* βS_n, $K(T_n) = T_n \cap K(\beta S_n)$, *each minimal right ideal and each minimal left ideal of* T_n *contains* 2^{2^κ} *idempotents, and each maximal group in* $K(T_n)$ *contains a free group on* 2^{2^κ} *generators.*

PROOF. [**15**, Theorems 2.3, 2.13, and 2.18] and [**45**, Theorem 1.65]. □

We shall not state here the infinitary generalization of Theorem 6.10 obtained using Theorem 6.12 because of the additional notation needed, stating instead the following consequence of that generalization.

THEOREM 6.14 (Carlson, Hindman, and Strauss). *Let* $u, v, k \in \mathbb{N}$, *let* M *be a* $u \times v$ *first entries matrix with entries from* $\mathbb{Z}$, *let* C *be a central subset of* $\mathbb{N}$, *and let* G *be a* K_k*-free graph on* $\mathbb{N}$. *There is a sequence* $\langle \vec{x_n}\rangle_{n=1}^{\infty}$ *in* $\mathbb{N}^v$ *such that for every* $F \in \mathcal{P}_f(\mathbb{N})$, *the set of entries of* $M(\sum_{n\in F}\vec{x_n})$ *is an independent subset of* C.

PROOF. [**17**, Theorem 4.8]. □

Another application of Theorem 6.12 is the following.

THEOREM 6.15 (Carlson, Hindman, and Strauss). *Let* $\mathbb{N}$ *be finitely colored. Then there is a color class* D *which is central in* $\mathbb{N}$ *and*

(1) *there exists a pairwise disjoint collection* $\{D_{i,j} : i, j \in \omega\}$ *of central subsets of* D *and for each* $i \in \omega$ *there exists a sequence* $\langle x_{i,n}\rangle_{n=i}^{\infty}$ *in* $D_{i,i}$ *such that whenever* F *is a finite nonempty subset of* ω *and* $f : F \to \{1, 2, \ldots, \min F\}$ *one has that* $\Sigma_{n\in F}\, x_{f(n),n} \in D_{i,j}$ *where* $i = f(\min F)$ *and* $j = f(\max F)$; *and*
(2) *at stage* n *when one is chosing* $(x_{0,n}, x_{1,n}, \ldots, x_{n,n})$, *each* $x_{i,n}$ *may be chosen as an arbitrary element of a certain central subset of* $D_{i,i}$, *with the choice of* $x_{i,n}$ *independent of the choice of* $x_{j,n}$.

PROOF. [**16**, Corollary 2.9]. □

In [**10**] several combined additive and multiplicative combinatorial structures were shown to exist in any multiplicatively large set. As an example consider the following.

THEOREM 6.16 (Beiglböck, Bergelson, Hindman, and Strauss). *Let* C *be a central subset of* $(\mathbb{N}, \cdot)$ *and let* $k \in \mathbb{N}$. *There exist* $a, b, d \in \mathbb{N}$ *such that*

$$\begin{aligned}\big\{b(a+id)^j : i,j \in \{0,1,\ldots,k\}\big\} \cup \big\{bd^j : j \in \{0,1,\ldots,k\}\big\}\\ \cup \big\{a+id : i \in \{0,1,\ldots,k\}\big\} \cup \{d\} \subseteq C\,.\end{aligned}$$

PROOF. [**10**, Corollary 4.3]. □

A significant portion of Theorem 6.16 has been recently derived by combinatorial methods in [**13**].

By way of contrast, comparatively little multiplicative structure is guaranteed to additively large sets as is demonstrated by several purely combinatorial results in [**9**]. (We do not address these here as we are concerned with applications of the algebra of βS.) However, if a set A is very large additively, there must be significant multiplicative structure.

THEOREM 6.17 (Beiglböck, Bergelson, Hindman, and Strauss). *Let $A \subseteq \mathbb{N}$ and assume that $K(\beta\mathbb{N},+) \subseteq \overline{A}$. Then for all $t \in \mathbb{Z}$, $c\ell K(\beta\mathbb{N},+) \subseteq \overline{(t+A)\cap\mathbb{N}}$ and in particular, $(t+A)\cap\mathbb{N}$ is central in $(\mathbb{N},+)$ and in $(\mathbb{N},\cdot)$.*

PROOF. [**9**, Theorem 3.11]. □

In [**21**, Theorem 2.5], Davenport presented a proof using the algebraic structure of $\beta(\mathbb{N}^k)$ of the *multidimensional van der Waerden theorem*, otherwise known as Grünwald's theorem or Gallai's theorem. (Grünwald and Gallai were the same person.)

THEOREM 6.18 (Grünwald=Gallai). *Let $k, r \in \mathbb{N}$ and asssume that $\mathbb{N}^k = \bigcup_{i=1}^r A_i$. Then there exist $l, d, a_1, a_2, \ldots, a_k \in \mathbb{N}$ and $j \in \{1, 2, \ldots, r\}$ such that $\big\{(a_1+n_1d, a_2+n_2d, \ldots, a_k+n_kd) : \{n_1, n_2, \ldots, n_k\} \subseteq \{1, 2, \ldots, l\}\big\} \subseteq A_j$.*

It is an old result of Bergelson [**11**] that if $\mathbb{N}$ is partitioned into finitely many cells, then one cell C satisfies $\overline{d}(\{n \in C : \overline{d}(C\cap(-n+C)) > 0\}) > 0$, where $\overline{d}$ is upper asymptotic density. In a similar vein we have the following, where d is the Følner density as defined in Definition 5.12.

THEOREM 6.19 (Hindman and Strauss). *Let $k \in \mathbb{N}$. Let S be a left cancellative semigroup satisfying SFC and assume that for all $x, y \in S$, $|\{s \in S : sx = y\}| \leq k$. Let $s \in S$ and let $A \subseteq S$. If $d(A) > 0$, then for each $l \in \mathbb{N}$ there exists $d \in \mathbb{N}$ such that $d(\{b \in S : \{s^d b, s^{2d} b, \ldots, s^{ld} b\} \subseteq A) > 0$.*

PROOF. [**47**, Theorems 4.16 and 5.5]. □

By [**47**, Theorems 2.4, 2.7, and 4.11] if S is as in Theorem 6.19 and A is a piecewise syndetic subset of S, then $d(A) > 0$, so the conclusion of Theorem 6.19 applies to A. One may wonder why the geometric progression is written in the form $\{s^d b, s^{2d} b, \ldots, s^{ld} b\}$ rather than $\{bs^d, bs^{2d}, \ldots, bs^{ld}\}$. It is shown in [**47**, Theorem 5.8] that there is a countable cancellative semigroup S satisfying SFC and a subset A such that $\overline{A}$ contains a left ideal of βS (so in particular A is central and hence piecewise syndetic) but there is some $s \in S$ such that $\{bs^n : n \in \mathbb{N}$ and $b \in S\} \cap A = \emptyset$.

We close with two simply stated questions about Ramsey Theory which have been open for decades.

QUESTION 6.20. Let $r \in \mathbb{N}$ and let $\mathbb{N} = \bigcup_{i=1}^r A_i$. Must there exist $i \in \{1, 2, \ldots, r\}$ and x and y such that $\{x, y, x+y, x\cdot y\} \subseteq A_i$?

The answer is known to be "yes" if $r = 2$. (This is a computer result of Ron Graham's, not published by him. Some details can be found in [**39**].) The first author of this survey firmly believes in the validity of the stronger version which requires for each $m \in \mathbb{N}$ a finite sequence $\langle x_t\rangle_{t=1}^m$ with $FS(\langle x_t\rangle_{t=1}^m) \cup FP(\langle x_t\rangle_{t=1}^m) \subseteq A_i$. (He has often stated it as a "fact" which has not been proved.)

It is a trivial consequence of the Finite Sums Theorem that if $r \in \mathbb{N}$ and $\mathbb{N} = \bigcup_{i=1}^{r} A_i$, then there exist $i \in \{1, 2, \ldots, r\}$ and a sequence $\langle x_t \rangle_{t=1}^{\infty}$ such that $\{x_t : t \in \mathbb{N}\} \cup \{x_t + x_s : t, s \in \mathbb{N} \text{ and } t \neq s\} \subseteq A_i$.

QUESTION 6.21. Is there a set $B \subseteq \mathbb{N}$ such that B does not contain $FS(\langle x_t \rangle_{t=1}^{\infty})$ for any sequence $\langle x_t \rangle_{t=1}^{\infty}$, but whenever B is finitely colored, there must be a sequence $\langle x_t \rangle_{t=1}^{\infty}$ with $\{x_t : t \in \mathbb{N}\} \cup \{x_t + x_s : t, s \in \mathbb{N} \text{ and } t \neq s\}$ monochromatic?

References

1. C. Adams, *Large finite sums sets with closure missing the smallest ideal of $\beta\mathbb{N}$*, Topology Proceedings **31** (2007), 403-418.
2. C. Adams, N. Hindman and D. Strauss, *Largeness of the set of finite products in a semigroup*, Semigroup Forum **76** (2008), 276-296[1].
3. P. Adams and D. Strauss, *Homomorphisms into $\mathbb{N}^*$*, Semigroup Forum **65** (2002), 460-470.
4. T. Alaste and M. Filali, *Slowly oscillating functions and closed left ideals of βS*, Topology and its Applications **156** (2009), 669-673.
5. R. Arens, *The adjoint of a bilinear operation*, Proc. Amer. Math. Soc. **2** (1951), 839-848.
6. L. Argabright and C. Wilde, *Semigroups satisfying a strong Følner condition*, Proc. Amer. Math. Soc. **17** (1967), 587-591.
7. J. Baumgartner, *Almost-disjoint sets, the dense set problem and the partition calculus*, Ann. Math. Logic **9** (1976), 401-439.
8. M. Beiglböck, V. Bergelson, T. Downarowicz, and A. Fish, *Solvability of Rado Systems in D-sets*, Topology and its Applications **156** (2009), 2565-2571.
9. M. Beiglböck, V. Bergelson, N. Hindman, and D. Strauss, *Multiplicative structures in additively large sets*, J. Comb. Theory (Series A) **113** (2006), 1219-1242. (MR 2007f: 05174)
10. M. Beiglböck, V. Bergelson, N. Hindman, and D. Strauss, *Some new results in multiplicative and additive Ramsey Theory*, Trans. Amer. Math. Soc. **360** (2008), 819-847.
11. V. Bergelson, *A density statement, generalizing Schur's Theorem*, J. Comb. Theory (Series A) **43** (1986), 338-343.
12. A. Blass and N. Hindman, *On strongly summable ultrafilters and union ultrafilters*, Trans. Amer. Math. Soc. **304** (1987), 83-99.
13. D. Bond and R. McCutcheon, *Reading "A variant of the Hales-Jewett theorem" on its anniversary*, Topology and its Applications **156** (2009), 2595-2599.
14. S. Burns and N. Hindman, *Quasi-central sets and their dynamical characterization*, Topology Proceedings **31** (2007), 445-455.
15. T. Carlson, N. Hindman, and D. Strauss, *The Graham-Rothschild Theorem and the algebra of βW*, Topology Proceedings **28** (2004), 361-399.
16. T. Carlson, N. Hindman, and D. Strauss, *Ramsey theoretic consequences of some new results about algebra in the Stone-Čech compactification*, Integers **5(2)** (2005) #A04 1-26.
17. T. Carlson, N. Hindman, and D. Strauss, *An infinitary extension of the Graham-Rothschild Parameter Sets Theorem*, Trans. Amer. Math. Soc. **358** (2006), 3239-3262.
18. T. Carlson, N. Hindman, J. McLeod, and D. Strauss, *Almost disjoint large subsets of semigroups*, Topology and its Applications **155** (2008), 433-444.
19. P. Civin and B. Yood, *The second conjugate space of a Banach algebra as an algebra*, Pacific J. Math. **11** (1961), 847-870.
20. H. Dales, A. Lau, and D. Strauss, *Banach algebras on semigroups and their compactifications*, Memoirs of the American Mathematical Society, to appear.
21. D. Davenport, *Ultrafilters and Ramsey Theory*, Contemp. Math. **275** (2001), 33-38.
22. M. Day, *Amenable semigroups*, Illinois J. Math. **1** (1957), 509-544.
23. D. De and N. Hindman, *Image partition regularity near zero*, Discrete Math. **309** (2009), 3219-3232.
24. D. De, N. Hindman, and D. Strauss, *A new and stronger Central Sets Theorem*, Fundamenta Mathematicae **199** (2008), 155-175.

[1] All of the items in this list of references that include Hindman as an author and have a publication date of 1995 or later are currently available at `http://mysite.verizon.net/nhindman/` except for item [**45**].

25. D. De, N. Hindman, and D. Strauss, *Sets central with respect to certain subsemigroups of* βS_d, Topology Proceedings, **33** (2009), 55-79.
26. M. DiNasso and M. Forti, *Ultrafilter semirings and nonstandard submodels of the Stone-Čech compactification of the natural numbers*, Contemp. Math. **380** (2005), 45-51.
27. R. Ellis, *Lectures on topological dynamics*, Benjamin, New York, 1969.
28. S. Ferri, N. Hindman, and D. Strauss, *Digital representation of semigroups and groups*, Semigroup Forum **77** (2008), 36-63.
29. S. Ferri and D. Strauss, *Homomorphisms into Stone-Čech remainders of countable groups*, Semigroup Forum **71** (2005) 428-438.
30. M. Filali, *t-sets and some algebraic properties in* βS *and in* $l_\infty(S)^*$, Semigroup Forum **65** (2002), 285-300.
31. M. Filali, E. Lutsenko, and I. Protasov, *Boolean group ideals and the ideal structure of* βG, Math. Stud. **30** (2008) 1-10.
32. H. Furstenberg, *Recurrence in ergodic theory and combinatorical number theory*, Princeton University Press, Princeton, 1981.
33. V. Gavrylkiv, *Right-Topological semigroup operations on inclusion hyperspaces*, Mat. Stud. **29** (2008), 18-34.
34. E. Glasner, *On two problems concerning topological centres*, Topology Proceedings **33** (2009), 29-39.
35. R. Graham and B. Rothschild, *Ramsey's Theorem for n-parameter sets*, Trans. Amer. Math. Soc. **159** (1971), 257-292.
36. A. Grainger, *Ultrafilters on the collection of finite subsets of an infinite set*, Semigroup Forum **67** (2003), 443-453.
37. A. Grainger, *Ideals of ultrafilters on the collection of finite subsets of an infinite set*, Semigroup Forum **73** (2006), 234-242.
38. N. Hindman, *Finite sums from sequences within cells of a partition of* $\mathbb{N}$, J. Comb. Theory (Series A) **17** (1974), 1-11.
39. N. Hindman, *Partitions and sums and products of integers*, Trans. Amer. Math. Soc. **247** (1979), 227-245.
40. N. Hindman, *Algebra in the Stone-Čech compactification and its applications to Ramsey Theory*, Sci. Math. Jpn. **62** (2005), 321-329.
41. N. Hindman, *Partition regularity of matrices*, in Combinatorial Number Theory, B. Landman, M. Nathanson, J. Nešetřil, R. Nowakowski, and C. Pomerance, editors, deGruyter, Berlin, 2007, 265-298. Also published as Integers **7(2)** (2007), A-18. `http://www.integers-ejcnt.org/vol7-2.html`
42. N. Hindman *Small sets satisfying the Central Sets Theorem* in Combinatorial Number Theory, B. Landman, M. Nathanson, J. Nešetřil, R. Nowakowski, C. Pomerance, and A. Robertson, editors, deGruyter, Berlin, 2009, 57-64. Also published as Integers **9**(Supplement) (2007), Article 5. `http://www.integers-ejcnt.org/vol9supp.html`
43. N. Hindman and I. Moshesh, *Image partition regularity of affine transformations*, J. Comb. Theory (Series A) **114** (2007), 1375-1393.
44. N. Hindman, J. Pym, and D. Strauss, *Multiplications in additive compactifications of* $\mathbb{N}$ *and* $\mathbb{Z}$, Topology and its Applications, **131** (2003), 149-176.
45. N. Hindman and D. Strauss, *Algebra in the Stone-Čech compactification: theory and applications*, de Gruyter, Berlin, 1998.
46. N. Hindman and D. Strauss, *Recent progress in the topological theory of semigroups and the algebra of* βS, in Recent Progress in General Topology, II, M. Husek and J. van Mill, eds., Elsevier, Amsterdam, (2002), 227-251.
47. N. Hindman and D. Strauss, *Density in arbitrary semigroups*, Semigroup Forum **73** (2006), 273-300.
48. N. Hindman and D. Strauss, *Discrete groups in* $\beta\mathbb{N}$, Topology and its Applications **154** (2007), 2099-2103.
49. N. Hindman and D. Strauss, *Characterization of simplicity and cancellativity in* βS, Semigroup Forum **75** (2007), 70-76.
50. N. Hindman and D. Strauss, *Subsemigroups of* βS *containing the idempotents*, Topology Proceedings **35** (2010), 233-246.
51. N. Hindman and D. Strauss, *Cartesian products of sets satisfying the Central Sets Theorem*, Topology Proceedings **35** (2010), 203-223.

52. N. Hindman and D. Strauss, *A simple characterization of sets satisfying the Central Sets Theorem*, New York J. Math. **15** (2009), 405-413.
53. N. Hindman and D. Strauss, *Sets satisfying the Central Sets Theorem*, Semigroup Forum, to appear.
54. N. Hindman and D. Strauss, *Density and invariant means in left cancellative left amenable semigroups*, Topology and its Applications, to appear.
55. N. Hindman, D. Strauss, and Y. Zelenyuk, *Large rectangular semigroups in Stone-Čech compactifications* Trans. Amer. Math. Soc. **355** (2003), 2795-2812.
56. R. Jin, *The sumset phenomenon* Proc. Amer. Math. Soc. **130** (2002), 855-861.
57. S. Koppelberg, *The Stone-Čech compactification of a semilattice*, Semigroup Forum **72** (2006), 63-74.
58. L. Legette, *Maximal groups in* βS *can be trivial*, Topology and its Applications **156** (2009), 2632-2641.
59. A. Maleki, *Solving equations in* $\beta\mathbb{N}$, Semigroup Forum **61** (2001), 373-384.
60. G. Moche, *The sizes of preimages of points under the natural map from* $K(\beta(\mathbb{N}\times\mathbb{N}))$ *to* $K(\beta\mathbb{N})\times K(\beta\mathbb{N})$, Dissertation, Howard University, 2002.
61. I. Moshesh, *Image partition regularity of affine transformations*, Dissertation, Howard University, 2006.
62. G. Polya, *Untersuchungen über Lücken und Singularitaten von Potenzreihen*, Math. Zeit. **29** (1929), 549-640.
63. H. Prömel and B. Voigt, *Graham-Rothschild parameter sets*, in Mathematics of Ramsey Theory, J. Nešetřil and V. Rödl, eds., Springer-Verlag, Berlin, 1990, 113-149.
64. I. Protasov, *Dynamical equivalences on* G^*, Topology and its Applications **155** (2008), 1394-1402.
65. I. Protasov, *Topologies on groups determined by right cancellable ultrafilters*, Comment. Math. Univ. Carolinae, to appear.
66. I. Protasov, *Finite groups in* βG, Mat. Stud. **10** (1998), 17-22.
67. I. Protasov and O. Protasova, *Sketch of group balleans*, Mat. Stud. **22** (2004), 10-20.
68. I. Protasov and O. Protasova, *On closed ideals of* βG, Semigroup Forum **75** (2007), 237-240.
69. I. Protasov and J. Pym, *Continuity of multiplication in the largest compactification of a locally compact group*, Bull. London Math. Soc. **33(3)** (2001), 279-282.
70. R. Rado, *Studien zur Kombinatorik*, Math. Zeit. **36** (1933), 242-280.
71. S. Shelah, *Proper forcing*, Springer-Verlag, Berlin, 1982.
72. H. Shi and H. Yang, *Nonmetrizable topological dynamical characterization of central sets*, Fundamenta Math. **150** (1996), 1-9.
73. D. Strauss, *The smallest ideals of* $\beta\mathbb{N}$ *under addition and multiplication*, Topology and its Applications **149** (2005), 289-292.
74. Y. Zelenyuk, *On the ultrafilter semigroup of a topological group*, Semigroup Forum **73** (2006), 301-307.
75. Y. Zelenyuk, *On the ultrafilter semigroup of an abelian topological group*, Semigroup Forum **74** (2007), 467-472.
76. Y. Zelenyuk, *Almost maximal spaces*, Topology and its Applications **154** (2007), 339-357.
77. Y. Zelenyuk, *Local homomorphisms of topological groups*, J. Aust. Math. Soc. **83** (2007), 1-14.
78. Y. Zelenyuk, *Finite groups in Stone-Čech compactifications*, Bull. London Math. Soc. **40** (2008), 337-346.
79. Y. Zelenyuk, *The number of minimal right ideals of* βG, Proc. Amer. Math. Soc. **137** (2009), 2483-2488.
80. Y. Zelenyuk, *Regular idempotents in* βS, Trans. Amer. Math. Soc., to appear.
81. Y. Zelenyuk, *The smallest ideal of* βS *is not closed*, Topology Proceedings, to appear.
82. Y. Zelenyuk and Y. Zelenyuk, *Free groups in the smallest ideal of* βG, Semigroup Forum **78** (2009), 360-367.

Department of Mathematics, Howard University, Washington, DC 20059, USA
E-mail address: nhindman@aol.com

Department of Pure Mathematics, University of Leeds, Leeds LS2 9J2, UK
E-mail address: d.strauss@hull.ac.uk

Contemporary Mathematics
Volume **530**, 2010

Ultrapower of $\mathbb{N}$ and Density Problems

Renling Jin

Abstract. We show how we can apply ultrapower methods to density problems in additive/combinatorial number theory.

1. Introduction and Background Information

In the past decade, the methods from nonstandard analysis have been successfully applied to density problems to obtain many results in additive/combinatorial number theory (cf. [**BJ, Ji1, Ji2, Ji3, Ji4, Ji5, Ji6, Ji7, Ji8, JK**]). Since a nonstandard universe can be constructed by taking an ultrapower of the standard universe, some of the methods used in these papers can also been developed "algebraically" through ultrapower methods without requiring too much knowledge from mathematical logic. In this article, we introduce some of these results through ultrapower methods. In particular, we present results related to Kneser's Theorem and Plünnecke's Theorem.

Let $\mathbb{N}$ be the set of all non-negative integers. To measure the size of a finite set $A \subseteq \mathbb{N}$, one can count the number of elements in A, known as the cardinality of A and denoted by $|A|$. If a is the least element in A and b is the greatest element in A, one can also measure the density of A relative to $[a,b]$ by the ratio $\frac{|A|}{b-a+1}$. When A is infinite, the cardinality of A is no longer useful for distinguishing the size of A from other infinite sets. But the density of a finite set can be extended to the density of A as the asymptotic trend of the densities of $A \cap [a_n, b_n]$ where $[a_n, b_n]$ is a sequence of finite intervals of non-negative integers with $\lim_{n\to\infty}(b_n - a_n) = \infty$. We can compare the "sizes" of two infinite subsets of $\mathbb{N}$ by comparing their densities.

Let $a, b \in \mathbb{N}$. We will write $[a,b]$ exclusively for the interval of integers between a and b including a and b. The following commonly used densities are under our consideration. Let $A \subseteq \mathbb{N}$. For $a, b \in \mathbb{N}$ let $A(a,b) = |A \cap [a,b]|$ and $A(b) = A(1,b)$. The *Shnirel'man density* $\sigma(A)$, the *lower asymptotic density* $\underline{d}(A)$, the *upper asymptotic density* $\overline{d}(A)$, and the *upper Banach density* $BD(A)$ of A are defined by

$$\sigma(A) = \inf_{n \geq 1} \frac{A(n)}{n},$$

1991 *Mathematics Subject Classification.* Primary 11B05, 11B13, 11U10, 03H15.

Key words and phrases. Shnirel'man density, lower asymptotic density, upper asymptotic density, upper Banach density, Plünnecke's inequality, Kneser's theorem, basis, nonstandard analysis.

The author is supported in part by NSF RUI grant DMS#0500671.

$$\underline{d}(A) = \liminf_{n\to\infty} \frac{A(n)}{n},$$

$$\overline{d}(A) = \limsup_{n\to\infty} \frac{A(n)}{n}, \text{ and}$$

$$BD(A) = \lim_{n\to\infty} \sup_{k\in\mathbb{N}} \frac{A(k, k+n-1)}{n}.$$

Clearly, these densities have the following order by their magnitude.

$$0 \leq \sigma(A) \leq \underline{d}(A) \leq \overline{d}(A) \leq BD(A) \leq 1$$

for every $A \subseteq \mathbb{N}$. The order of these densities by their popularity among many number theorists seems to be opposite of their magnitudes. Upper Banach density is especially unfamiliar to some number theorists. However, the concept of upper Banach density bears the most resemblance among these densities to the concept of a probability measure space. For example, in [**Fu**, Lemma 3.17] Furstenberg established a correspondence principle between upper Banach density and probability measure and used it to prove many interesting number theoretic results. The author has also developed a general scheme, which establishes a connection between upper Banach density and Shnirel'man density/lower asymptotic density through Loeb probability measure spaces, which enable us to generate many new theorems about upper Banach density (cf. [**Ji2**]).

Shnirel'man density is probably the most popular density among many number theorists. There have been many important classical theorems about Shnirel'man density.

1.1. Kneser's Theorem. Shnirel'man in 1930 proved a theorem that for any $A, B \subseteq \mathbb{N}$, if $0 \in A$ and $1 \in B$, then

$$\sigma(A+B) \geq \sigma(A) + \sigma(B) - \sigma(A)\sigma(B)$$

where $A + B = \{a + b : a \in A \text{ and } b \in B\}$ (cf. [**HR**, Theorem 1 on page 3] or [**Na1**, Theorem 7.5 on page 193]). Let P be the set of all prime numbers and $A = P \cup \{0, 1\}$. By the theorem above Shnirel'man showed that there is a positive integer h such that $\sigma(hA) = 1$ where

$$hA = \underbrace{A + A + \cdots + A}_{h}.$$

This result is the first significant advancement on the famous Goldbach Conjecture. It shows that there is a fixed positive integer h such that every positive integer greater than 1 is the sum of at most h prime numbers.

Mann in 1942 improved Shnirel'man's Theorem by showing that if $0 \in A \cap B$, then $\sigma(A+B) \geq \min\{1, \sigma(A) + \sigma(B)\}$ (cf. [**HR**, Theorem 3 on page 5]). Mann's theorem was included in Khinchin's little book "Three pearls of number theory" as one of the three pearls (cf. [**Kh**]).

It is often the case that after a theorem about Shnirel'man density is proven, people want to know whether it can be generalized to a theorem about lower asymptotic density. However, one cannot replace σ by $\underline{d}$ in either Shnirel'man's Theorem or in Mann's Theorem. Let $2k^2 < g$ and $A = [0, k-1] + \{gn : n \in \mathbb{N}\}$. A is the union of k arithmetic progressions with a common difference g. Clearly, $\underline{d}(A) = \frac{k}{g}$ and

$$\underline{d}(A+A) = \frac{2k-1}{g} = 2\underline{d}(A) - \frac{1}{g} < 2\underline{d}(A) - \left(\frac{k}{g}\right)^2 = 2\underline{d}(A) - \underline{d}(A)^2 < 2\underline{d}(A) \leq 1.$$

However, this counterexample is essentially the only reason why σ cannot be replaced by $\underline{d}$ in Shnirel'man's Theorem or in Mann's Theorem. In 1953 Kneser proved the following theorem.

THEOREM 1.1 (M. Kneser, 1953). *Let $A, B \subseteq \mathbb{N}$ such that $\underline{d}(A+B) < \underline{d}(A) + \underline{d}(B)$. Then there exist positive integer d and $G \subseteq [0, d-1]$ such that*

(1) $\underline{d}(A+B) \geq \underline{d}(A) + \underline{d}(B) - \frac{1}{d}$,
(2) $A + B \subseteq G + \{dn : n \in \mathbb{N}\}$, *and*
(3) $(G + \{dn : n \in \mathbb{N}\}) \setminus (A+B)$ *is finite.*

The proof of Theorem 1.1 can be found in [**HR**, page 51–75][1]. It is not difficult to prove that Theorem 1.1 is equivalent to the following theorem.

THEOREM 1.2. *Let $A, B \subseteq \mathbb{N}$ such that $\underline{d}(A+B) < \underline{d}(A) + \underline{d}(B)$. Then there exist positive integer d and sets $F, F' \subseteq [0, d-1]$ such that*

(1) $A \subseteq F + \{dn : n \in \mathbb{N}\}$, $B \subseteq F' + \{dn : n \in \mathbb{N}\}$, *and*
(2) $\underline{d}(A) + \underline{d}(B) > \frac{|F|+|F'|-1}{d}$.

Theorem 1.2 clearly shows that if $\underline{d}(A+B) < \underline{d}(A) + \underline{d}(B)$, then each of A and B must be large subsets of the union of arithmetic progressions with a common difference d. This formulation of Kneser's Theorem is in the same style as in the so called Freiman's inverse problem for finite sets, which says that if $A + B$ is small, then A and B must have some arithmetic structure (cf. [**Na2**]).

We will present theorems about upper Banach density parallel to Kneser's Theorem in §2 and §3.

1.2. Plünnecke's Theorem. A set $B \subseteq \mathbb{N}$ is called an *essential component* if $\sigma(A+B) > \sigma(A)$ whenever $A \subseteq \mathbb{N}$ and $0 < \sigma(A) < 1$. Since the early time of the last century people have been interested in finding which sets $B \subseteq \mathbb{N}$ can be an essential component (cf. [**HR**]). By Shnirel'man's Theorem [**HR**, page 3] it can easily be seen that if $0 \in B$ and $\sigma(B) > 0$, then B is an essential component. However, even if $\sigma(B) = 0$, B can still be an essential component. A set $B \subseteq \mathbb{N}$ is called a *basis* of order h if $hB = \mathbb{N}$. Let $h > 1$. Note that B is a basis of order h iff $\sigma(hB) = 1$. If B is a basis of some finite order, then B is an essential component although such B may have Shnirel'man density 0. For example, $B = \{n^2 : n \in \mathbb{N}\}$ is a basis of order 4 by Lagrange's Theorem and $\sigma(B) = 0$. In 1937 Erdös proved that if B is a basis of order h, then

$$\sigma(A+B) \geq \sigma(A) + \frac{1}{2h} \cdot \sigma(A)\left(1 - \sigma(A)\right). \tag{1.1}$$

A short time later, Landau noticed that in Erdös' proof h can be replaced by average order h^* (cf. [**HR**, page 10]). Let $B \subseteq \mathbb{N}$ be a basis of order h. For each $m \in \mathbb{N}$ let

[1] Kneser's Theorem actually deals with the sum of multiple sets. We state only the version for the sum of two sets here for simplicity.

$h_B(m) = \min\{h' \in \mathbb{N} : m \in h'B\}$. The *average order* h^* of B is defined by

$$h^* = \sup_{n\geq 1} \frac{1}{n} \sum_{m=1}^{n} h_B(m).$$

It is easy to see that $h^* \leq h \leq 2h^*$ (cf. [**HR**, page 12]). In 1938 Rohrbach proved a theorem for lower asymptotic density parallel to Erdös–Landau's result. A set $B \subseteq \mathbb{N}$ is called an *asymptotic basis* of order h if hB contains all sufficiently large positive integers. The *average asymptotic order* h^* of an asymptotic basis is defined by

$$h^* = \limsup_{n\to\infty} \frac{1}{n} \sum_{m=1}^{n} h_B(m)$$

where $h_B(m)$ is defined to be 0 if $m \notin hB$. Rohrbach proved (cf. [**HR**, page 45]) that if B is an asymptotic basis of average asymptotic order h^*, then

$$\underline{d}(A+B) \geq \underline{d}(A) + \frac{1}{2h^*} \cdot \underline{d}(A)\left(1 - \underline{d}(A)\right). \tag{1.2}$$

In 1970 Plünnecke obtained the following significant improvement of Erdös–Landau's result (cf. [**Pl**] or [**Na2**, page 225]).

THEOREM 1.3 (Plünnecke, 1970). *If B is a basis of order h, then for every $A \subseteq \mathbb{N}$*

$$\sigma(A+B) \geq \sigma(A)^{1-\frac{1}{h}}. \tag{1.3}$$

Since it is easy to show that $x^{1-\frac{1}{h}} \geq x + \frac{1}{h} \cdot x(1-x)$ for any $h > 0$ and $0 \leq x \leq 1$ by elementary calculus, Plünnecke's Theorem implies (1.1) even when $\frac{1}{2h}$ is replaced by $\frac{1}{h}$. Since the average order h^* of a basis B is less than or equal to 2 times the order h of B, Plünnecke's Theorem also implies (1.1) when h is replaced by h^*.

However, we cannot replace σ by $\underline{d}$ and replace the order h by the average asymptotic order h^* in (1.3) as Rohrbach did in (1.2) to Erdös Theorem (1.1). For example, let $A = \{1 + 3n : n \in \mathbb{N}\}$ and $B = \{i + 3n : i = 0, 1 \text{ and } n \in \mathbb{N}\}$. Then $A + B = \{i + 3n : i = 1, 2 \text{ and } n \in \mathbb{N}\}$, $\sigma(A) = \underline{d}(A) = \frac{1}{3}$ and $\sigma(A+B) = \underline{d}(A+B) = \frac{2}{3}$. It is easy to check that B is a basis of order $h = 2$, average order $h^* = \frac{3}{2}$, and average asymptotic order $h^{**} = \frac{4}{3}$. Note that

$$\underline{d}(A)^{1-\frac{1}{h^{**}}} = \left(\frac{1}{3}\right)^{\frac{1}{4}} > \sigma(A)^{1-\frac{1}{h^*}} = \underline{d}(A)^{1-\frac{1}{h^*}}$$

$$= \left(\frac{1}{3}\right)^{\frac{1}{3}} > \frac{2}{3} = \sigma(A+B) = \underline{d}(A+B).$$

We will discuss the generalization of Plünnecke's Theorem to other densities in §2 and §3.

2. Level One Applications

In this section we will introduce ultrapower methods and develop a scheme, which allows us to obtain a theorem about upper Banach density parallel to every existing theorem about Shnirel'man density or lower asymptotic density without making too much effort.

2.1. Ultrapower of $\mathbb{R}$.

DEFINITION 2.1. A collection $\mathcal{F}$ of subsets of $\mathbb{N}$ is called a **filter** if the following are true:

(1) $\emptyset \notin \mathcal{F}$ and $\mathbb{N} \in \mathcal{F}$,
(2) $A \cap B \in \mathcal{F}$ for any $A, B \in \mathcal{F}$,
(3) $A \in \mathcal{F}$ and $A \subseteq B$ imply $B \in \mathcal{F}$ for any $A, B \subseteq \mathbb{N}$.

The filter $\mathcal{F}$ is called a **non-principal ultrafilter** if

(4) $\{n\} \notin \mathcal{F}$ for each $n \in \mathbb{N}$,
(5) for every $A \subseteq \mathbb{N}$, either $A \in \mathcal{F}$ or $\mathbb{N} \setminus A \in \mathcal{F}$.

The existence of a non-principal ultrafilter on $\mathbb{N}$ is guaranteed by the axiom of choice. From now on we fix a non-principal ultrafilter $\mathcal{F}$ on $\mathbb{N}$. We assume that the reader knows the basic properties of ultrafilters. For example, we assume the reader knows that the intersection of finitely many sets $A_i \in \mathcal{F}$ is again in $\mathcal{F}$, every cofinite subset of $\mathbb{N}$ is in $\mathcal{F}$, and if the union of finitely many sets is in $\mathcal{F}$, then so is one of those sets. We write $\mathbb{R}^{\mathbb{N}}$ for the set of all functions f from some $F_f \in \mathcal{F}$ to $\mathbb{R}$. Note that the term $\mathbb{R}^{\mathbb{N}}$ often denotes the set of all functions from $\mathbb{N}$ to $\mathbb{R}$. We change this convention for technical convenience.

DEFINITION 2.2. Given any $f, g \in \mathbb{R}^{\mathbb{N}}$, let

(1) $f \sim g$ iff $\{n : f(n), g(n)$ are defined and $f(n) = g(n)\} \in \mathcal{F}$,
(2) $[f] = \{g \in \mathbb{R}^{\mathbb{N}} : g \sim f\}$, and
(3) ${}^*\mathbb{R} = \mathbb{R}^{\mathbb{N}}/\mathcal{F} = \{[f] : f \in \mathbb{R}^{\mathbb{N}}\}$.

It is easy to check that "$\sim$" defined in (1) above is an equivalence relation. Hence $[f]$ is an equivalence class in (2) above and ${}^*\mathbb{R}$ is the set of all equivalence classes. For each $A \subseteq \mathbb{R}$ let ${}^*A = A^{\mathbb{N}}/\mathcal{F} = \{[f] : f \in A^{\mathbb{N}}\}$ where $A^{\mathbb{N}}$ is the set of all functions from some $F_f \in \mathcal{F}$ to A. A set $B \subseteq {}^*\mathbb{R}$ is called internal if B is an ultraproduct of a sequence of sets $B_n \in \mathbb{R}$, i.e., $B = \Pi_{n\in\mathbb{N}} B_n/\mathcal{F} = \{[f] : f \in \mathbb{R}^{\mathbb{N}}$ and $f(n) \in B_n$ for every $n \in \mathbb{N}$ such as $f(n)$ is defined and $B_n \neq \emptyset\}$. For each $a \in \mathbb{R}$ let f_a be the constant function with value a. If we identify each $a \in \mathbb{R}$ with $[f_a] \in {}^*\mathbb{R}$, we can view $\mathbb{R}$ as a subset of ${}^*\mathbb{R}$. We can extend $\leq$, $+$, $\cdot$, etc. from $\mathbb{R}$ to ${}^*\mathbb{R}$.

DEFINITION 2.3. Given any $[f], [g] \in {}^*\mathbb{R}$, let

(1) $[f] \leq [g]$ iff $\{n : f(n) \leq g(n)\} \in \mathcal{F}$,
(2) $[f] + [g] = [f + g]$, and
(3) $[f] \cdot [g] = [f \cdot g]$

where $f + g$ and $f \cdot g$ are ordinary addition and multiplication of two functions. By the same idea we can extend any relation or function on $\mathbb{R}$ to a relation or function on ${}^*\mathbb{R}$. For example, if $F(x_1, x_2, \ldots, x_k)$ is a k–dimensional function from $\mathbb{R}^k$ to $\mathbb{R}$, then we can extend F to a function from ${}^*\mathbb{R}^k$ to ${}^*\mathbb{R}$ by letting $F([g_1], [g_2], \ldots, [g_k])$ be the equivalence class $[\bar{f}]$ of the function $\bar{f}$ where $\bar{f}(n) = F(g_1(n), g_2(n), \ldots, g_k(n))$. Note that if B is the ultraproduct of a sequence of finite sets $B_n \subseteq \mathbb{R}$, then the cardinality of B is defined by $|B| = [f]$ where $f(n) = |B_n|$.

We can also extend the boolean operations among subsets of $\mathbb{R}$ to internal subsets of ${}^*\mathbb{R}$. The following proposition is easy to prove.

PROPOSITION 2.4. Let $A = \Pi_{n\in\mathbb{N}} A_n/\mathcal{F}$ and $B = \Pi_{n\in\mathbb{N}} B_n/\mathcal{F}$ be two internal subsets of ${}^*\mathbb{R}$. Then

(1) $A \subseteq B$ if $\{n \in \mathbb{N} : A_n \subseteq B_n\} \in \mathcal{F}$,
(2) $A \cap B = \Pi_{n\in\mathbb{N}}(A_n \cap B_n)/\mathcal{F}$,
(3) $A \cup B = \Pi_{n\in\mathbb{N}}(A_n \cup B_n)/\mathcal{F}$, and
(4) $A \setminus B = \Pi_{n\in\mathbb{N}}(A_n \setminus B_n)/\mathcal{F}$.

PROPOSITION 2.5. The relation $\leq$ is a linear order on ${}^*\mathbb{R}$.

PROOF. Let $[f], [g] \in {}^*\mathbb{R}$. Then either $\{n \in \mathbb{N} : f(n) \leq g(n)\} \in \mathcal{F}$ or $\{n \in \mathbb{N} : f(n) > g(n)\} \in \mathcal{F}$ by (5) of Definition 2.1. Hence either $[f] \leq [g]$ or $[f] > [g]$ by (1) of Definition 2.3. □

PROPOSITION 2.6. If $[f], [g] \in {}^*\mathbb{Z}$ such that $[f] \leq [g] \leq [f] + k$ for some $k \in \mathbb{N}$ (k is the equivalence class of a constant function with value k), then there is $m \in \mathbb{N}$ such that $m \leq k$ and $[g] = [f] + m$.

PROOF. Let

$$\begin{aligned} X &= \{n \in \mathbb{N} : f(n) \leq g(n) \leq f(n) + k\} \\ &= \{n \in \mathbb{N} : f(n) \leq g(n)\} \cap \{n \in \mathbb{N} : g(n) \leq f(n) + k\} \in \mathcal{F}. \end{aligned}$$

Then $X \in \mathcal{F}$. Since X is the union of $k + 1$ sets $X_i = \{n \in \mathbb{N} : g(n) = f(n) + i\}$ for $i = 0, 1, \ldots, k$, then one of these X_i's must be in $\mathcal{F}$. Let $X_m \in \mathcal{F}$ for some $0 \leq m \leq k$. Thus $[g] = [f] + m$. □

PROPOSITION 2.7. If $A, B \subseteq \mathbb{R}$, then ${}^*(A + B) = {}^*A + {}^*B$.

PROOF. Let $[f] \in {}^*(A + B)$. For each $n \in \mathbb{N}$ choose $a_n \in A$ and $b_n \in B$ such that $a_n + b_n = f(n)$. Define $g_1(n) = a_n$ and $g_2(n) = b_n$. Then $f = g_1 + g_2$. Hence $[f] = [g_1] + [g_2] \in {}^*A + {}^*B$. For each $[g_1] + [g_2] \in {}^*A + {}^*B$ let $f(n) = g_1(n) + g_2(n)$. Then $\{n : f(n) \in A + B\} \in \mathcal{F}$. Hence $[g_1] + [g_2] = [f] \in {}^*(A + B)$. □

From Proposition 2.6 it is not hard to check that $({}^*\mathbb{Z}; \leq, +, \cdot, 0, 1)$ is a discrete ordered ring containing $(\mathbb{Z}; \leq, +, \cdot, 0, 1)$ as a subring. By Proposition 2.6 again we have that if $[f] \in {}^*\mathbb{N}$ and $[f] \neq k$ for every $k \in \mathbb{N}$, then $[f] > k$ for every $k \in \mathbb{N}$. We call $[f] \in {}^*\mathbb{N} \setminus \mathbb{N}$ a hyperfinite integer. For example, if Id is the identity function $Id(n) = n$ for every $n \in \mathbb{N}$, then $[Id]$ is a hyperfinite integer. For each $[f] \in {}^*\mathbb{Z}$, the map $k \mapsto [f] + k$ is an order–isomorphic embedding from $\mathbb{Z}$ onto $[f] + \mathbb{Z} \subseteq {}^*\mathbb{Z}$. Therefore, for a set $A \subseteq \mathbb{N}$ and $[f] \in {}^*\mathbb{N}$ we can define the densities of ${}^*A$ on $[f] + \mathbb{N}$.

2.2. Densities in a (possibly remote) copy of $\mathbb{N}$.

DEFINITION 2.8. Let $A \subseteq \mathbb{N}$ and $[f] \in {}^*\mathbb{N}$. The Shnirel'man density of ${}^*A$ in $[f] + \mathbb{N}$ is defined by

$$\sigma_{[f]}({}^*A) = \inf_{i\in\mathbb{N}, i\geq 1} \frac{{}^*A([f] + 1, [f] + i)}{i}$$

and the lower asymptotic density of ${}^*A$ in $[f] + \mathbb{N}$ is defined by

$$\underline{d}_{[f]}({}^*A) = \liminf_{i\in\mathbb{N}\,\&\, i\to\infty} \frac{{}^*A([f] + 1, [f] + i)}{i}.$$

Notice that $\sigma_{[f]}({}^*A)$ is the Shnirel'man density of the set $\{n \in \mathbb{N} : [f] + n \in {}^*A\}$ and $\underline{d}_{[f]}({}^*A)$ is the lower asymptotic density of the set $\{n \in \mathbb{N} : [f] + n \in {}^*A\}$. Recall that ${}^*A([f], [g]) = \left|\prod_{n\in\mathbb{N}}(A \cap [f(n), g(n)])/\mathcal{F}\right| = [h]$ where $h(n) = A(f(n), g(n))$ for any $[f], [g] \in {}^*\mathbb{N}$. Note that $\sigma_{[f]}({}^*A) = \sigma(A)$ and $\underline{d}_{[f]}({}^*A) = \underline{d}(A)$ if $[f] = 0$. If $0 \leq [g] - [f] \in \mathbb{N}$, then ${}^*A([f], [g])$ is also in $\mathbb{N}$. The next theorem, although straightforward, is the main tool in this section.

THEOREM 2.9. *Let $A \subseteq \mathbb{N}$ and $\alpha \in \mathbb{R}$. Then the following are equivalent.*

(1) $BD(A) \geq \alpha$.

(2) *There is* $[f] \in {}^*\mathbb{N}$ *such that* $\sigma_{[f]}({}^*A) \geq \alpha$.

(3) *There is* $[f] \in {}^*\mathbb{N}$ *such that* $\underline{d}_{[f]}({}^*A) \geq \alpha$.

PROOF. We first prove that (1) implies (2). Let $[a_n, b_n] \subseteq \mathbb{N}$ be such that

$$\lim_{n\to\infty}(b_n - a_n) = \infty \text{ and } \lim_{n\to\infty}\frac{A(a_n, b_n)}{b_n - a_n + 1} = BD(A) \geq \alpha.$$

We intend to find $\{c_k, d_k \in \mathbb{N} : k \in \mathbb{N}\}$ such that

$$\lim_{k\to\infty}(d_k - c_k) = \infty \text{ and } \frac{A(c_k + 1, c_k + i)}{i} > \alpha - \frac{1}{k}$$

for every $i \in [1, d_k - c_k]$, which will imply that $\sigma_{[f]}({}^*A) \geq \alpha$ where f is the function with $f(k) = c_k$.

For positive integer $m, k \in \mathbb{N}$ let

$$l_{m,k} = \max\left\{x \in \mathbb{N} : \exists c \in [a_m, b_m]\, \forall i \in [1, x]\, \left(\frac{A(c+1, c+i)}{i} > \alpha - \frac{1}{k}\right)\right\}.$$

Claim For every positive integer k, the sequence $\{l_{m,k} : m \in \mathbb{N}\}$ is upper unbounded in $\mathbb{N}$.

Proof of Claim Suppose the claim is not true. We derive a contradiction. Let k_0 be a positive integer such that l_{m,k_0}'s are bounded by a positive integer $L \in \mathbb{N}$. Let m be large enough so that

$$\frac{A(a_m, b_m)}{b_m - a_m + 1} > \alpha - \frac{1}{2k_0} \text{ and } \frac{L}{b_m - a_m + 1} < \frac{1}{2k_0}.$$

We define a finite sequence $a_m - 1 = d_0 < d_1 < \cdots < d_t \leq b_m$ such that

$$\frac{A(d_i + 1, d_{i+1})}{d_{i+1} - d_i} \leq \alpha - \frac{1}{k_0} \text{ and } b_m - d_t \leq L.$$

Suppose we have found d_i with $b_m - d_i > L$. Let

$$S = \left\{x \in [d_i + 1, b_m] : \frac{A(d_i + 1, x)}{x - d_i} \leq \alpha - \frac{1}{k_0}\right\}.$$

$S \neq \emptyset$ because $l_{m,k_0} < L$. Let $d_{i+1} = \max S$. By induction we can define d_i's until the last term $d_t > b_m - L$. Since

$$\begin{aligned}\frac{A(a_m, b_m)}{b_m - a_m + 1} &\leq \frac{\sum_{i=0}^{t-1} A(d_i + 1, d_{i+1}) + A(d_t + 1, b_m)}{b_m - a_m + 1}\\ &\leq \left(\alpha - \frac{1}{k_0}\right)\frac{\sum_{i=0}^{t-1}(d_{i+1} - d_i)}{b_m - a_m + 1} + \frac{L}{b_m - a_m + 1}\\ &\leq \alpha - \frac{1}{k_0} + \frac{1}{2k_0} = \alpha - \frac{1}{2k_0},\end{aligned}$$

which contradicts the assumption that $\frac{A(a_m, b_m)}{b_m - a_m + 1} > \alpha - \frac{1}{2k_0}$. This completes the proof of the claim.

By the claim we can choose a positive integer m_k for each positive integer k such that $\lim_{k\to\infty} l_{m_k,k} = \infty$ ($d_k = c_k + l_{m_k,k}$ is the number mentioned in the beginning of this proof). Let $f(k) = c_k$ be such that

$$\frac{A(c_k+1, c_k+i)}{i} > \alpha - \frac{1}{k}$$

for every $i \in [1, l_{m_k,k}]$. We need to show that $\sigma_{[f]}({}^*A) \geq \alpha$. Given $i \in \mathbb{N}$ with $i \geq 1$. Since $A(f(k)+1, f(k)+i)$ for all $k \in \mathbb{N}$ has at most $i+1$ possible values and

$$\frac{A(f(k)+1, f(k)+i)}{i} > \alpha - \frac{1}{k}$$

whenever k is large enough such that $i < l_{m_k,k}$, then there is $K_i \in \mathbb{N}$ such that

$$\frac{A(f(k)+1, f(k)+i)}{i} \geq \alpha$$

for every $k > K_i$. Hence

$$\left\{k \in \mathbb{N} : \frac{A(f(k)+1, f(k)+i)}{i} \geq \alpha\right\} \in \mathcal{F}.$$

This shows that

$$\frac{{}^*A([f]+1, [f]+i)}{i} \geq \alpha$$

is true for any positive integer $i \in \mathbb{N}$. Therefore, $\sigma_{[f]}({}^*A) \geq \alpha$.

It is trivial that (2) implies (3).

We now show that (3) implies (1). To prove $BD(A) \geq \alpha$ it suffices to show that for any positive integer $k \in \mathbb{N}$ there exists an interval $[a,b] \subseteq \mathbb{N}$ such that $b - a > k$ and

$$\frac{A(a,b)}{b-a+1} > \alpha - \frac{1}{k}.$$

Fix a positive integer k. Since $\underline{d}_{[f]}({}^*A) \geq \alpha$, then there is a positive integer $m > k$ such that

$$\frac{{}^*A([f]+1, [f]+m)}{m} > \alpha - \frac{1}{k}.$$

This implies that

$$S = \left\{n \in \mathbb{N} : \frac{A(f(n)+1, f(n)+m)}{m} > \alpha - \frac{1}{k}\right\} \in \mathcal{F}.$$

In particular, $S \neq \emptyset$. Let $n \in S$, $a = f(n)+1$, and $b = f(n)+m$. The interval $[a,b]$ is what we are looking for. This completes the proof of the theorem □

Theorem 2.9 is a bridge connecting upper Banach density with Shnirel'man density and lower asymptotic density through ultrapower methods. By this connection we have found many theorems about upper Banach density, each of which is parallel to an existing theorem about Shnirel'man density or lower asymptotic density (cf. [**Ji2, Ji3**]). Next we derive two theorems about upper Banach density to demonstrate the idea.

2.3. When Kneser and Plünnecke meet Banach. The first theorem appeared in [**Ji2**] and is parallel to Plünnecke's Theorem.

DEFINITION 2.10. Let $B \subseteq \mathbb{N}$. B is called a piecewise basis of order h if there exists a sequence $\{c_k : k \in \mathbb{N}\}$ of positive integers such that

$$h\cdot(B \cap [c_k, c_k + k]) \supseteq [hc_k, hc_k + k].$$

Note that if B is a basis of order h, then B is a piecewise basis of order at most h because we can choose $c_k = 0$. Note also that $h\cdot(B \cap [c_k, c_k + k]) \supseteq [hc_k, hc_k + k]$ is equivalent to

$$h\cdot((B \cap [c_k, c_k + k]) - c_k) \supseteq [0, k].$$

THEOREM 2.11. *Let $A, B \subseteq \mathbb{N}$. If B is a piecewise basis of order h, then*

$$BD(A + B) \geq BD(A)^{1-\frac{1}{h}}.$$

PROOF. Let $\{c_k : k \in \mathbb{N}\}$ be the sequence associated with B in Definition 2.10. Let $g(k) = c_k$ and Id be the identity function on $\mathbb{N}$. Then we have that

$$h\cdot({}^*B \cap [[g], [g] + [Id]]) \supseteq [h[g], h[g] + [Id]].$$

In particular, we have $h\cdot(({}^*B \cap ([g] + \mathbb{N})) - [g]) \supseteq \mathbb{N}$. This shows that the set $({}^*B \cap ([g] + \mathbb{N})) - [g]$ is a basis of order h.

Let $BD(A) = \alpha$. By Theorem 2.9 there is $[f] \in {}^*\mathbb{N}$ such that $\sigma_{[f]}({}^*A) = \alpha$. This is equivalent to the condition that $\sigma(({}^*A \cap ([f] + \mathbb{N})) - [f]) = \alpha$.

By Plünnecke's Theorem we have that

$$\begin{aligned}\sigma_{[f]+[g]}({}^*(A + B)) &\geq \sigma((({}^*A + {}^*B) \cap ([f] + [g] + \mathbb{N})) - ([f] + [g]))\\ &\geq \sigma((({}^*A \cap ([f] + \mathbb{N})) - [f]) + (({}^*B \cap ([g] + \mathbb{N})) - [g])) \geq \alpha^{1-\frac{1}{h}}.\end{aligned}$$

By Theorem 2.9 again we have $BD(A+B) \geq \alpha^{1-\frac{1}{h}}$. This completes the proof. □

The second theorem appeared in [**Ji3**] and is parallel to Kneser's Theorem.

THEOREM 2.12. *Let $A, B \subseteq \mathbb{N}$. If $BD(A + B) < BD(A) + BD(B)$, then there exist a positive integer d, a set $G \subseteq [0, d-1]$, and a sequence $\{[a_k, b_k] \subseteq \mathbb{N} : k \in \mathbb{N}\}$ of intervals such that*

(1) $BD(A + B) \geq \frac{|G|}{d} \geq BD(A) + BD(B) - \frac{1}{d}$,
(2) $\lim_{k\to\infty}(b_k - a_k) = \infty$, *and*
(3) $(A + B) \cap [a_k, b_k] \supseteq (a_k + G + \{dn : n \in \mathbb{N}\}) \cap [a_k, b_k]$.

PROOF. Let $BD(A) = \alpha$ and $BD(B) = \beta$. By Theorem 2.9 there are $[f], [g] \in {}^*\mathbb{N}$ such that $\underline{d}_{[f]}({}^*A) = \alpha$ and $\underline{d}_{[g]}({}^*B) = \beta$. By Theorem 2.9 again, if $\underline{d}_{[f]+[g]}({}^*A + {}^*B) \geq \alpha + \beta$, then $BD(A + B) \geq \alpha + \beta$, contradicting the assumption, so we have that $\underline{d}_{[f]+[g]}({}^*A + {}^*B) < \alpha + \beta$. By Kneser's Theorem we can find a positive integer d and a set $G \subseteq [0, d-1]$ such that

$$\begin{aligned}\underline{d}_{[f]+[g]}({}^*(A + B)) &= \underline{d}_{[f]+[g]}({}^*A + {}^*B)\\ &\geq \underline{d}((({}^*A - [f]) \cap \mathbb{N}) + (({}^*B - [g]) \cap \mathbb{N}))\\ &= \frac{|G|}{d} \geq \alpha + \beta - \frac{1}{d},\end{aligned}$$

which implies $BD(A+B) \geq \frac{|G|}{d} \geq \alpha+\beta-\frac{1}{d}$, and

$$\begin{aligned}({}^*A+{}^*B)\cap([f]+[g]+\mathbb{N}) &\supseteq ({}^*A\cap([f]+\mathbb{N}))+({}^*B\cap([g]+\mathbb{N}))\\ &\supseteq (({}^*A\cap([f]+\mathbb{N}))+({}^*B\cap([g]+\mathbb{N})))\cap([f]+[g]+m+\mathbb{N})\\ &= ([f]+[g]+G+\{dn : n\in\mathbb{N}\})\cap([f]+[g]+m+\mathbb{N})\end{aligned}$$

for some $m\in\mathbb{N}$. Let $k\in\mathbb{N}$. Since

$$\begin{aligned}&({}^*A+{}^*B)\cap[[f]+[g]+m,[f]+[g]+m+k]\\ &\quad\supseteq ([f]+[g]+G+\{dn : n\in\mathbb{N}\})\cap[[f]+[g]+m,[f]+[g]+m+k],\end{aligned}$$

there exists $n\in\mathbb{N}$ such that

$$\begin{aligned}&(A+B)\cap[f(n)+g(n)+m, f(n)+g(n)+m+k]\\ &\quad\supseteq (f(n)+g(n)+G+\{dn : n\in\mathbb{N}\})\cap[f(n)+g(n)+m, f(n)+g(n)+m+k].\end{aligned}$$

Let $a_k = f(n)+g(n)+m$ and $b_k = f(n)+g(n)+m+k$. Clearly, the sequence $\{[a_k,b_k] : k\in\mathbb{N}\}$ is the sequence we desired. This completes the proof. □

3. Level Two Applications

In §2 we developed a general way of deriving a theorem about upper Banach density parallel to each existing theorem about Shnirel'man density or lower asymptotic density via Theorem 2.9. However, a simple application of Theorem 2.9 often results in a theorem, which in some sense is not optimal. For example in Theorem 2.12 the structure of $A+B$ is characterized only on a small portion of $\mathbb{N}$. The characterization of the structure of $A+B$ can be made on a much larger set. In Theorem 2.11 the definition of a piecewise basis seems artificial. In this section we will discuss whether Theorem 1.3 can be generalized to lower asymptotic density, upper asymptotic density, and upper Banach density. The theorem about upper Banach density is in fact a significant improvement of Theorem 2.11. The proofs of the results in this section can be found in [**Ji7**] and [**Ji8**].

Although ultrapower methods introduced in §2 works fine, it is more convenient to work under the full strength of nonstandard analysis. In ultrapower methods, one might view $[f]$ not as a number but as a function (more precisely, as an equivalence class containing f). But from a model theoretic point of view $[f]$ is just a single point in the extension ${}^*\mathbb{R}$ of $\mathbb{R}$. People probably do not consider a real number as a sequence of rational numbers when working on real analysis problems although the Cauchy definition of a real is an equivalence class of a Cauchy sequence of rational numbers. If the reader is interested in mathematical logic, the Łoś Theorem should also be a great help.

For constructing a nonstandard universe we first take $\mathbb{R}$ as a set of atoms. Then let $V_0=\mathbb{R}$, $V_{n+1}=V_n\cup\mathcal{P}(V_n)$, and $V=\bigcup_{n=0}^{\bar{N}} V_n$, where $\mathcal{P}$ is the power set operator, for some sufficiently large positive integer $\bar{N}$. We call $(V,\in)$ the standard universe. The standard universe is large enough to contain every possible mathematical object involved in a standard mathematical argument. For example $\leq$ on $\mathbb{R}$ is a set of ordered pairs. Hence $\leq$ is an element in V_3. The nonstandard universe ${}^*V$ is the ultrapower of V modulo $\mathcal{F}$. For each $[f],[g]\in{}^*V$ define $[f]\in[g]$ if the set $\{n\in\mathbb{N} : f(n)\in g(n)\}$ is in $\mathcal{F}$. Let $i : V\mapsto{}^*V$ be such that $i(a)=[f_a]$ where f_a is the constant function on $\mathbb{N}$ with value a. Łos' Theorem says that for any first–order formula $\varphi(x_1,x_2,\ldots,x_k)$ in the language of one binary relation $\in$

and any $[f^{(1)}], [f^{(2)}], \ldots, [f^{(k)}] \in {}^*V$, $\varphi([f^{(1)}], [f^{(2)}], \ldots, [f^{(k)}])$ is true in ${}^*V$ if and only if

$$\{n \in \mathbb{N} : \varphi(f^{(1)}(n), f^{(2)}(n), \ldots, f^{(k)}(n)) \text{ is true in } V\} \in \mathcal{F}.$$

Łos' Theorem implies the famous transfer principle, which says that for any first–order formula $\varphi(x_1, x_2, \ldots, x_k)$ in the language of one binary relation $\in$ and for any $a_1, a_2, \ldots, a_n \in V$ the sentence $\varphi(a_1, a_2, \ldots, a_k)$ is true in $(V, \in)$ if and only if $\varphi([f_{a_1}], [f_{a_2}], \ldots, [f_{a_k}])$ is true in $({}^*V, \in)$.

The proofs in [**Ji7**] and [**Ji8**] heavily use nonstandard analysis techniques. We do not intend to include those proofs here. Instead we will explain the general ideas of the proofs.

3.1. Kneser meets Banach again. The following theorem improves Theorem 2.12. The proof of the following theorem can be found in [**Ji7**].

THEOREM 3.1. *Let $A, B \subseteq \mathbb{N}$ be such that $BD(A) = \alpha$, $BD(B) = \beta$, and $BD(A+B) < \alpha + \beta$. Then there are positive $g \in \mathbb{N}$ and $G \subseteq [0, g-1]$ such that*

(1) $BD(A+B) \geq \alpha + \beta - \frac{1}{g}$,

(2) $A + B \subseteq G + g\mathbb{N}$,

(3) *if* $\left\{\left[a_n^{(i)}, b_n^{(i)}\right] : n \in \mathbb{N}\right\}$ *for $i = 1, 2$ are two sequences of intervals such that*

$$\lim_{n\to\infty} \left(b_n^{(i)} - a_n^{(i)}\right) = \infty,$$

$$\lim_{n\to\infty} \frac{A\left(a_n^{(1)}, b_n^{(1)}\right)}{b_n^{(1)} - a_n^{(1)} + 1} = \alpha, \ \lim_{n\to\infty} \frac{B\left(a_n^{(2)}, b_n^{(2)}\right)}{b_n^{(2)} - a_n^{(2)} + 1} = \beta,$$

and

$$0 < \liminf_{n\to\infty} \frac{b_n^{(1)} - a_n^{(1)}}{b_n^{(2)} - a_n^{(2)}} \leq \limsup_{n\to\infty} \frac{b_n^{(1)} - a_n^{(1)}}{b_n^{(2)} - a_n^{(2)}} < \infty,$$

then there exist $\left[c_n^{(i)}, d_n^{(i)}\right] \subseteq \left[a_n^{(i)}, b_n^{(i)}\right]$ *for each $n \in \mathbb{N}$ and $i = 1, 2$ such that*

$$\lim_{n\to\infty} \frac{d_n^{(i)} - c^{(i)}}{b_n^{(i)} - a_n^{(i)}} = 1$$

and

$$(A+B) \cap \left[c_n^{(1)} + c_n^{(2)}, d_n^{(1)} + d_n^{(2)}\right] = (G + g\mathbb{N}) \cap \left[c_n^{(1)} + c_n^{(2)}, d_n^{(1)} + d_n^{(2)}\right].$$

REMARK 3.2. (1) The condition

$$0 < \liminf_{n\to\infty} \frac{b_n^{(1)} - a_n^{(1)}}{b_n^{(2)} - a_n^{(2)}} \leq \limsup_{n\to\infty} \frac{b_n^{(1)} - a_n^{(1)}}{b_n^{(2)} - a_n^{(2)}} < \infty,$$

in (3) of Theorem 3.1 is necessary because otherwise one can choose

$$A = \bigcup_{n=1}^{\infty} \left(\left[2^{(2n)^2}, 1.5 \times 2^{(2n)^2} - 2^{(2n-1)^2+1}\right] \right.$$
$$\left. \cup \left[1.5 \times 2^{(2n)^2} + 2^{(2n-1)^2+1}, 2 \times 2^{(2n)^2}\right]\right),$$

$$B = \bigcup_{n=1}^{\infty} \left(\left[2^{(2n+1)^2}, 1.5 \times 2^{(2n+1)^2} - 2^{(2n)^2+1} \right] \right.$$
$$\left. \cup \left[1.5 \times 2^{(2n+1)^2} + 2^{(2n)^2+1}, 2 \times 2^{(2n+1)^2} \right] \right).$$

Then $BD(A) = \alpha = BD(B) = \beta = 1$, which trivially implies $BD(A+B) < BD(A) + BD(B)$. On the other hand, let $a_n^{(1)} = 2^{(2n)^2}$, $b_n^{(1)} = 2 \times 2^{(2n)^2}$, $a_n^{(2)} = 2^{(2n+1)^2}$, and $b_n^{(2)} = 2 \times 2^{(2n+1)^2}$. Then all conditions of Theorem 3.1 except the one mentioned above are true. However, the structure described in the last line of (3) in Theorem 3.1 cannot be true because $(A+B) \cap \left[a_n^{(1)} + a_n^{(2)}, b_n^{(1)} + b_n^{(2)}\right]$ has large gaps in the middle of the interval.

(2) Let A achieve its upper Banach density on a sequence of intervals $\left[a_n^{(1)}, b_n^{(1)}\right]$ and let B achieve its upper Banach density on a sequence of intervals $\left[a_n^{(2)}, b_n^{(2)}\right]$. We probably shouldn't hope to characterize the structure of $A + B$ outside of the intervals $\left[a_n^{(1)} + a_n^{(2)}, b_n^{(1)} + b_n^{(2)}\right]$ because the upper Banach densities of A and B would not change if we delete all elements of A outside the union of the intervals $\left[a_n^{(1)}, b_n^{(1)}\right]$ and delete all elements of B outside the union of the intervals $\left[a_n^{(2)}, b_n^{(2)}\right]$. On the other hand, we cannot hope to replace $c^{(i)}, d^{(i)}$ by $a^{(i)}, b^{(i)}$ in the last line of Theorem 3.1 for the same reason as in the Kneser's Theorem where the structure of $A+B$ is characterized not in $\mathbb{N}$ but in $\mathbb{N} \setminus [0, m]$ for some $m \in \mathbb{N}$. This is why in Theorem 3.1 the structure of $A+B$ is characterized on $\left[c_n^{(1)} + c_n^{(2)}, d_n^{(1)} + d_n^{(2)}\right]$ instead.

(3) The proof of Theorem 3.1 is much more complicated than the proof of Theorem 2.12. In order to prove Theorem 3.1 one should improve Theorem 2.9 first. In fact $BD(A) \geq \alpha$ implies $\underline{d}_{[f]}({}^*A) \geq \alpha$ for many f's.

Let $[f] < [g]$ be in ${}^*\mathbb{Z}$ such that $[g] - [f]$ is a hyperfinite integer. For each internal subset C of $[[f], [g]]$ the cardinality of C is an element in ${}^*\mathbb{N}$. Define $\mu(C) = \frac{|C|}{[g]-[f]+1}$. Then $0 \leq \mu(C) \leq 1$ and $\mu(C) \in {}^*\mathbb{R}$. Note that for every $r \in {}^*\mathbb{R}$, $0 \leq r \leq 1$ the set S_r of all standard reals $s \in \mathbb{R}$ with $s < r$ has the least upper bound β in $\mathbb{R}$. It is not hard to see that $|r - \beta| < \frac{1}{n}$ for every $n \in \mathbb{N}$. We say that r and β are infinitesimally close. Note also that such β is unique. We call β the standard part of r and write $st(r) = \beta$. Hence $st \circ \mu$ maps every internal subset $C \subseteq [[f], [g]]$ to a standard real number between 0 and 1. In fact $st \circ \mu$ is a finitely additive probability measure on the algebra of all internal subsets of $[[f], [g]]$. For any $X \subseteq [[f], [g]]$ we can use $st \circ \mu$ to define lower measure and upper measure of X and call X measurable if the lower measure and upper measure of X coincide. By measure–completion process $st \circ \mu$ can be extended to a countably additive, complete, atom-less probability measure μ_L on the σ–algebra of all measurable subsets of $[[f], [g]]$. This probability space is called Loeb space. With the idea of Loeb space together with Birkhoff Ergodic Theorem we can improve Theorem 2.9 in the following theorem (cf. [**Ji2**]).

THEOREM 3.3. *Let $A \subseteq \mathbb{N}$ be such that $BD(A) = \alpha > 0$. Suppose $[a_n, b_n] \subseteq \mathbb{N}$ such that*

$$\lim_{n\to\infty}(b_n - a_n) = \infty \text{ and } \lim_{n\to\infty}\frac{A(a_n, b_n)}{b_n - a_n + 1} = \alpha.$$

Let $[f]$ be a hyperfinite integer. Then $[a_{[f]}, b_{[f]}]$ is an interval of hyperfinite length and for μ_L-almost all $x \in [a_{[f]}, b_{[f]}]$ we have $\underline{d}_x({}^*A) = \alpha$ where μ_L is the Loeb probability measure on $[a_{[f]}, b_{[f]}]$. Note that $a_{[f]}$ is the equivalence class $[g]$ where $g(n) = a_{f(n)}$.

By combining Theorem 3.3 and Kneser's Theorem we can pin down the structure of $A + B$ in the interval $\left[a_n^{(1)} + a_n^{(2)}, b_n^{(1)} + b_n^{(2)}\right]$ for each hyperfinite integer n. It takes a small trick to show that the structures of $A + B$ in the intervals $\left[a_n^{(1)} + a_n^{(2)}, b_n^{(1)} + b_n^{(2)}\right]$ for all hyperfinite integers n are actually coherent. Hence we can now characterize the structure of $A+B$ in the union of all these hyperfinite intervals. Finally, by the transfer principle we pull down the nonstandard result to the standard world and obtain Theorem 3.1.

3.2. Plünnecke meets Banach again. It is not hard to show that if B is a basis of order h, then $\underline{d}(A+B) \geq \underline{d}(A)^{1-\frac{1}{h}}$. However, $\underline{d}(A+B) \geq \underline{d}(A)^{1-\frac{1}{h}}$ for B being an asymptotic basis of order h, although true, is not a direct consequence of Theorem 1.3. It is in fact a corollary of Theorem 3.4 which we will present below.

Recall that in Theorem 1.3 we cannot replace σ by $\underline{d}$ and replace a basis B of order h by an asymptotic basis of average asymptotic order h^*. What kind of generalization of Theorem 1.3 to lower asymptotic density can we have?

Let $B \subseteq \mathbb{N}$. The set B is called a *lower asymptotic basis* of order h if

$$\underline{d}(hB) = 1,$$

the set B is called an *upper asymptotic basis* of order h if

$$\overline{d}(hB) = 1,$$

and the set B is called an *upper Banach basis* of order h if

$$BD(hB) = 1.$$

Recall that if $h > 1$, then B is a basis of order h iff $\sigma(hB) = 1$. Hence the style of our definition of the three asymptotic bases above is consistent with that of B being a basis. Note that if B is an asymptotic basis of asymptotic order h_0 and of average asymptotic order h_1, then B is a lower asymptotic basis of order h_2 with $h_1 \leq h_2 \leq h_0$. Note that if B is a piecewise basis of piecewise order h_0, then B is an upper Banach basis of order $h_1 \leq h_0$.

Let P again be the set of all prime numbers. P is not a basis because P does not contain 0 and 1. If $A = P \cup \{0, 1\}$, then A is a basis of order h for some $h \in \mathbb{N}$. However, the order h may be larger than 4 (h is at most 7 by a result of Olivier Ramaré). By Vinogradov's Three-Prime Theorem P is an asymptotic basis of asymptotic order order 4. By a result in [**Es**] P is a lower asymptotic basis of order 3.[2] Of course, P would be an asymptotic basis of asymptotic order 3 if the famous Goldbach Conjecture for the sum of two prime numbers has a positive answer. This makes Theorem 3.4 below interesting.

---

[2]This result was discovered independently by Nikolai Chudakov, Johannes van der Corput, and Theodor Estermann at about the same time.

The three theorems below are the results of effort for generalizing Plünnecke's Theorem to the three asymptotic densities.

THEOREM 3.4. *Let $A, B \subseteq \mathbb{N}$ and B be a lower asymptotic basis of order h. Then*

$$\underline{d}(A+B) \geq \underline{d}(A)^{1-\frac{1}{h}}.$$

THEOREM 3.5. *There are $A, B \subseteq \mathbb{N}$ with $\overline{d}(A) = \frac{1}{2}$ and B an upper asymptotic basis of order 2 such that*

$$\overline{d}(A+B) = \overline{d}(A).$$

THEOREM 3.6. *Let $A, B \subseteq \mathbb{N}$ and B be a upper Banach basis of order h. Then*

$$BD(A+B) \geq BD(A)^{1-\frac{1}{h}}.$$

As a corollary of Theorem 3.4, we have that for any set $A \subseteq \mathbb{N}$, $\underline{d}(A+P) \geq \underline{d}(A)^{2/3}$ where P is the set of all prime numbers. It is interesting to see whether the lower bound $\underline{d}(A)^{2/3}$ of $\underline{d}(A+P)$ in this corollary can be improved.

It is not a surprise that the behavior of upper asymptotic density is different from the behavior of lower asymptotic density or the behavior of upper Banach density. We have discovered many instances of that phenomenon.

Theorem 3.6 is a significant improvement of Theorem 2.11 because a piecewise basis of piecewise order h is clearly an upper Banach basis of order at most h and the definition of upper Banach basis seems more natural than the definition of piecewise basis.

The proof of Theorem 3.5 does not involve nonstandard methods. The upper asymptotic basis B for Theorem 3.5 constructed in [**Ji7**] is a modification of the thin basis constructed by Cassels (cf. [**HR**, Theorem 12 on page 39]).

The reader can see that the proof of Theorem 2.11 does not get into Plünnecke's original idea of Plünnecke's graph. Plünnecke's graph was used to obtain a powerful inequality and that inequality leads to Theorem 1.3. The proof of Theorem 3.4 and Theorem 3.6 combines the strength of Plünnecke's inequality and nonstandard methods. In the proof of Theorem 3.4, we apply Plünnecke's inequality to the segments of $^*A$ in a hyperfinite interval $[0, n]$. Since the set $A_n = {}^*A \cap [0, n]$ is hyperfinite, it is easy for us to make small adjustments in order to fit the condition required by Plünnecke's inequality. Because of this, a would–be long ϵ–δ argument becomes a very straightforward argument.

The proof of Theorem 3.6 requires the use of Theorem 3.3. Let $BD(A) = \alpha$. Theorem 3.3 shows that there are arbitrarily long intervals $[a, b]$ such that the set $^*A \cap [a, b]$ for each such interval $[a, b]$ is homogeneously distributed. Now Theorem 3.6 follows from this homogeneity and Plünnecke's inequality.

References

[BJ] P. Bihani and R. Jin, *Kneser's theorem for upper Banach density*, Journal de théorie des nombres de Bordeaux, vol 18 (2006), no 2, pp. 323–343.

[Es] T. Estermann, *On Goldbach's problem: proof that almost all even positive integers are sums of two primes*, Proceedings of London Mathematical Society, 44 (1938), pp. 307–314.

[Fu] H. Furstenberg, *Recurrence in Ergodic Theory and Combinatorial Number Theory*, Princeton University Press, 1981.

[HR] H. Halberstam and K. F. Roth, *Sequences*, Oxford University Press, 1966

[Ji1] R. Jin, *Sumset phenomenon*, Proceedings of American Mathematical Society, Vol. 130, No. 3 (2002), pp. 855–861.

[Ji2] R. Jin, *Nonstandard methods for upper Banach density problems*, Journal of Number Theory, 91 (2001), pp. 20–38.

[Ji3] R. Jin, *Standardizing nonstandard methods for upper Banach density problems*, in the DIMACS series *Unusual Applications of Number Theory*, edited by M. Nathanson, Vol. 64 (2004)

[Ji4] R. Jin, *Inverse problem for upper asymptotic density*, Transactions of American Mathematical Society, 355 (2003), No. 1, pp. 57–78.

[Ji5] R. Jin, *Solution to the inverse problem for upper asymptotic density*, Journal für die reine und angewandte Mathematik (Crelle's Journal), 595 (2006), pp. 121–166.

[Ji6] R. Jin, *Inverse problem for cuts*, Logic and Analysis, 1, No. 1 (2007), pp. 61–89.

[Ji7] R. Jin, *Pinning down the structure of $A+B$ when $A+B$ has small upper Banach density*, http://jinr.people.cofc.edu/research/publication.html

[Ji8] R. Jin, *Plünnecke's Theorem for asymptotic densities*, http://jinr.people.cofc.edu/research/publication.html

[JK] R. Jin and H. J. Keisler, *Abelian group with layered tiles and the sumset phenomenon*, Transactions of American Mathematical Society, Vol. 355 (2003), No. 1, pp. 79–97.

[Kh] A. I. Khinchin, *Three Pearls of Number Theory*, Translated from the 2d (1948) rev. Russian ed. by F. Bagemihl, H. Komm, and W. Seidel, Rochester, N.Y., Graylock Press, 1952.

[Na1] M. B. Nathanson, *Additive Number Theory–The Classical Bases*, Springer, 1996.

[Na2] M. B. Nathanson, *Additive Number Theory–Inverse Problems and the Geometry of Sumsets*, Springer, 1996.

[Pl] H. Plünnecke, *Eine zahlentheoretische Anwendung der Graphentheorie*, Journal für die reine und angewandte Mathematik, **234** (1970), 171–183.

Department of Mathematics, College of Charleston, Charleston, South Carolina, 29424

E-mail address: jinr@cofc.edu

Contemporary Mathematics
Volume **530**, 2010

The Ultraproduct Construction

H. Jerome Keisler

ABSTRACT. This is a brief survey of the ultraproduct construction, which is meant to provide background material for the readers of this volume.

1. Introduction

The ultraproduct construction is a uniform method of building models of first order theories which has applications in many areas of mathematics. It is attractive because it is algebraic in nature, but preserves all properties expressible in first order logic. The idea goes back to the construction of nonstandard models of arithmetic by Skolem [**51**] in 1934. In 1948, Hewitt [**16**] studied ultraproducts of fields. For first order structures in general, the ultraproduct construction was defined by Łoś [**37**] in 1955. The subject developed rapidly beginning in 1958 with a series of abstracts by Frayne, Morel, Scott, and Tarski, which led to the 1962 paper [**14**]. Other early papers are [**31**] by Kochen, and [**18**] by the author. The groundwork for the application of ultraproducts to mathematics was laid in the late 1950's through the 1960's. The purpose of this article is to give a survey of the classical results on ultraproducts of first order structures in order to provide some background for the papers in this volume. Over the years, many generalizations of the ultraproduct construction, as well as applications of ultraproducts to non-first order structures, have appeared in the literature. To keep this paper of reasonable length, we will not include such generalizations in this survey. For earlier surveys of ultraproducts see [**7**], [**12**], [**24**]. For much more about ultraproducts see the books [**9**], [**10**], [**49**], and [**54**].

We assume familiarity with a few basic concepts from model theory. For the convenience of the reader we give a crash course here. The cardinality of a set X is denoted by $|X|$. The cardinality of $\mathbb{N}$ is denoted by ω. The set of all subsets of a set I is denoted by $\mathcal{P}(I)$, and the set of finite subsets of I by $\mathcal{P}_\omega(I)$. Given mappings $f : X \to Y$ and $g : Y \to Z$, the composition $g \circ f : X \to Z$ is the mapping $x \mapsto g(f(x))$. A **first order vocabulary** L consists of a set of finitary relation symbols, function symbols, and constant symbols. We use $\mathcal{A}, \mathcal{B}, \ldots$ to denote L-structures with universe sets $A, B, \ldots$. By the cardinality of $\mathcal{A}$ we mean the cardinality of its universe set A. The notation $\mathcal{A} \models \varphi(a_1, \ldots, a_n)$ means that the formula $\varphi(x_1, \ldots, x_n)$ is true in $\mathcal{A}$ when each x_i is interpreted by the corresponding

2010 *Mathematics Subject Classification.* Primary 03C20, Secondary 03H05, 54D80.
Key words and phrases. Ultraproducts.

a_i. The notation $h : \mathcal{A} \to \mathcal{B}$ means that h is a homomorphism of $\mathcal{A}$ into $\mathcal{B}$, that is, h maps A into B and each atomic formula which is true for a tuple in $\mathcal{A}$ is true for the h-image of the tuple in $\mathcal{B}$. The notation $h : \mathcal{A} \subseteq \mathcal{B}$ means that h is an (isomorphic) embedding of $\mathcal{A}$ into $\mathcal{B}$, that is, h maps A into B and each quantifier-free formula of L which is true for a tuple in $\mathcal{A}$ is true for the h-image of the tuple in $\mathcal{B}$. $h : \mathcal{A} \cong \mathcal{B}$ means that h is an isomorphism from $\mathcal{A}$ onto $\mathcal{B}$, and $\mathcal{A} \cong \mathcal{B}$ means that $\mathcal{A}$ and $\mathcal{B}$ are isomorphic. The set of all sentences true in $\mathcal{A}$ is called the **complete theory** of $\mathcal{A}$. $\mathcal{A}$ and $\mathcal{B}$ are called **elementarily equivalent**, in symbols $\mathcal{A} \equiv \mathcal{B}$, if they have the same complete theory. The notation $h : \mathcal{A} \prec \mathcal{B}$ means that h is an **elementary embedding** from $\mathcal{A}$ into $\mathcal{B}$, that is, h maps A into B and each formula of L which is true for a tuple in $\mathcal{A}$ is true for the h-image of the tuple in $\mathcal{B}$. Clearly, $h : \mathcal{A} \prec \mathcal{B}$ implies that $\mathcal{A} \equiv \mathcal{B}$. We say that $\mathcal{B}$ is an **elementary extension** of $\mathcal{A}$ and write $\mathcal{A} \prec \mathcal{B}$ if $A \subseteq B$ and the identity map is an elementary embedding of $\mathcal{A}$ into $\mathcal{B}$. It is easy to see that if $h : \mathcal{A} \prec \mathcal{B}$, then $\mathcal{B}$ is isomorphic to some elementary extension of $\mathcal{A}$.

A fundamental result that is used very often in model theory is the **compactness theorem**, which says that if every finite subset of a set T of sentences has a model, then T has a model. One application of compactness is the construction of extremely rich models called saturated models. An L-structure $\mathcal{A}$ is said to be κ-**saturated** if every set of first order formulas with fewer than κ parameters from A which is finitely satisfied in $\mathcal{A}$ is satisfied in $\mathcal{A}$. $\mathcal{A}$ is **saturated** if it is $|A|$-saturated. Morley and Vaught [**39**] proved that any two elementarily equivalent saturated structures of the same cardinality are isomorphic, that each infinite structure $\mathcal{A}$ has a saturated elementary extension in each inaccessible cardinal $\kappa \geq |A| + |L|$, and has a κ^+-saturated elementary extension of cardinality 2^κ whenever $2^\kappa \geq |A|$ and $\kappa \geq |L|$.

Given two vocabularies $L_1 \subseteq L_2$, the **reduct** of an L_2-structure $\mathcal{A}_2$ to L_1 is the L_1-structure $\mathcal{A}_1$ obtained by forgetting the interpretation of each symbol of $L_2 \setminus L_1$. An **expansion** of an L_1-structure $\mathcal{A}_1$ to L_2 is an L_2-structure formed by adding interpretations of the symbols of $L_2 \setminus L_1$, that is, an L_2-structure whose reduct to L_1 is $\mathcal{A}_1$.

2. Ultraproducts and ultrapowers

We begin with the definition of an ultrafilter over an index set I. An ultrafilter over I can be defined as the collection of all sets of measure 1 with respect to a finitely additive measure $\mu : \mathcal{P}(I) \to \{0,1\}$. Here is an equivalent definition in more primitive terms.

DEFINITION 2.1. Let I be a non-empty set. A **proper filter** U over I is a set of subsets of I such that:

(i) U is closed under supersets; if $X \in U$ and $X \subseteq Y \subseteq I$ then $Y \in U$.
(ii) U is closed under finite intersections; if $X \in U$ and $Y \in U$ then $X \cap Y \in U$.
(iii) $I \in U$ but $\emptyset \notin U$.

An **ultrafilter** over I is a proper filter U over I such that:
(iv) For each $X \subseteq I$, exactly one of the sets $X, I \setminus X$ belongs to U.

THEOREM 2.2. *(Tarski [**53**]) Every proper filter over a set I can be extended to an ultrafilter over I.*

For an infinite set I, an important example of a proper filter over I is the **Fréchet filter**, which is the set of all cofinite (complements of finite) subsets of I. An ultrafilter that contains the Fréchet filter is called a **free ultrafilter**. By Theorem 2.2, the Fréchet filter can be extended to an ultrafilter over I, so free ultrafilters over I exist.

The only ultrafilters over I which are not free are the **principal ultrafilters**, which are of the form $U = \{X \subseteq I : i_0 \in X\}$ for some $i_0 \in I$. For a set I of finite cardinality n, every ultrafilter over I is principal, and thus there are only n ultrafilters over I.

The following result of Pospíšil [**41**] shows that there are as many ultrafilters over an infinite set I as there are sets of subsets of I.

THEOREM 2.3. *For each set I of infinite cardinality κ, the set of ultrafilters over I has cardinality 2^{2^κ}.*

We now define the ultraproduct operation on sets. Let U be an ultrafilter over I, and for each $i \in I$ let A_i be a nonempty set. The ultraproduct $\prod_U A_i$ is obtained by first taking the cartesian product $\Pi_{i\in I} A_i$ and then identifying two elements which are equal for U-almost all $i \in I$. Here is the formal definition.

DEFINITION 2.4. Let U be an ultrafilter over I. Two elements f, g of the cartesian product $\prod_{i\in I} A_i$ are said to be **U-equivalent**, in symbols $f =_U g$, if the set $\{i : f(i) = g(i)\}$ belongs to U. The U-equivalence class of f is the set $f_U = \{g : f =_U g\}$. The **ultraproduct** $\prod_U A_i$ is defined as the set of U-equivalence classes

$$\prod_U A_i = \{f_U : f \in \prod_{i\in I} A_i\}.$$

In the above definition, it is easily checked that $=_U$ is an equivalence relation on $\prod_{i\in I} A_i$. Given a nonempty set A, the **ultrapower** of A modulo U is the defined as the ultraproduct $\prod_U A = \prod_U A_i$ where $A_i = A$ for each $i \in I$. The **natural embedding** is the mapping $d : A \to \prod_U A$ such that $d(a)$ is the U-equivalence class of the constant function with value a. It is easily seen that d is injective.

We now introduce the ultraproduct operation on first order structures. For each $i \in I$, let $\mathcal{A}_i$ be an L-structure with universe set A_i. Briefly, the **ultraproduct** $\prod_U \mathcal{A}_i$ is the unique L-structure with universe $\prod_U A_i$ such that each basic formula holds in the ultraproduct if and only if it holds in $\mathcal{A}_i$ for U-almost all i. Here is the formal definition.

DEFINITION 2.5. Given an ultrafilter U over I and L-structures $\mathcal{A}_i, i \in I$, the **ultraproduct** $\prod_U \mathcal{A}_i$ is the unique L-structure $\mathcal{B}$ such that:

- The universe of $\mathcal{B}$ is the set $B = \prod_U A_i$.
- For each atomic formula $\varphi(x_1, \ldots, x_k)$ which has at most one symbol from the vocabulary L, and each $f_1, \ldots, f_k \in \prod_{i\in I} A_i$,

$$\mathcal{B} \models \varphi(f_{1U}, \ldots, f_{kU}) \text{ iff } \{i : \mathcal{A} \models \varphi(f_1(i), \ldots, f_k(i))\} \in U.$$

Using the properties of ultrafilters, one can check that there is a unique L-structure $\mathcal{B}$ with the above properties, so the ultraproduct is well-defined. The details are tedious but routine. As with sets, the **ultrapower** of an L-structure $\mathcal{A}$ modulo U is defined as the ultraproduct $\prod_U \mathcal{A} = \prod_U \mathcal{A}_i$ where $\mathcal{A}_i = \mathcal{A}$ for each $i \in I$.

3. The theorem of Łoś

We now prove the fundamental theorem of Łoś, which makes ultraproducts useful in model theory. It shows that a formula holds in an ultraproduct $\prod_U \mathcal{A}_i$ if and only if it holds in $\mathcal{A}_i$ for U-almost all i.

THEOREM 3.1. *(Łoś* [**37**]*) Let U be an ultrafilter over I, and let $\mathcal{A}_i$ be an L-structure for each $i \in I$. Then for each formula $\varphi(x_1, \ldots, x_n)$ of L and each $f_1, \ldots, f_n \in \prod_{i\in I} A_i$, we have*

$$\prod_U \mathcal{A}_i \models \varphi(f_{1U}, \ldots, f_{nU}) \text{ iff } \{i : \mathcal{A}_i \models \varphi(f_1(i), \ldots, f_n(i))\} \in U.$$

PROOF. We argue by induction on the complexity of φ. The definition of ultraproduct gives the result when φ is an atomic formula of the form $F(x_1, \ldots, x_n) = y$. An induction on the complexity of terms gives the result for atomic formulas of the form $t(x_1, \ldots, x_n) = y$, and then the definition of ultraproduct gives the result for arbitrary atomic formulas of L. The steps for logical connectives are easy.

To complete the proof we give the step for existential quantifiers. Suppose the result holds for the formula $\varphi(f_{1U}, \ldots, f_{nU}, g_U)$ where $f_1, \ldots, f_n, g \in \prod_{i\in I} A_i$. We prove the result for the formula $\exists y\, \varphi(f_{1U}, \ldots, f_{nU}, y)$. Using the inductive hypothesis and the fact that U is closed under supersets, we see that the following are equivalent:

$$\prod_U \mathcal{A}_i \models \exists y\, \varphi(f_{1U}, \ldots, f_{nU}, y)$$

$$(\exists g) \prod_U \mathcal{A}_i \models \varphi(f_{1U}, \ldots, f_{nU}, g_U)$$

$$(\exists g)\, \{i : \mathcal{A}_i \models \varphi(f_1(i), \ldots, f_n(i), g(i))\} \in U$$

$$\{i : \mathcal{A}_i \models \exists y\, \varphi(f_1(i), \ldots, f_n(i), y)\} \in U.$$

This completes the induction. □

COROLLARY 3.2. *For each set of sentences T in L, every ultraproduct of models of T is a model of T.*

COROLLARY 3.3. *For each L-structure $\mathcal{A}$ and ultrafilter U over I, $d : \mathcal{A} \prec \prod_U \mathcal{A}$. If $\mathcal{A}$ is finite, $d : \mathcal{A} \cong \prod_U \mathcal{A}$.*

In applications, it is often convenient to rename the elements of an ultrapower. We say that an isomorphic embedding $h : \mathcal{A} \to \mathcal{B}$ is an **ultrapower embedding** if $h = j \circ d$ for some isomorphism $j : \prod_U \mathcal{A} \cong \mathcal{B}$. The natural embedding $d : \mathcal{A} \to \prod_U \mathcal{A}$ is an ultrapower embedding. We say that $\mathcal{B}$ is an **ultrapower extension** of $\mathcal{A}$ if $A \subseteq B$ and the identity map $\iota : \mathcal{A} \to \mathcal{B}$ is an ultrapower embedding. Note that every ultrapower embedding is an elementary embedding, every ultrapower extension is an elementary extension, and every ultrapower of $\mathcal{A}$ is isomorphic to an ultrapower extension of $\mathcal{A}$.

4. Some consequences of Łoś' Theorem

An important property of the ultraproduct construction is that it behaves well when new symbols are added to the vocabulary. The following simple observation is quite powerful when combined with Łoś' Theorem.

PROPOSITION 4.1. *(Expansion Property) Suppose $L_1 \subseteq L_2$, and for each $i \in I$, $\mathcal{A}_i$ is an L_1-structure and $\mathcal{B}_i$ is an expansion of $\mathcal{A}_i$ to L_2. Then for every ultrafilter U over I, $\prod_U \mathcal{B}_i$ is an expansion of $\prod_U \mathcal{A}_i$.*

Given an ultrafilter U over a set I and a mapping $h : A \to B$, define $\prod_U h$ to be the mapping $f_U \mapsto (h \circ f)_U$ from $\prod_U A$ into $\prod_U B$. The next result is a consequence of the Expansion Property and Łoś' Theorem.

PROPOSITION 4.2. *Let U be an ultrafilter over I. The mapping $h \mapsto \prod_U h$ is a functor on the category of all homomorphisms $h : \mathcal{A} \to \mathcal{B}$ between L-structures. If $h : \mathcal{A} \to \mathcal{B}$ then $\prod_U h : \prod_U \mathcal{A} \to \prod_U \mathcal{B}$. If h is surjective, then so is $\prod_U h$. If h is an isomorphic embedding, then so is $\prod_U h$. If h is an elementary embedding, then so is $\prod_U h$.*

The initial interest in ultraproducts in the late 1950's was sparked by the discovery of a proof of the Compactness Theorem for first order logic via ultraproducts (see [**14**]). This proof was attractive because it gave a direct algebraic construction of the required model.

THEOREM 4.3. *(Ultraproduct Compactness) Let S be an infinite set of sentences of L and let I be the set of all finite subsets of S. For each $i \in I$ let $\mathcal{A}_i$ be a model of i. Then there is an ultrafilter U over I such that the ultraproduct $\prod_U \mathcal{A}_i$ is a model of S.*

PROOF. For each $i \in I$, let X_i be the set of all $j \in I$ such that $i \subseteq j$. Let F be the set of all $X \subseteq I$ such that $X \supseteq X_i$ for some $i \in I$. Note that $i \in X_i$, and $X_{i \cup j} = X_i \cap X_j$. It follows that F is a proper filter over I. By Theorem 2.2, F can be extended to an ultrafilter U over I. For each $\varphi \in S$ and $j \in X_{\{\varphi\}}$, $\mathcal{A}_j$ is a model of φ. Moreover, $X_{\{\varphi\}} \in U$. Therefore by Łoś' Theorem, $\prod_U \mathcal{A}_i$ is a model of φ. Hence $\prod_U \mathcal{A}_i$ is a model of S as required. □

The compactness theorem is an easy corollary of this result. For this reason, the ultraproduct construction can be used as a substitute for the compactness theorem with an algebraic flavor.

Another important property of ultraproducts is that an ultraproduct of ultraproducts is isomorphic to a single ultraproduct. This property was also proved in [**14**] by applying Łoś' Theorem. To avoid complicated notation, we will state the result only for ultrapowers.

DEFINITION 4.4. Let U, V be ultrafilters over sets I, J. The **product** $U \times V$ is the set

$$U \times V = \{Y \subseteq I \times J : \{j \in J : \{i \in I : \langle i, j \rangle \in Y\} \in U\} \in V\}.$$

The following result shows that the product of two ultrafilters produces an ultrapower of an ultrapower.

PROPOSITION 4.5. *(See [**14**]) Let U be an ultrafilter over I and V be an ultrafilter over J, and let $\mathcal{A}$ be any L-structure. Then:*

(i) $U \times V$ *is an ultrafilter over* $I \times J$.

(ii) $\prod_{U \times V} \mathcal{A} \cong \prod_V (\prod_U \mathcal{A})$.

(iii) Each of the ultrapowers $\prod_U \mathcal{A}$ *and* $\prod_V \mathcal{A}$ *is elementarily embeddable in* $\prod_{U \times V} \mathcal{A}$.

The order in the product $U \times V$ matters. See [**9**], Exercise 6.1.19, for examples where $\prod_{U \times V} \mathcal{A}$ is not isomorphic to $\prod_{V \times U} \mathcal{A}$.

5. Uniform and countably incomplete ultrafilters

From now on, we will confine our attention to ultrafilters which are uniform and countably incomplete. In this section we explain why.

DEFINITION 5.1. An ultrafilter U over I is **uniform** if every $X \in U$ has cardinality $|X| = |I|$.

If I is a singleton $I = \{i_0\}$, then $\{I\}$ is a uniform ultrafilter over I. But if I is a finite set of cardinality $|I| > 1$, then every ultrafilter over I is principal, so there is no uniform ultrafilter over I.

If I is infinite, then the set $F = \{X \subseteq I : |I \setminus X| < |I|\}$ of subsets with small complements is a proper filter over I, and an ultrafilter U over I is uniform if and only if U contains F. By Theorem 2.2, F can be extended to an ultrafilter over I, so there exist uniform ultrafilters over I.

For ultraproducts, we can always replace a non-uniform ultrafilter by a uniform ultrafilter. Suppose U is a non-uniform ultrafilter over I, and let J be an element of U of minimum cardinality. Then the set $V = U \cap \mathcal{P}(J)$ is a uniform ultrafilter over J, and every ultraproduct $\prod_U \mathcal{A}_i$ is isomorphic to the ultraproduct $\prod_V \mathcal{A}_j$ by the mapping $g_U \mapsto (g \restriction J)_V$.

DEFINITION 5.2. An ultrafilter U is **countably complete** if U is closed under countable intersections. U is **countably incomplete** if U has a countable subset V such that $\bigcap V = \emptyset$.

It is an easy exercise to show that an ultrafilter U is countably incomplete if and only if it is not countably complete.

Every principal ultrafilter is countably complete. However, the hypothesis that there exists a non-principal countably complete ultrafilter is a very strong axiom of infinity that is not provable from ZFC. The first cardinal κ such that there is a non-principal countably complete ultrafilter over a set of cardinality κ is called the first measurable cardinal. This cardinal, if it exists, is exceedingly large (for example, κ must be the κ-th inaccessible cardinal, and even the κ-th Ramsey cardinal; see [**29**]). Countably complete ultraproducts satisfy an analogue of Łoś' Theorem for the infinitary logic with conjunctions and quantifiers of length $< \kappa$ (see [**19**]). It follows that when U is a countably complete ultrafilter and the cardinality of $\mathcal{A}$ is less than the first measurable cardinal, the ultrapower $\prod_U \mathcal{A}$ is trivial, that is, $d : \mathcal{A} \cong \prod_U \mathcal{A}$. For this reason, the study of countably complete ultrapowers belongs to the theory of large cardinals. It is an large and active area of research, but is outside the scope of this article.

We conclude this section with some results which hold for all countably incomplete ultrafilters. The following easy result shows that countably incomplete ultrapowers of infinite structures are always non-trivial.

PROPOSITION 5.3. *Let U be a countably incomplete ultrafilter over I and let $\mathcal{A}$ be infinite. Then d maps A properly into the ultrapower $\prod_U A$, and hence $\prod_U \mathcal{A}$ is isomorphic to a proper elementary extension of $\mathcal{A}$.*

Here are some results about cardinalities of ultraproducts.

THEOREM 5.4. *(Frayne, Morel and Scott [**14**]) Let U be a countably incomplete ultrafilter. Then $\prod_U A_i$ is either finite or of cardinality $\geq 2^\omega$. Thus an ultraproduct $\prod_U A_i$ is never countably infinite.*

The following improvement was given in Keisler [**21**] for ultraproducts of infinite sets, and in Shelah [**46**] for ultraproducts of finite sets.

THEOREM 5.5. *Let U be a countably incomplete ultraproduct. If $\prod_I A_i$ is infinite, then $|\prod_U \mathcal{A}| = |\prod_U \mathcal{A}|^\omega$.*

Here is a property of countably incomplete ultraproducts which is used in many applications, such as the Loeb measure in probability theory, and the nonstandard hull of a Banach space.

THEOREM 5.6. *(Keisler [**18**]). Suppose L is countable and U is a countably incomplete ultrafilter over I. Then every ultraproduct $\prod_U \mathcal{A}_i$ is ω_1-saturated.*

6. Complete embeddings

One advantage of ultrapowers is that they always produce complete embeddings in the following sense.

DEFINITION 6.1. We say that mapping $h : \mathcal{A} \to \mathcal{B}$ is a **complete embedding** of $\mathcal{A}$ into $\mathcal{B}$ if for every expansion of $\mathcal{A}'$ of $\mathcal{A}$ there is an expansion $\mathcal{B}'$ of $\mathcal{B}$ such that $h : \mathcal{A}' \prec \mathcal{B}'$. $\mathcal{B}$ is a **complete extension** of $\mathcal{A}$ if $A \subseteq B$ and the identity mapping $\iota : A \to B$ is a complete embedding.

Note that every complete embedding is an elementary embedding. By Proposition 4.1, the natural embedding $d : \mathcal{A} \to \prod_U \mathcal{A}$ is a complete embedding, and hence every ultrapower embedding is a complete embedding. The converse of this fact is false—there are complete embeddings which are not ultrapower embeddings (see [**9**], Exercise 6.4.6). However, the next result shows that each complete embedding is locally an ultrapower embedding.

THEOREM 6.2. *Suppose $h : \mathcal{A} \to \mathcal{B}$ is a complete embedding. Then for each finite subset S of $\mathcal{B}$ there is a $\mathcal{C} \prec \mathcal{B}$ such that $S \subseteq C$ and $h : \mathcal{A} \to \mathcal{C}$ is an ultrapower embedding.*

This is a consequence of a stronger result in [**20**], which states that $h : \mathcal{A} \to \mathcal{B}$ is a complete embedding if and only if it is a limit ultrapower embedding (we will not define limit ultrapowers here, but mention only that they are generalizations of ultrapowers which share many of their properties).

The following two results do not mention ultrapowers but are proved using ultrapowers.

THEOREM 6.3. *(Rabin [**43**] and Keisler [**20**]; see also [**9**]) Suppose κ is infinite and less than the first measurable cardinal. Then the following are equivalent*

(i) $\kappa = \kappa^\omega$.

(ii) Every structure of cardinality κ (with any number of relations) has a proper elementary extension of cardinality κ.

(iii) Every structure of cardinality κ has a proper complete extension of cardinality κ.

The next result improves the classical upward Löwenheim-Skolem-Tarski theorem when the vocabulary L is large.

THEOREM 6.4. *([**9**], Corollary 6.5.12) Suppose κ is infinite and less than the first measurable cardinal. Then every structure of cardinality κ (with any number of relations) has an elementary extension of cardinality λ if and only if $\lambda \geq \kappa^\omega$.*

7. Nonstandard universes

In applications of the ultrapower, one often picks an ultrafilter U and simultaneously takes the ultrapower of everything in sight modulo U. An efficient way to do this is to begin with a superstructure and use the ultrapower to build a nonstandard universe. We will briefly sketch how this is done, and then point out a connection between nonstandard universes and complete embeddings. For more details, see [**9**], or Chapter 15 of [**28**]. Ultrapowers are also used to construct models of various nonstandard set theories, such as Nelson's internal set theory and Hrbaček set theory, showing that they are conservative over ZFC (see [**33**] for a full treatment and references).

Given a set X, the n-th **cumulative power set** of X is defined recursively by

$$V_0(X) = X, \quad V_{n+1}(X) = V_n(X) \cup \mathcal{P}(V_n(X)).$$

The **superstructure** over X is the union of the cumulative power sets and is denoted by $V(X)$,

$$V(X) = \bigcup_{n=0}^{\infty} V_n(X).$$

The superstructure $V(X)$ has a membership relation $\in$ between elements of $V_n(X)$ and $V_{n+1}(X)$, $n = 0, 1, 2, \ldots$. We treat the elements of X as atoms, and always assume that $\emptyset \notin X$ and that no $x \in X$ contains any elements of $V(X)$. We then consider the structure $\mathcal{V}(X) = \langle V(X), \in \rangle$ whose vocabulary has the single binary relation $\in$.

DEFINITION 7.1. A function $f : I \to V(X)$ is called **bounded** if $f : I \to V_n(X)$ for some n, so $\bigcup_n (V_n(X))^I$ is the set of all bounded functions. Given an ultrafilter U over I, the **bounded ultrapower** $\prod_U^b \mathcal{V}(X)$ of $\mathcal{V}(X)$ modulo U is the substructure of the ordinary ultrapower $\prod_U \mathcal{V}(X)$ whose universe is the set

$$\prod_U^b V(X) = \{g_U : g \in \bigcup_n (V_n(X))^I\}$$

of U-equivalence classes of bounded functions. The interpretation of $\in$ in $\prod_U^b \mathcal{V}(X)$ is denoted by $\in_U$.

A **bounded quantifier formula** is a first order formula in which each quantifier has the form $(\forall u \in v)$ or $(\exists u \in v)$.

DEFINITION 7.2. A **nonstandard universe** is a triple $(V(X), V({}^*X), *)$ such that:

- $V(X)$ and $V(^*X)$ are superstructures,
- $* : V(X) \to V(^*X)$,
- $\mathbb{N} \subseteq X$,
- $*$ maps $\mathbb{N}$ properly into $^*\mathbb{N}$, and
- (Transfer Principle) For each bounded quantifier formula $\varphi(v_1, \ldots, v_k)$ and $a_1, \ldots, a_k \in V(X)$,

$$\mathcal{V}(X) \models \varphi(a_1, \ldots, a_k) \text{ if and only if } \mathcal{V}(^*X) \models \varphi(^*a_1, \ldots, {}^*a_k).$$

The following basic result converts a bounded ultrapower of a superstructure into a nonstandard universe in a canonical way. The advantage of doing this is that it replaces the relation $\in_U$ by the ordinary membership relation $\in$. The Transfer Principle is proved using Łoś' Theorem.

THEOREM 7.3. *(Mostowski Collapse) For each superstructure $\mathcal{V}(X)$ and countably incomplete ultrafilter U, there is a unique nonstandard universe $(V(X), V(^*X), *)$ and mapping $h : \prod_U^b \mathcal{V}(X) \to \mathcal{V}(^*X)$ such that:*

- *$^*X = \prod_U X$, and $h(g_U) = g_U$ for each $g_U \in \prod_U X$, and*
- *For each $g_U \in \prod_U^b V(X) \setminus \prod_U X$, $h(g_U) = \{h(f_U) : f_U \in_U g_U\}$.*

We now observe that each nonstandard universe harbors a whole tower of complete embeddings.

PROPOSITION 7.4. *Let $(V(X), V(^*X), *)$ be a nonstandard universe. For each n, let $\mathcal{V}_n(X) = \langle V_n(X) \in\rangle$ and $^*(\mathcal{V}_n(X)) = \langle ^*(V_n(X)), \in\rangle$. Then $* : \mathcal{V}_n(X) \to {}^*(\mathcal{V}_n(X))$ is a complete embedding.*

PROOF. Since $V_n(X)$ is an element of $V(X) \setminus X$, $^*(V_n(X))$ is an element of $V(^*X) \setminus {}^*X$. Therefore $^*(V_n(X))$ is also a subset of $V(^*X)$. We have $* : \mathcal{V}_n(X) \prec {}^*(\mathcal{V}_n(X))$ because $*$ preserves bounded formulas in $\mathcal{V}(X)$. But any finitary function or relation on $V_n(X)$ is an element of $V_m(X)$ for some m, and we also have $* : \mathcal{V}_m(X) \prec {}^*(\mathcal{V}_m(X))$. This shows that $* : \mathcal{V}_n(X) \to {}^*(\mathcal{V}_n(X))$ is a complete embedding. □

Combining this with Theorem 6.2, we see that every nonstandard universe is locally an ultrapower embedding.

COROLLARY 7.5. *Let $(V(X), V(^*X), *)$ be a nonstandard universe. For each n and each finite set $S \subseteq {}^*(V_n(X))$ there is a $\mathcal{C} \prec \mathcal{V}_n(^*X)$ such that $S \subseteq C$ and $* : \mathcal{V}_n(X) \to \mathcal{C}$ is an ultrapower embedding.*

Benci constructed a nonstandard universe with the nice property that $^*X = X$, so that only one superstructure in needed instead of two.

THEOREM 7.6. *(Benci [**3**]) For each set X such that $\mathbb{N} \subseteq X$ and $|X|^\omega = |X|$, there is a nonstandard universe $(V(X), V(X), *)$.*

The first step in constructing $(V(X), V(X), *)$ is to take a free ultrafilter U over $\mathbb{N}$ and form the bounded ultrapower $\prod_U^b \mathcal{V}(X)$. Then choose a bijection j from $\prod_U X$ onto X, and for $x \in X$ define $^*x = j(d(x))$. Finally, extend $*$ to a mapping from $V(X)$ into itself using the Mostowski collapse.

8. The Rudin-Keisler ordering

The ultrapower construction was the motivation for the so-called Rudin-Keisler ordering (introduced by M.E. Rudin [**44**] and the author [**25**]). It is a pre-ordering on the class of all ultrafilters. Heuristically, higher ultrafilters in the ordering produce larger ultrapowers with respect to elementary embeddings. This ordering has been extensively studied in the literature, particulary for ultrafilters over $\mathbb{N}$, because it has a rich structure and leads to many attractive problems. The minimal ultrafilters over $\mathbb{N}$ in this ordering ordering offer promising possibilities for applications, because they give the "smallest" nontrivial ultrapowers.

Given a function $f : I \to J$ and an ultrafilter U over I, we define

$$f[U] = \{Y \subseteq J : f^{-1}(Y) \in U\}.$$

It is easy to see that $f[U]$ is an ultrafilter over J.

DEFINITION 8.1. Given ultrafilters U over I and V over J, $V \leq_{RK} U$ means that there exists a function $f : I \to J$ such that $V = f[U]$. We write $U \equiv_{RK} V$ if $[U \leq_{RK} V$ and $V \leq_{RK} U]$, and $U <_{RK} V$ if $[U \leq_{RK} V$ but not $U \equiv_{RK} V]$.

Note that if $V \leq_{RK} U$ then $\min\{|Y| : Y \in V\} \leq \min\{|X| : X \in U\}$. It is clear that $\leq_{RK}$ is transitive and symmetric. It is also upward directed— one can check that whenever V, W are ultrafilters and $U = V \times W$, we have $V \leq_{RK} U$ and $W \leq_{RK} U$. It is not hard to see that an ultrafilter U is countably incomplete if and only if there is a free ultrafilter V over $\mathbb{N}$ such that $V \leq_{RK} U$.

The following result was found independently by several people (for a proof see [**10**], Theorem 9.2).

THEOREM 8.2. *Let U be an ultrafilter over I and let $f : I \to J$. Then we have $f[U] \equiv_{RK} U$ if and only if the restriction of f to some $X \in U$ is one to one.*

The next result gives the connection between the pre-ordering $\leq_{RK}$ and ultrapowers. It shows that higher ultrafilters with respect to $\leq_{RK}$ give bigger ultrapowers with respect to elementary embeddability.

PROPOSITION 8.3. *([**25**]; see also [**4**] and [**9**], Exercise 4.3.41). Let U be an ultrafilter over I and V be an ultrafilter over J. Then $V \leq_{RK} U$ if and only if for every $\mathcal{A}$, $\prod_V \mathcal{A}$ is elementarily embeddable in $\prod_U \mathcal{A}$. Also, $V \equiv_{RK} U$ if and only if for every $\mathcal{A}$, $\prod_V \mathcal{A} \cong \prod_U \mathcal{A}$.*

A recurrent theme in the literature is to exploit the interplay between ultrafilters $V \leq_{RK} U$ and elements of an ultrapower modulo U (see, for example, [**3**], [**4**], [**5**],[**6**],[**36**],[**40**],[**42**]).

To explain the idea, we introduce some notation and state a result. Let U be an ultrafilter over I and let $\mathcal{B} = \prod_U \mathcal{A}$. For each function $f : I \to A$, let $\mathcal{B}[f]$ be the set of all elements $(g \circ f)_U \in \mathcal{B}$ where $g : A \to A$.

We make several observations: If $f : I \to A$ then $f[U] \leq_{RK} U$. If $f, h : I \to A$ and $h[U] \leq_{RK} f[U]$ then $h_U \in \mathcal{B}[f]$. If f is a constant function, then $f[U]$ is principal and $\mathcal{B}[f] = d(\mathcal{A})$. If the structure $\mathcal{A}$ has a function symbol for every $g : A \to A$, then $\mathcal{B}[f]$ is just the substructure of $\mathcal{B}$ generated by f_U.

PROPOSITION 8.4. *Suppose U is an ultrafilter over I, $\mathcal{B} = \prod_U \mathcal{A}$, and f is a function from I into A.*

(i) $\mathcal{B}[f] \prec \mathcal{B}$.

(ii) $\mathcal{B}[f] \cong \prod_{f[U]} \mathcal{A}$.
(iii) If $f[U] \equiv_{RK} U$ *then* $\mathcal{B}[f] = \mathcal{B}$.
(iv) If $|I| \leq |A|$ *and* $\mathcal{B}[f] = \mathcal{B}$, *then* $f[U] \equiv_{RK} U$.

PROOF. (i) is proved by induction on complexity of formulas.

(ii) $f[U]$ is an ultrafilter over A, and the isomorphism is given by the mapping $(g \circ f)_U \mapsto g_{f[U]}$.

(iii) Suppose $f[U] \equiv_{RK} U$. By Theorem 8.2, there is an $X \in U$ such that the restriction of f to X is one to one. Then for any $h : I \to A$ there exists $g : A \to A$ such that $(g \circ f)_U = h_U$, and hence $\mathcal{B}[f] = \mathcal{B}$.

(iv) Suppose $|I| \leq |A|$ and $\mathcal{B}[f] = \mathcal{B}$. Then there is a one to one function $h : I \to A$. Since $\mathcal{B}[f] = \mathcal{B}$, $h_U = (g \circ f)_U$ for some $g : A \to A$. Then $h[U] = (g \circ f)[U] = g[f[U]]$, so $h[U] \leq_{RK} f[U]$. Since h is one to one we have $U \equiv_{RK} h[U]$, and $f[U] \leq_{RK} U$ by definition, so $f[U] \equiv_{RK} U$. □

One can apply Proposition 8.4 to show that if V, W are free ultrafilters over J, K and $U = V \times W$, then $V <_{RK} U$ and $W <_{RK} U$. (Hint: We have already observed that $V \leq_{RK} U$. To show $V <_{RK} U$, take $A = J$, let f be the projection from $J \times K$ onto J, prove that $\mathcal{B}[f] \neq \mathcal{B}$, and apply Proposition 8.4 (iii)).

We now turn to the ultrafilters over $\mathbb{N}$.

The **Stone-Čech compactification** of the discrete topology on $\mathbb{N}$ is denoted by $\beta(\mathbb{N})$, and can be defined as the topology on the set of all ultrafilters over $\mathbb{N}$ which has as a closed base the family of all sets $\{U \in \beta(\mathbb{N}) : X \in U\}$ where $X \in \mathcal{P}(\mathbb{N})$. The Stone representation theorem ([**52**]) shows that $\beta(\mathbb{N})$ is a compact totally disconnected Hausdorff space. One can identify each $n \in \mathbb{N}$ with the principal ultrafilter which contains the singleton $\{n\}$, so that $\beta(\mathbb{N}) \setminus \mathbb{N}$ is the space of all free ultrafilters over $\mathbb{N}$.

W. Rudin [**45**] first discovered that there are free ultrafilters over $\mathbb{N}$ with different topological properties. Assuming the continuum hypothesis, he proved that the space $\beta(\mathbb{N}) \setminus \mathbb{N}$ is not point homogeneous. Frolík [**15**] later proved this fact in ZFC.

Ng and Render [**40**] used the Rudin-Keisler ordering on $\beta\mathbb{N}$ to prove the following theorem about ultrapowers indexed by $\mathbb{N}$.

THEOREM 8.5. *Let* U *be a free ultrafilter over* $\mathbb{N}$, *let* A *be infinite, and let* S *be the topology on* $\prod_U A$ *whose closed basis is the family of all sets of the form* $\{f_U : f \in C^{\mathbb{N}}\}$ *where* $C \subseteq A$. *Then* S *is not countably compact.*

We say that a free ultrafilter U over $\mathbb{N}$ is **minimal** if there is no free ultrafilter V over $\mathbb{N}$ such that $V <_{RK} U$. Note that if U is minimal then there is no free ultrafilter V over any set J such that $V <_{RK} U$. Assuming Martin's axiom, Booth [**6**] proved that minimal ultrafilters exist, and Blass [**4**] proved that there are 2^{2^ω} minimal ultrafilters. On the other hand, Kunen [**35**] proved that the existence of minimal ultrafilters is independent of ZFC. There are several natural equivalent conditions for a minimal ultrafilter.

THEOREM 8.6. *(Kunen, Rowbottom; see* [**10**]*) Let* U *be a free ultrafilter over* $\mathbb{N}$. *The following are equivalent:*

(i) U *is minimal;*

(ii) U *is selective (i.e.,* U *contains a choice set for every partition of* $\mathbb{N}$ *into infinitely many classes which are not elements of* U*);*

(iii) U is Ramsey (i.e., U contains a homogeneous set for every partition of $[\mathbb{N}]^k$ into two classes, where $[\mathbb{N}]^k$ is the set of unordered k-tuples from $\mathbb{N}$)

There are also nice conditions which involve ultrapowers. Here is a condition which is a variant of a theorem of Benci and Di Nasso [**2**]. It can be proved using Theorem 8.6 (iii).

THEOREM 8.7. *A free ultrafilter U over $\mathbb{N}$ is minimal if and only if for each $x \in \prod_U \mathbb{R}$ there is a function $g : \mathbb{N} \to \mathbb{R}$ such that $g_U = x$ and g is either constant, strictly increasing, or strictly decreasing.*

The following theorem is a variant of results of Puritz [**42**] and Ng and Render [**40**], and is a consequence of Proposition 8.4.

THEOREM 8.8. *Let $\mathcal{A}$ be an infinite structure, U be a free ultrafilter over $\mathbb{N}$, and $\mathcal{B} = \prod_U \mathcal{A}$.*

(i) U is minimal if and only if for every $f : \mathbb{N} \to A$, either $\mathcal{B}[f] = d(\mathcal{A})$ or $\mathcal{B}[f] = \mathcal{B}$.

(ii) Suppose $\mathcal{A}$ has symbols for every function $g : A \to A$. Then U is minimal if and only if the only substructures of $\mathcal{B}$ are $d(\mathcal{A})$ and $\mathcal{B}$ itself.

9. Regular ultrafilters

In this section we introduce the regular ultrafilters, which behave especially well with respect to the ultrapower construction.

DEFINITION 9.1. A free ultrafilter U over I is called **regular** if there is a set $E \subseteq U$ such that $|E| = |I|$ and each $i \in I$ belongs to only finitely many $X \in E$.

We begin with some easy facts about regular ultrafilters.

PROPOSITION 9.2. *(i) If J is infinite, $I = \mathcal{P}_\omega(J)$, and $\{i \in I : j \in i\} \in U$ for each $j \in J$, then U is regular.*

(ii) There exist regular ultrafilters over each infinite set I.

(iii) Every regular ultrafilter is countably incomplete and uniform.

(iv) Suppose U is a regular ultrafilter over I and V is an ultrafilter over J. If $|I| = |J|$ and $U \le_{RK} V$, then V is regular. If $|J| \le |I|$ then $U \times V$ and $V \times U$ are regular.

It is obvious that if U is an ultrafilter over I, then any ultrapower $\prod_U A$ has cardinality at most $|A^I|$. The following result shows that when U is regular, this maximum cardinality is attained.

THEOREM 9.3. *(Frayne, Morel and Scott [**14**]) If U is a regular ultrafilter over I and A is infinite, then $|\prod_U A| = |A^I|$.*

It follows that each infinite set has ultrapowers of arbitrarily large cardinality. Here is another consequence, whose statement does not mention ultrapowers. Suppose L has at least a unary predicate symbol P, and let $P^{\mathcal{A}}$ be the interpretation of P in $\mathcal{A}$. By a (κ, λ)-structure we mean a structure $\mathcal{A}$ such that $|A| = \kappa$ and $|P^{\mathcal{A}}| = \lambda$.

COROLLARY 9.4. *Suppose κ, λ, and μ are cardinals such that $\omega \le \lambda \le \kappa$. Then every (κ, λ)-structure has an elementary extension which is a $(\kappa^\mu, \lambda^\mu)$-structure.*

While Theorem 9.3 gives a simple formula for the cardinality of a regular ultrapower, cardinalities of regular ultraproducts are much more complicated. Given a regular ultrafilter U over I, let $Fin(U)$ be the set of infinite cardinalities of ultraproducts $\prod_U A_i$ where each A_i is finite. S. Koppelberg [**32**] showed that $|2^I| \in Fin(U)$. Shelah ([**49**], page 357) showed that for each finite set of C cardinals κ such that $\kappa = \kappa^\omega < |2^I|$, there is a regular ultrafilter U over I with $Fin(U) = C \cup \{|2^I|\}$. Using the methods of [**27**] and [**32**] one can get various examples where $Fin(U)$ is infinite.

The next result shows that regular ultrapowers are large in a model-theoretic sense. A structure $\mathcal{A}$ is called κ-**universal** if every structure $\mathcal{B}$ of cardinality $|B| < \kappa$ which is elementarily equivalent to $\mathcal{A}$ is elementarily embeddable in $\mathcal{A}$. Morley and Vaught [**39**] showed that every κ-saturated structure is κ^+-universal.

THEOREM 9.5. *(Frayne, Morel and Scott [**14**]; Keisler [**26**]) Suppose U is an ultrafilter over a set of cardinality κ. Then U is regular if and only if whenever $|L| \le \kappa$, every ultrapower $\prod_U \mathcal{A}$ is κ^+-universal.*

It is natural to ask: When is $\prod_U \mathcal{A}$ κ^+-saturated? The next result shows that the answer depends only on the complete theory of $\mathcal{A}$.

THEOREM 9.6. *(Keisler [**26**]) Let U be a regular ultrafilter over a set of cardinality κ. If $|L| \le \kappa$ and $\mathcal{A} \equiv \mathcal{B}$, then $\prod_U \mathcal{A}$ is κ^+-saturated if and only if $\prod_U \mathcal{B}$ is κ^+-saturated.*

Let us say that a regular ultrafilter U over a set of cardinality κ **saturates** a complete theory T if for every model $\mathcal{A}$ of T, $\prod_U \mathcal{A}$ is κ^+-saturated. By Theorem 9.6, it does not matter which model of T we take. Given two complete theories S, T with countable vocabularies, we write $S \triangleleft T$ if every regular ultrafilter which saturates T saturates S. This relation can be used to classify complete theories. Intuitively, higher theories in this ordering are more complex than lower ones. The $\triangleleft$-**class** of T is the set of all S such that $S \triangleleft T$ and $T \triangleleft S$. It is clear that $\triangleleft$ is is reflexive and transitive, so it induces a partial order on the $\triangleleft$-classes. The paper [**26**] showed that there are at least two $\triangleleft$-classes, including a lowest and a highest $\triangleleft$-class, and posed several questions which are still open, including: Is this partial order linear? How many $\triangleleft$-classes are there? Is there a syntactical characterization of the $\triangleleft$-classes? The following theorem gives some partial results.

THEOREM 9.7. *(Shelah [**47**]) There are at least four $\triangleleft$-classes, including a lowest, second lowest, and highest. There are syntactical conditions for the lowest (stable theories without the finite cover property) and second lowest (stable theories with the finite cover property).*

Recently, Malliaris [**38**] has made further progress on this problem.

We conclude this section with a discussion of ultrafilters which are uniform and countably incomplete but not regular. If there is a first measurable cardinal κ, one can give an easy example of such an ultrafilter over κ. Any countably complete free ultrafilter U over κ is uniform and not regular. Let V be a free ultrafilter over $\mathbb{N}$. Then both $U \times V$ and $V \times U$ are uniform and countably incomplete, and $\prod_{U \times V} \mathcal{A} \cong \prod_{V \times U} \mathcal{A} \cong \prod_V \mathcal{A}$ whenever $|A| < \kappa$. Then in view of Theorem 9.3, $U \times V$ and $V \times U$ are both non-regular.

In [**9**] we asked whether there are any uniform non-regular ultrafilters on an infinite set of cardinality smaller than the first measurable cardinal. A related question is whether a set A can have ultrapowers whose cardinality is not a power of $|A|$. These questions have prompted a great deal of research. Ketonen [**30**] showed that if there is a uniform non-regular ultrafilter over ω_1, then $O^\#$ exists, which implies the consistency of various large cardinal axioms with ZFC. This suggested that one may need large cardinal assumptions to get uniform non-regular ultrafilters. Donder [**11**] proved that the statement "*Every uniform ultrafilter is regular*" is consistent relative to ZFC. Foreman, Magidor and Shelah [**13**] have shown that the statements "*There exist uniform non-regular ultrafilters over each successor of a regular cardinal*" and "*There is a uniform ultrafilter U over ω_1 such that* $|\prod_U \omega_1| = \omega_1$" are consistent relative to ZFC plus a large cardinal axiom. Jin and Shelah [**17**] have shown that the statements "*There is a uniform ultrafilter U such that the cardinality of $\prod_U \mathbb{N}$ is (1) inaccessible, (2) a singular strong limit cardinal*" are consistent relative to ZFC plus a large cardinal axiom. Such ultrafilters must be non-regular by Theorem 9.3.

10. Good ultrafilters and isomorphic ultrapowers

In this last section we introduce good ultrafilters, which are of interest because they produce ultrapowers which are as saturated as possible. We then discuss the well-known result that two elementarily equivalent structures have isomorphic ultrapowers.

DEFINITION 10.1. An ultrafilter U over I is **good** if for each $f : \mathcal{P}_\omega(I) \to U$ such that $a \subseteq b$ implies $f(a) \supseteq f(b)$, there exists $g : \mathcal{P}_\omega(I) \to U$ such that $a, b \in \mathcal{P}_\omega(I)$ implies $g(a) \subseteq f(a)$ and $g(a \cup b) = g(a) \cap g(b)$.

It is easily seen that every free ultrafilter over a countable set is good.

PROPOSITION 10.2. *Every countably incomplete good ultrafilter is regular.*

PROOF. This is stated as Exercise 6.1.3 in [**9**]. We give a proof here. Suppose U is a countably incomplete good ultrafilter over I. Take sets $X_0 \supseteq X_1 \supseteq \cdots$ in U such that $\bigcap_n X_n = \emptyset$. Define $f : \mathcal{P}_\omega(I) \to U$ by $f(a) = X_{|a|}$. Since U is good there exists $g : \mathcal{P}_\omega(I) \to U$ such that f and g are as in Definition 10.1. Then the family $E = \{g(\{i\}) : i \in I\}$ makes U regular. □

The converse of Proposition 10.2 is not true. For example, let U be a regular ultrafilter over I, V be a regular ultrafilter over J, and $|J| \le |I|$. Then $V \times U$ is good if and only if U is good. If either $|J| < |I|$ or V is not good, then $U \times V$ is regular but not good.

Good ultrafilters were introduced in Keisler [**18**] and [**23**], where their existence was proved under the assumption of the generalized continuum hypothesis. Kunen later proved their existence outright.

THEOREM 10.3. *(Kunen [**34**]) For every infinite set I, there exist $2^{2^{|I|}}$ countably incomplete good ultrafilters over I.*

For good ultrafilters, unlike regular ultrafilters in general, there is a simple formula for the cardinality of an ultraproduct of finite sets.

THEOREM 10.4. *(Shelah [**49**], page 343). Let U be a good ultrafilter over I, and suppose each A_i is finite. Then the cardinality of $\prod_U A_i$ is either finite or $|2^I|$.*

The next result gives the main model-theoretic property of good ultrafilters. Theorem 5.6 is the special case where $\kappa = \omega$.

THEOREM 10.5. *(Keisler [**22**], [**26**]) Let U be an ultrafilter over a set of cardinality κ. The following are equivalent:*

(i) U is good.

(ii) Every ultrapower $\prod_U \mathcal{A}$ with a vocabulary of cardinality $\leq \kappa$ is κ^+-saturated.

(iii) Every ultraproduct $\prod_U \mathcal{A}_i$ with a vocabulary of cardinality $\leq \kappa$ is κ^+-saturated.

(iv) Every ultraproduct $\prod_U \mathcal{A}_i$ with a vocabulary of cardinality $\leq \kappa$ is κ^+-universal.

Under the generalized continuum hypothesis, this gives a first version of the result that elementarily equivalent structures have isomorphic ultrapowers.

THEOREM 10.6. *(Keisler [**18**]) Assume that $2^\kappa = \kappa^+$. Suppose that $|L| \leq \kappa$, $\mathcal{A}, \mathcal{B}$ have cardinality $\leq 2^\kappa$, and $\mathcal{A} \equiv \mathcal{B}$. Then $\prod_U \mathcal{A} \cong \prod_U \mathcal{B}$ for some ultrafilter U over a set of cardinality κ.*

PROOF. We give a proof from Theorem 10.5. Let U be a good ultrafilter. By Theorem 10.5 and $2^\kappa = \kappa^+$, the ultrapowers $\prod_U \mathcal{A}$ and $\prod_U \mathcal{B}$ are saturated structures of cardinality κ^+. By Corollary 3.3 they are also elementarily equivalent, so they are isomorphic by the result of Morley and Vaught. □

A decade later, Shelah eliminated the generalized continuum hypothesis from this theorem, but with an ultrafilter over a set of cardinality 2^κ instead of κ.

THEOREM 10.7. *(Shelah [**48**]) Suppose that $\mathcal{A}, \mathcal{B}$ have cardinality $\leq \kappa$, and $\mathcal{A} \equiv \mathcal{B}$. Then $\prod_U \mathcal{A} \cong \prod_U \mathcal{B}$ for some ultrafilter U over a set of cardinality 2^κ.*

Shelah [**50**] showed that it is not provable in ZFC that whenever $\mathcal{A}, \mathcal{B}$ are countable and $\mathcal{A} \equiv \mathcal{B}$, there is an ultrafilter U over a countable set such that $\prod_U \mathcal{A} \cong \prod_U \mathcal{B}$. Thus when $\kappa = \omega$, Theorem 10.6 really needs the continuum hypothesis, and in Theorem 10.7 one cannot always take U over a countable set.

Combining Theorem 10.7 with Łoś's Theorem 3.1, we get algebraic characterizations of elementary equivalence, elementary embeddings, and elementary classes.

COROLLARY 10.8. *(Isomorphism theorem) $\mathcal{A} \equiv \mathcal{B}$ if and only if there is an ultrafilter U such that $\prod_U \mathcal{A} \cong \prod_U \mathcal{B}$.*

COROLLARY 10.9. *$h : \mathcal{A} \prec \mathcal{B}$ if and only if there is an ultrafilter U and an isomorphism $j : \prod_U \mathcal{A} \cong \prod_U \mathcal{B}$ such that $e \circ h = j \circ d$, where $d : \mathcal{A} \prec \prod_U \mathcal{A}$ and $e : \mathcal{B} \prec \prod_U \mathcal{B}$ are the natural embeddings.*

COROLLARY 10.10. *A class K of L-structures is the class of all models of some first order theory if and only if K is closed under ultraproducts and isomorphisms, and the complement of K is closed under ultrapowers.*

COROLLARY 10.11. *A class K of L-structures is the class of all models of some first order sentence if and only if both K and its complement are closed under ultraproducts and isomorphisms.*

References

1. V. Benci, *A construction of a nonstandard universe*, pp. 471-478 in *Advances in Dynamical Systems and Quantum Physics*, ed. by S. Albeverio et. al., World Scientific 1995.
2. V. Benci and M. Di Nasso, *Alpha-theory: an elementary axiomatics for nonstandard analysis*, Expositiones Math. **21** (2003), 355-386.
3. M. Benedikt, *Ultrafilters which extend measures*, J. Symb. Logic **63** (1998), 638-662.
4. A. Blass, *The Rudin-Keisler ordering of P-points*, Trans. Amer. Math. Soc. **179** (1973), 145-166.
5. A. Blass, *A model-theoretic view of some special ultrafilters*, pp. 79-90 in *Logic Colloquium 77*, ed. by A. Macintyre et. al., North-Holland 1978.
6. D. Booth, *Ultrafilters on a countable set*, Annals of Math. Logic **2** (1971), 1-24.
7. C.C. Chang, *Ultraproducts and other methods of constructing models*, pp. 85-121 in *Sets, Models and Reecursion Theory*, ed. by J.N. Crossley, North-Holland 1967.
8. C.C. Chang and H.J. Keisler, *Applications of ultrproducts of pairs of cardinals to the theory of models*, Pacific J. Math. **12** (1962), 835-845.
9. C.C. Chang and H.J. Keisler, *Model Theory, Third Edition*, North Holland Elsevier, 1990.
10. W.W.Comfort and S. Negrepontis, *The Theory of Ultrafilters*, Springer-Verlag 1974.
11. H.-D. Donder, *Regularity of ultrafilters and the core model*, Israel J. Math. **63** (1988), 289-322.
12. P. Eklof, *Ultraproducts for algebraists*, pp. 105-138 in *Handbook of Mathematical Logic*, ed. by J. Barwise, North-Holland 1977.
13. M. Foreman, M. Magidor, and S. Shelah, *Martin's maximum, saturated ideals, and non-regular ultrafilters. Part II*, Annals of Math. **127** (1988), 521-545.
14. T.E. Frayne, A.C. Morel, and D.S. Scott, *Reduced direct products*, Fund. Math. **51** (1962), 195-228 (Abstract: Notices Amer. Math. Soc. **5** (1958), p. 674).
15. Z. Frolík, *Sums of ultrafilters*, Bull. Amer. Math. Soc. **73** (1967), 87-91.
16. E. Hewitt, *Rings of real-valued continuous functions*, Trans. Amer. Math. Soc. **64** (1948), 45-99.
17. R. Jin and S. Shelah, *Possible size of an ultrapower of ω*, Archive for Math. Logic **38** (1999), 61-77.
18. H.J. Keisler, *Ultraproducts and elementary classes*, Koninkl. Ned. Akad. Wetensch. Proc. Ser. A **64** (= Indag. Math. **23**) (1961), 477.495.
19. H.J. Keisler, *Some applications of the theory of models to set theory*, pp. 80-86 in *Logic, Methodology and Philosophy of Science*, ed. by E. Nagel et. al., Stanford Univ. Press 1962.
20. H.J. Keisler, *Limit ultrapowers*, Trans. Amer. Math. Soc. **107** (1963), 383-408.
21. H.J. Keisler, *On cardinalities of ultraproducts*, Bull. Amer. Math. Soc. **70** (1964), p. 644-647.
22. H.J. Keisler, *Ultraproducts and saturated models*, Koninkl. Ned. Akad. Wetensch. Proc. Ser. A **67** (= Indag. Math. **26**) (1964), 178-186.
23. H.J. Keisler, *Good ideals in fields of sets*, Annals of Math. **79** (1964), 338-359.
24. H.J. Keisler, *A survey of ultraproducts*, pp. 112-126 in *Logic, Methodology and Philosophy of Science*, ed. by Y. Bar-Hillel, North-Holland 1965.
25. H.J. Keisler, Mimeographed Lecture Notes, U.C.L.A. 1967.
26. H.J. Keisler, *Ultraproducts which are not saturated*, J. Symbolic Logic **32** (1967), 23-46.
27. H.J. Keisler, *Ultraproducts of finite sets*, J. Symbolic Logic **32** (1967), 47-57.
28. H.J. Keisler, *Foundations of Infinitesimal Calculus* (online editiion) (2007), available at www.math.wisc.edu/~keisler.
29. H.J. Keisler and A. Tarski, *From accessible to inaccessible cardinals*, Fund. Math. **53** (1964), p. 225-308.
30. J. Ketonen, *Nonregular ultrafilters and large cardinals*, Trans. Amer. Math. Soc. **224** (1976), 61-73.
31. S.B. Kochen, *Ultraproducts in the theory of models*, Ann. Math. Ser. 2, **74** (1961), 221-261.
32. S. Koppelberg, *Cardinalities of ultraproducts of finite sets*, J. Symbolic Logic **45** (1980), 574-584.
33. V. Kanovei and M. Reeken, *Nonstandard Analysis, Axiomatically*, Springer 2004.
34. K. Kunen, *Ultrafilters and independent sets*, Trans. Amer. Math. Soc. *172*, (1972), 199-206.
35. K. Kunen, *Some points in $\beta(\mathbb{N})$*, Math. Proc. Cambridge Philosophical Soc. **80(3)** (1976), 385-398.
36. D. Lascar, *Stability in Model Theory*, Longman 1987.

37. J. Łoś, *Quelques remarques, théorèmes et problèmes sur les classes définissables d'algèbres*, pp. 98-113 in *Mathematical Interpretations of Formal Systems*, North-Holland 1955.
38. M. Malliaris, *Realization of φ-types and Keisler's order*, Annals of Pure and Applied Logic, in press.
39. M. Morley and R. Vaught, *Homogeneous universal models*, Math. Scand. **11** (1962), 37-57.
40. S-A Ng and H. Render, *The Puritz order and its relationship to the Rudin-Keisler order*, pp. 157-166 in *Reuniting the Antipodes*, ed. by P. Schuster et. al., Kluwer 2001.
41. B. Pospíšil, *Remark on bicompact spaces*, Ann. Math. **38** (1937), 845-846.
42. C. Puritz, *Ultrafilters and standard functions in non-standard arithmetic*, Proc. London Math. Soc. **22** (1971), 705-733.
43. M.O. Rabin, *Arithmetical extensions with prescribed cardinality*, Koninkl. Ned. Akad. Wetensch. Proc. Ser. A **62** (= Indag. Math. **21**) (1959), 439-446.
44. M.E. Rudin, *Types of ultrafilters*, pp. 147-151 in *Topology Seminar, Wisconsin 1965*, Annals of Math. Studies no. 60 (1966).
45. W. Rudin, *Homogeneity problems in the theory of Čech compactifications*, Duke Math. Journal **23** (1956), 409-419.
46. S. Shelah, *On the cardinality of ultraproducts of finite sets*, J. Symbolic Logic **35** (1970), 83-84.
47. S. Shelah, *Saturation of ultrapowers and Keisler's order*, Ann. Math. Logic **4** (1972), 75-114.
48. S. Shelah, *Every two elementarily equivalent models have isomorphic ultrapowers*, Israel J. Math. **10** (1972), 224-233.
49. S. Shelah, *Classification Theory, Revised Edition*, North-Holland 1990.
50. S. Shelah, *Vive la différence I: nonisomorphism of ultrapowers of countable models*, pp. 357-405 in *Set Theory of the Continuum*, Springer-Verlag 1992.
51. T. Skolem, *Über die Nicht-Charakterisierbarkeit der Zahlenreihe mittels endlich oder abzählbar unendlich vieler Aussagen mit ausschliesslich Zahlenvariablen*, Fund. Math. **23** (1934), 150-161.
52. M.H. Stone, *The representation theorem for Boolean algebras*, Trans. Amer. Math. Soc. **40** (1936), 37-111.
53. A. Tarski, *Une contribution à la théorie de la mesure*, Fund. Math. **15** (1930), 42-50.
54. G. Weaver, *Henkin-Keisler Models*, Springer 1997.

DEPARTMENT OF MATHEMATICS, UNIVERSITY OF WISCONSIN, 480 LINCOLN DRIVE, MADISON WI 53706

E-mail address: `keisler@math.wisc.edu`

Contemporary Mathematics
Volume **530**, 2010

Ultrafilters and Large Cardinals

Itay Neeman

Abstract. This paper is a survey of basic large cardinal notions, and applications of large cardinal ultrafilters in forcing. The main application presented is the consistent failure of the singular cardinals hypothesis. Other applications are mentioned that involve variants of Prikry forcing, over models of choice and models of determinacy.

My talk at the Ultramath conference was about ultrafilters and large cardinals. As we shall see shortly, many large cardinal axioms can be viewed as asserting the existence of ultrafilters with specific properties. In a sense then one can say that all theorems that use these large cardinals are theorems about ultrafilters. But the study of these large cardinal axioms is far too wide a subject for a single talk, or even for many. My talk concentrated on one aspect, the use of the large cardinal ultrafilters in forcing. This paper follows a similar path.

The paper is intended for non-specialists, and the material is presented in a way that minimizes any prerequisites. Forcing is explained in Section 2, and the basic definitions of large cardinals in Section 1. Section 3 includes one of the most celebrated results combining these two topics, namely the consistent failure of the singular cardinals hypothesis. The proof illustrates most vividly how large cardinal ultrafilters are used in forcing constructions. Finally Section 4 gives some concluding remarks, on later uses of ultrafilters in forcing, both under the axiom of choice and under the axiom of determinacy.

The paper is expository, and the results presented, with the exception of some theorems in the last section, are not due to the author. It differs in some aspects from my talk at the Ultramath conference. In particular Theorem 4.2 was not yet known at the time of the conference.

1. Large cardinals

Recall that V denotes the universe of all sets. It is the union of the levels of the von Neumann hierarchy, V_α, $\alpha \in \text{On}$.

This material is based upon work supported by the National Science Foundation under Grant No. DMS-0556223.

Keywords: Large cardinals, ultrafilters, forcing, measurable cardinals, supercompact cardinals, Prikry forcing, singular cardinals hypothesis.

MSC 2010: 03E55, 03E35.

An embedding $\pi\colon V \to M$ is *elementary* if it preserves truth, meaning that for any formula φ and any $a_1, \ldots, a_n \in V$, $\varphi(\pi(a_1), \ldots, \pi(a_n))$ holds in M iff $\varphi(a_1, \ldots, a_n)$ holds in V.

Of course the identity map from V to V is elementary, but it is of no interest, and all maps below are assumed to be non-trivial. In all embeddings $\pi\colon V \to M$ below, π is a definable class, and $M \subseteq V$, as V is the universe of all sets. We often do not mention this explicitly. (There is another type of embeddings, called *generic* elementary embeddings, with M contained in a generic extension of V, and π definable in the extension.) Again in all embeddings $\pi\colon V \to M$ below, M is assumed to be transitive, though we do not mention this assumption explicitly.

The large cardinal axioms we work with in this paper assert the existence of (non-trivial) elementary embeddings $\pi\colon V \to M$. The strength of an axiom depends on the amount of agreement it gives between M and V. For example: the embedding is *δ-strong* if all bounded subsets of δ belong to M, namely $\bigcup_{\xi<\delta} \mathcal{P}(\xi) \subseteq M$. It is δ-supercompact if all subsets of M of size δ belong to M, namely $M^\delta \subseteq M$.

Axioms asserting the existence of such embeddings are called large cardinal axioms because they can be phrased as making assertions about very large cardinals. The next few claims isolate the most relevant cardinal to an embedding π.

CLAIM 1.1. *Let α be an ordinal. If $\pi\restriction\alpha$ is the identity, then $\pi\restriction V_\alpha$ is the identity.*

PROOF. An easy transfinite induction using elementarity. The limit case is clear. Suppose then that $\pi\restriction\alpha+1$ is the identity. Fix $x \in V_{\alpha+1}$, namely $x \subseteq V_\alpha$. Then by elementarity, $w \in x$ iff $\pi(w) \in \pi(x)$. Since $\pi\restriction V_\alpha$ is the identity by induction, $w \in x$ iff $w \in \pi(x)$ for all $w \in V_\alpha$. To get $\pi(x) = x$, it is enough then to show that $\pi(x) \subseteq V_\alpha$. Since $x \subseteq V_\alpha$, by elementarity $\pi(x) \subseteq V^M_{\pi(\alpha)}$, where $V^M_{\pi(\alpha)}$ is the $\pi(\alpha)$-th level of the von Neumann hierarchy computed in M. Using now the assumption that $\pi(\alpha) = \alpha$ we get $\pi(x) \subseteq V^M_\alpha$. An absoluteness argument shows V^M_α is equal to $V_\alpha \cap M$, hence $\pi(x) \subseteq V_\alpha \cap M \subseteq V_\alpha$. □

It follows that for any (non-trivial) elementary $\pi\colon V \to M$, there exists some ordinal α so that $\pi(\alpha) \neq \alpha$. The least such ordinal is called the *critical point* of π, denoted $\mathrm{Crit}(\pi)$.

CLAIM 1.2. *For any ordinal α, $\pi(\alpha) \geq \alpha$. In particular $\pi(\mathrm{Crit}(\pi)) > \mathrm{Crit}(\pi)$.*

PROOF. Again an induction. If α is least so that $\pi(\alpha) < \alpha$, then setting $\beta = \pi(\alpha)$ we have $\pi(\beta) \geq \beta = \pi(\alpha)$. But then $\beta < \alpha$ while $\pi(\beta) \geq \pi(\alpha)$, contradicting elementarity. □

Let $\kappa = \mathrm{Crit}(\pi)$. It is easy to see that κ must be a cardinal: If $f\colon \lambda \to \kappa$ is a surjection with $\lambda < \kappa$, then by elementarity $\pi(f)$ is a surjection of $\lambda = \pi(\lambda)$ onto $\pi(\kappa) > \kappa$. There is then $\alpha < \lambda$ so that $\pi(f)(\alpha) = \kappa$. But by elementarity and since $\alpha = \pi(\alpha)$, $\pi(f)(\alpha) = \pi(f(\alpha))$. Since $f(\alpha) < \kappa$, $\pi(f(\alpha)) = f(\alpha) < \kappa$, contradiction.

It is also easy to see that (1) $\kappa = \mathrm{Crit}(\pi)$ is not the (cardinal) successor of any $\lambda < \kappa$, hence it must be a limit cardinal; (2) κ does not have cofinality $\lambda < \kappa$, hence it must be regular; and (3) there is no bijection of κ and the powerset of λ for any $\lambda < \kappa$. The critical point of an elementary embedding must therefore be (strongly) inaccessible. In particular, ZFC does not prove the existence of non-trivial elementary embeddings. In fact, by elementarity, the critical point cannot even be the first inaccessible cardinal: κ is inaccessible both in V and in M. Thus in M there is an inaccessible cardinal below $\pi(\kappa)$, and by elementarity there is an inaccessible cardinal below κ in V.

Indeed κ cannot be the first cardinal with any property whose definition involves only quantifiers on subsets of κ. Any such definition is absolute between V and M. (V and M have the same subsets of κ, since for any $x \subseteq \kappa$, $x = \pi(x) \cap \kappa$, and $\pi(x) \cap \kappa$ belongs to M.) If the property holds of κ in V then it holds of $\kappa < \pi(\kappa)$ in M, and hence by elementarity it holds of some $\alpha < \kappa$ in V.

The critical point of an elementary embedding must therefore be a very large cardinal. Axioms asserting the existence of elementary embeddings in effect assert the existence of cardinals with sufficient reflection properties that they can serve as critical points. The properties can be captured using ultrafilters.

Consider an elementary $\pi \colon V \to M$, and let $\kappa = \mathrm{Crit}(\pi)$. Define a collection U of subsets of κ setting $X \in U$ iff $\kappa \in \pi(X)$.

CLAIM 1.3. *U is an ultrafilter on $\mathcal{P}(\kappa)$. U is non-principal, meaning that no singleton set belongs to U. U is $<\kappa$ complete, meaning that the intersection of any family of fewer than κ elements of U is itself in U. U is normal, meaning that any function $f \colon \kappa \to \kappa$ which is regressive on a set in U, is constant on a set in U. (f is* regressive *on $A \subseteq \kappa$ if $f(\alpha) < \alpha$ for all $\alpha \in A$.)*

PROOF. It is easy to verify that U is a filter using the elementarity of π. For example the formula $(\forall\alpha)(\alpha \in X \wedge \alpha \in Y \to \alpha \in X \cap Y)$ is true in V. Hence by elementarity $(\forall\alpha)(\alpha \in \pi(X) \wedge \alpha \in \pi(Y) \to \alpha \in \pi(X \cap Y))$. Applying this with $\alpha = \kappa$ it follows that if X and Y both belong to U, then so does $X \cap Y$. A similar argument replacing the pair $\langle X, Y \rangle$ with a sequence $\langle X_\xi \mid \xi < \tau \rangle$ for $\tau < \kappa$ establishes that U is $<\kappa$ complete.

We leave most other clauses to the reader, and continue only to show that U is an ultrafilter. Let $X \subseteq \kappa$. Then $X \cup (\kappa - X) = \kappa$. Hence by elementarity $\pi(X) \cup \pi(\kappa - X) = \pi(\kappa)$. Since $\pi(\kappa) > \kappa$ it follows that κ belongs to $\pi(X) \cup \pi(\kappa - X)$, hence one of X, $\kappa - X$ is in U. □

CLAIM 1.4. *Let U be a normal, $<\kappa$ complete non-principal ultrafilter on $\mathcal{P}(\kappa)$. Then there is an elementary embedding $\pi \colon V \to M$, with $\mathrm{Crit}(\pi) = \kappa$ and $X \in U$ iff $\kappa \in \pi(X)$ for all $X \subseteq \kappa$.*

PROOF SKETCH. Let $M^* = V^\kappa/U$; elements of M^* are equivalence classes of functions $f \colon \kappa \to V$ under the relation $f \sim g$ iff $\{\alpha \mid f(\alpha) = g(\alpha)\}$ belongs to U, and M^* is endowed with one binary relation $\in^*$ given by $f \in^* g$ iff $\{\alpha \mid f(\alpha) \in g(\alpha)\}$ belongs to U. Define $\pi^* \colon V \to M^*$ by $\pi(x) = [\mathrm{Cnst}_x]$, where $\mathrm{Cnst}_x \in V^\kappa$ is the function with constant value x. A standard result, called Łos's theorem, states that formula $\varphi([f_1], \ldots, [f_n])$ holds in M^* iff the set $\{\alpha \mid \varphi(f_1(\alpha), \ldots, f_n(\alpha))\}$ belongs to U. It follows from this that π^* is elementary from V to M^*. (In particular, M^* satisfies the axiom of extensionality.)

Using $<\kappa$ completeness of U, indeed countable completeness is enough, one can reduce the wellfoundedness of $\in^*$ to the wellfoundedness of $\in$, proving that $\in^*$ is wellfounded. It is easy to see that $\in^*$ is set-like, meaning that every element of M^* has only a set (not a proper class) of inequivalent $\in^*$ predecessors; indeed, all $\in^*$ predecessors of f have representatives whose rank in the von Neumann hierarchy is no larger than the rank of f. As a wellfounded set-like structure that satisfies the axiom of extensionality, M^* can be put in isomorphism with a transitive structure

M. Precisely, by $\in^*$ induction define $k(x) = \{k(z) \mid z \in^* x\}$, and let M be the image of k. Then $k\colon M^* \to M$ is an isomorphism, and M is transitive.

We now have an elementary embedding $\pi = k \circ \pi^*\colon V \to M$. The model M is called the *ultrapower* of V by the ultrafilter U, and $k \circ \pi^*$ is the *ultrapower embedding*.

Using the $<\kappa$ completeness of U and Łos's theorem, one can check that for any $\alpha < \kappa$, $f \in^* \mathrm{Cnst}_\alpha$ iff there exists $\beta < \alpha$ with $f \sim \mathrm{Cnst}_\beta$. It follows that $k([\mathrm{Cnst}_\alpha]) = \alpha$ for each $\alpha < \kappa$, and hence $k \circ \pi^* \restriction \kappa$ is the identity.

Let Id denote the identity function on κ. Then $\mathrm{Cnst}_\alpha \in^* \mathrm{Id}$ for every $\alpha < \kappa$, using completeness and the fact that U is non-principal. So $k([\mathrm{Id}]) \geq \kappa$. On the other hand, using normality any $f \in^* \mathrm{Id}$ is equivalent to Cnst_α for some $\alpha < \kappa$. So $k([\mathrm{Id}]) = \kappa$.

Note that $\mathrm{Id} \in^* \mathrm{Cnst}_\kappa$, so $k\circ\pi^*(\kappa) = k([\mathrm{Cnst}_\kappa]) > k([\mathrm{Id}]) = \kappa$. Since $k\circ\pi^* \restriction \kappa$ is the identity, it follows that $\mathrm{Crit}(k\circ\pi^*) = \kappa$. Finally, for any $X \subseteq \kappa$, $\kappa \in k\circ\pi^*(X)$ iff $\mathrm{Id} \in^* \mathrm{Cnst}_X$ iff $\{\alpha \mid \mathrm{Id}(\alpha) \in \mathrm{Cnst}_X(\alpha)\}$ belongs to U iff $X = \{\alpha \mid \alpha \in X\} \in U$. □

A cardinal κ is *measurable* if it is the critical point of an elementary embedding. Equivalently by the last two claims, κ is measurable iff there is a normal, $<\kappa$ complete non-principal ultrafilter on $\mathcal{P}(\kappa)$. The characteristic function of such an ultrafilter is called a *measure* on κ.

The definition of measurable cardinals is due to Ulam (see [**43**]) who characterized them using measures. The modern approach through elementary embeddings which we followed above is largely due to Scott, who used it in [**34**] to prove that there are no measurable cardinals in the constructible universe L.

In modern view, measures extract a certain *canonical* content from an elementary embedding. The first indication of this was a theorem of Kunen [**19**], that if U and U' are two measures on κ, then $L[U] = L[U']$ and $U \cap L[U] = U' \cap L[U']$. This result was the seed for a development of a theory of canonical inner models for large cardinal axioms. These are built from sequences of measures, and more generally objects called extenders that extract more from an embedding than a single ultrafilter. Using results similar to Kunen's theorem one can show that the inner models constructed are canonical, in much the same way that L is. One then obtains canonical inner models for large cardinal axioms. The use of ultrafilters (and more generally extenders) in characterizing the large cardinals embeddings is essential for canonicity, and indeed Kunen's theorem and its generalizations to stronger large cardinal axioms rely on iterated constructions of ultrapowers, much like the one in Claim 1.4

It is still open whether all large cardinal axioms admit canonical inner models. To give the reader an impression of the reach of current constructions, we list a few more axioms in the large cardinal hierarchy.

A cardinal κ is *δ-strong* if there is a δ-strong elementary embedding $\pi\colon V \to M$ with critical point κ and $\pi(\kappa) > \delta$. (Recall π is δ-strong if all bounded subsets of δ belong to M.) κ is δ-strong *with respect to* H if in addition $\pi(H) \cap \delta = H \cap \delta$. As $\pi(\kappa) > \delta$ this is equivalent to $\pi(H \cap \kappa) \cap \delta = H \cap \delta$, so the condition asserts that π stretches $H \cap \kappa$ to agree with $H \cap \delta$.

A cardinal τ is *Woodin* if for every $H \subseteq \tau$, there exists $\kappa < \tau$ which is $<\tau$-strong with respect to H (meaning δ-strong with respect to H for all $\delta < \tau$).

A cardinal κ is λ-*supercompact* if there is a λ-supercompactness elementary embedding $\pi: V \to M$ with critical point κ and $\pi(\kappa) > \lambda$. (Recall π is λ-supercompact if $M^\lambda \subseteq M$.) κ is *supercompact* if it is λ-supercompact for all λ.

Supercompactness can be characterized using ultrafilters. Let $\mathcal{P}_\kappa(\lambda)$ denote the set $\{a \subseteq \lambda \mid \text{Card}(a) < \kappa\}$. A function $f: \mathcal{P}_\kappa(\lambda) \to \lambda$ is *regressive* on X if for any $a \in X$, $f(a) \in a$. A filter U on $\mathcal{P}_\kappa(\lambda)$ is *normal* if any function regressive on a set in U is constant on a set in U. The filter is *fine* if for any $\xi \in \lambda$ the set $\{a \in \mathcal{P}_\kappa(\lambda) \mid \xi \in a\}$ belongs to U.

It is easy to see that a λ-supercompactness embedding $\pi: V \to M$ with critical point κ and $\pi(\kappa) > \lambda$ gives rise to a $<\kappa$ complete, fine, normal ultrafilter U on $\mathcal{P}_\kappa(\lambda)$. To see this note that by supercompactness, $\pi''\lambda$ belongs to M. Let $s = \pi''\lambda$, and define U by setting $X \in U$ iff $s \in \pi(X)$ for each $X \subseteq \mathcal{P}_\kappa(\lambda)$. Arguments similar to the ones in the proof of Claim 1.3 show that U is a $<\kappa$ complete, fine, normal ultrafilter. It is also true that such an ultrafilter gives rise to a λ-supercompactness embedding. The arguments for this direction are similar to these in the proof of Claim 1.4. This characterization of supercompactness embeddings using ultrafilters is due to Solovay. The ultrafilters, namely $<\kappa$ complete, fine, normal ultrafilters measuring subsets of $\mathcal{P}_\kappa(\lambda)$, are called *supercompactness* ultrafilters.

Measurable, Woodin, and supercompact are three of the most important large cardinal notions. Measurable cardinals are the starting point for the hierarchy of elementary embeddings, Woodin cardinals are closely connected with axioms of determinacy, and supercompact cardinals are used in proving the consistency of various forcing axioms.

Currently the construction of canonical inner models reaches levels that involve Woodin cardinals and interactions between Woodin cardinals and strong embeddings. A recent theorem of Woodin shows that once the construction is brought to a specific level somewhat beyond supercompact cardinals, the resulting inner model in fact captures practically all large cardinal properties in V. But the construction of canonical inner models for supercompact cardinals is still very much an open question. It is one of the major research projects involving large cardinals.

Other research projects deal with uses of large cardinal axioms. Some of the earliest uses involved partition properties. For example:

CLAIM 1.5. *Suppose κ is measurable, and let U be a normal, $<\kappa$ complete non-principal ultrafilter on $\mathcal{P}(\kappa)$. Let c be a $<\kappa$-coloring of finite subsets of κ. (Precisely, c is a function defined on finite subsets of κ, with range of cardinality $\lambda < \kappa$.) Then there is a set $H \in U$ which is* homogeneous *for c, meaning that for each n, the restriction of c to subsets of H of size n is constant.*

PROOF. We adapt the standard proof of the infinite Ramsey's theorem on ω, replacing the largeness notion used there, namely "unbounded in ω", with "element of U".

For each $n < \omega$ let c_n be the restriction of c to sets of size n. It is enough to find a homogeneous set $H_n \in U$ for each c_n. Then $H = \bigcap_{n<\omega} H_n$ belongs to U by countable completeness, and is homogeneous for c by definition.

We prove by induction on n that there is a homogeneous set in U for any $<\kappa$-coloring c_n of subsets of κ of size n. The case of $n = 1$ is immediate using the $<\kappa$-completeness of U. So let us handle the inductive case.

Fix c_{n+1}. For each $s \subseteq \kappa$ of size n, there is $H_s \in U$ so that c_{n+1} is constant on $\{s \cup \{\alpha\} \mid \alpha \in H_s\}$. Let q_s be the color taken by c_{n+1} on this set. By induction, there is a homogeneous set $\bar{H} \in U$ for the coloring $s \mapsto q_s$. Let q be the constant value of q_s on $s \subseteq \bar{H}$.

Set $H = \{\alpha \in \bar{H} \mid \alpha \in H_s \text{ for all } s \subseteq \alpha \cap \bar{H}\}$. Clearly c_{n+1} is homogeneous on H, taking only the value q. And H belongs to U using normality: For every $\alpha \in \bar{H} - H$ there is $s \subseteq \alpha \cap \bar{H}$ so that $\alpha \notin H_s$. Let $f_1(\alpha), \dots, f_n(\alpha)$ list the elements of such an s. Then the functions f_i are regressive on $\bar{H} - H$. If H does not belong to U then $\bar{H} - H$ belongs to U, and by normality there is a fixed s so that $\{\alpha \mid \{f_1(\alpha), \dots, f_n(\alpha)\} = s\}$ belongs to U. But this is impossible, since this set is disjoint from H_s which belongs to U. □

The property of κ given by the claim, restricted to 2-colorings, is called the *Ramsey property.* The claim establishes that measurable cardinals are Ramsey. In fact measurability is stronger, for example giving a set $A \in U$ so that all δ in A are themselves Ramsey. For more on infinitary partition properties see Kanamori [**18**, Chapter 2]. Here let us mention only one more property, the *tree property.* A cardinal κ has this property if every tree T of height κ and width $< \kappa$ (meaning that every level of the tree has size $< \kappa$), has a cofinal branch. Note that the tree property for ω is simply König's lemma: every finitely branching tree of infinite height has an infinite branch.

CLAIM 1.6. *If κ is measurable then κ has the tree property.*

PROOF. Let T be a tree of height κ and width $< \kappa$. Let A_α consist of the nodes on level α of T. Let $\pi \colon V \to M$ be elementary with $\text{Crit}(\pi) = \kappa$. Re-naming the nodes of T, we may assume that $T \subseteq \kappa$, and A_α is a bounded subset of κ. Since $\pi \restriction \kappa$ is the identity, it follows that $\pi(A_\alpha) = A_\alpha$.

The height of $\pi(T)$ is $\pi(\kappa) > \kappa$, so we may fix a node u of level κ in $\pi(T)$. Consider the branch of $\pi(T)$ leading to u, namely $Z = \{x \mid x <_{\pi(T)} u\}$. Each $x \in Z$ belongs to $\pi(A_\alpha)$ for some $\alpha < \kappa$, and since $\pi(A_\alpha) = A_\alpha$ it follows that $Z \subseteq \kappa$ and hence $<_T \restriction Z = <_{\pi(T)} \restriction Z$. From this and the fact that Z is a branch of $\pi(T)$ of height κ, it follows that Z is also a branch of T of height κ. □

The tree property fails at the successor of κ, for any regular κ so that $2^{<\kappa} = \kappa$, and hence certainly for any measurable cardinal. For singular κ, the tree property may hold at κ^+, but the known examples require substantial large cardinal strength. For example it is shown in Magidor–Shelah [**24**] that if $\text{Cof}(\kappa) = \omega$ and κ is a limit of supercompact cardinals, then the tree property holds at κ^+.

2. Forcing

Let M be a transitive model of ZFC. *Forcing* is a technique introduced by Paul Cohen [**5, 6**], for adjoining an external set G to M, to obtain an extension $M[G]$ of the initial model M. Our exposition of the technique follows the treatment in Shoenfield [**39**].

The new set G is a subset of some $\mathbb{P} \in M$. $M[G]$ consists of all sets which can be constructed from elements of M using the new set G. One way to make this precise is the following. For each $\tau \in M$ define $\tau[G] = \{\sigma[G] \mid (\exists p \in \mathbb{P}) \langle \sigma, p\rangle \in \tau \wedge p \in G\}$. This definition is made by induction on the rank of τ. Then define

$M[G] = \{\tau[G] \mid \tau \in M\}$. In forcing lingo, elements p of $\mathbb{P}$ are called *conditions*, and $\tau \in M$ is said to *name* $\tau[G]$.

CLAIM 2.1. *$M \cup \{G\} \subseteq M[G]$. In particular, if $G \notin M$, then $M[G]$ is a proper extension of M.*

PROOF. By induction on rank define $\check{x} = \{\langle \check{y}, p\rangle \mid y \in x \wedge p \in \mathbb{P}\}$. Then for every $x \in M$, $\check{x} \in M$ and $\check{x}[G] = x$. So $M \subseteq M[G]$.

Let $\dot{G} = \{\langle \check{p}, p\rangle \mid p \in \mathbb{P}\}$. Then $\dot{G}[G] = \{\check{p}[G] \mid p \in G\} = \{p \mid p \in G\} = G$, and therefore $G \in M[G]$. □

The fundamental theorem of forcing allows reasoning about $M[G]$ from within M. Precisely, there is a relation $\Vdash^M_{\mathbb{P}}$ on $\mathbb{P} \times$ {formulas with parameters}, so that (with a certain restriction on $\mathbb{P}$ and G, see below):

(1) If $M[G] \models \varphi(\tau_1[G], \dots, \tau_n[G])$, then there exists $p \in G$ so that $p \Vdash^M_{\mathbb{P}} \varphi(\tau_1, \dots, \tau_n)$.
(2) If $p \Vdash^M_{\mathbb{P}} \varphi(\tau_1, \dots, \tau_n)$, then $M[G] \models \varphi(\tau_1[G], \dots, \tau_n[G])$ for all G with $p \in G$.
(3) For each formula φ, the relation $p \Vdash^M_{\mathbb{P}} \varphi(\tau_1, \dots, \tau_n)$ is uniformly definable from $\mathbb{P}$ over M.

$p \Vdash^M_{\mathbb{P}} \varphi(\tau_1, \dots, \tau_n)$ is read *p forces $\varphi(\tau_1, \dots, \tau_n)$* in $\mathbb{P}$ over M.

Because of the definability requirement (3) one cannot expect the conditions to hold without some restriction on $\mathbb{P}$ and G. After all, conditions (1) and (2) involve quantification over G, which is not an element of M, and at face value may lead to a relation which is not definable over M. One of Cohen's great insights is that nonetheless, conditions (1) and (2) *do* hold for a relation with the definability in (3), provided one places the restriction that $\mathbb{P}$ is a partially ordered set (*poset*), and G ranges over M-generic filters on $\mathbb{P}$.

DEFINITION 2.2. A subset G of a poset $\mathbb{P}$ is a *filter* on $\mathbb{P}$ if (a) it is closed upward under the order of $\mathbb{P}$, and (b) every two elements p_1, p_2 of G are compatible within G, meaning that there is $r \in G$ which is below both p_1 and p_2 in the order of $\mathbb{P}$.

An *antichain* of $\mathbb{P}$ is a set of pairwise incompatible elements. G is *generic* over M if it meets every maximal antichain of $\mathbb{P}$ that belongs to M.

A subset D of $\mathbb{P}$ is *dense* if $(\forall p \in P)(\exists r < p) r \in D$. (By $r < p$ we mean that r is below p in the order of $\mathbb{P}$. We also say that r is *stronger* than p.) Genericity over M is equivalent to meeting every dense subset of $\mathbb{P}$ that belongs to M.

In general there is no guarantee that generic subsets of $\mathbb{P}$ must exist. But if M is countable then M generic sets can be constructed by a simple induction, using an enumeration in order type ω of the dense sets which belong to M. Typically they cannot exist inside M, and indeed this is one of the points of the method of forcing: to adjoin a set G that does not belong to M.

We assume for the rest of the section that $\mathbb{P}$ is a poset in M. We also assume that for every $p \in \mathbb{P}$, there is a generic G with $p \in G$. (This holds for example if M is countable.)

THEOREM 2.3 (The fundamental theorem of forcing). *For every formula φ there is a relation $p \Vdash^M_{\mathbb{P}} \varphi(\tau_1, \dots, \tau_n)$, uniformly definable from $\mathbb{P}$ over M, so that conditions (1) and (2) above hold when G is restricted to range over M generic filters on $\mathbb{P}$.*

We refer the reader to Kunen [**20**] and Jech [**16**] for a proof of the fundamental theorem. The theorem allows for a very powerful analysis of the extension model $M[G]$. For example, if the poset $\mathbb{P}$ is countable in M, then using the theorem one can show that every cardinal of M remains a cardinal in $M[G]$. This is Claim 2.8 below.

REMARK 2.4. By extending the relation $\Vdash_{\mathbb{P}}^{M}$ if needed, we may assume that if p forces $\varphi(\tau_1, \dots, \tau_n)$ and $q < p$, then q also forces $\varphi(\tau_1, \dots, \tau_n)$. The extended relation still satisfies (1) and (2). That (1) continues to hold is trivial. The proof of (2) for the extended relation uses the upward closure of filters.

REMARK 2.5. Again by extending $\Vdash_{\mathbb{P}}^{M}$ if needed, we may assume that if no $q < p$ forces the negation of $\varphi(\tau_1, \dots, \tau_n)$, then p forces $\varphi(\tau_1, \dots, \tau_n)$. The extended relation trivially continues to satisfy (1). To see that (2) continues to hold note that if there is a generic G with $p \in G$ such that $M[G] \not\models \varphi(\tau_1[G], \dots, \tau_n[G])$, then there is $q < p$ forcing $\neg\varphi(\tau_1, \dots, \tau_n)$. (There is some $\bar{q} \in G$ forcing $\neg\varphi(\tau_1, \dots, \tau_n)$ by (1). Since p and $\bar{q}$ both belong to G, they are compatible in G, so there is $q \in G$ stronger than both. By the previous remark and since $q < \bar{q}$, q forces $\neg\varphi(\tau_1, \dots, \tau_n)$.)

CLAIM 2.6. *For any $p \in \mathbb{P}$ and any $\varphi(\tau_1, \dots, \tau_n)$, there exists $q < p$ which either forces $\varphi(\tau_1, \dots, \tau_n)$, or forces $\neg\varphi(\tau_1, \dots, \tau_n)$. (Such q is said to* decide *$\varphi(\tau_1, \dots, \tau_n)$.)*

PROOF. This is immediate from the previous remark, but for clarity we give a direct proof.

Let G be M-generic on $\mathbb{P}$. Then $\varphi(\tau_1[G], \dots, \tau_n[G])$ either holds or fails in $M[G]$. If it holds, then by condition (2) there is $q \in G$ forcing $\varphi(\tau_1, \dots, \tau_n)$, and using Remark 2.4 and compatibility of conditions in G we may take $q < p$. A similar argument produces a condition forcing the negation of $\varphi(\tau_1, \dots, \tau_n)$ in the case that $\varphi(\tau_1[G], \dots, \tau_n[G])$ fails in $M[G]$. □

REMARK 2.7. Suppose p forces "τ is a function with domain $\check{\lambda}$". Then for every $\alpha \in \lambda$ there is β and $q < p$ so that q forces "τ applied to $\check{\alpha}$ is equal to $\check{\beta}$". (We write this simply as $q \Vdash_{\mathbb{P}}^{M} \tau(\check{\alpha}) = \check{\beta}$.) The proof is similar to that of the previous claim. Let G be M-generic with $p \in G$. Then $f = \tau[G]$ is a function. Let $\beta = f(\alpha)$, and find $q \in G$ below p forcing $\tau(\check{\alpha}) = \check{\beta}$.

CLAIM 2.8. *Suppose $\mathbb{P}$ is countable in M, and G is an M-generic filter on $\mathbb{P}$. Then all cardinals of M remain cardinals in $M[G]$.*

PROOF. Let κ be a cardinal of M and suppose for contradiction that there is $\lambda < \kappa$ and a surjective function $f\colon \lambda \to \kappa$ in $M[G]$. Let $\tau \in M$ be such that $f = \tau[G]$.

Fix $p_0 \in G$ forcing that τ is a function. Such p_0 exists using (1). By Remark 2.4, every $p < p_0$ forces τ to be a function. For each $p < p_0$ and each $\alpha < \lambda$, there exists at most one ordinal β so that $p \Vdash_{\mathbb{P}}^{M} \tau(\check{\alpha}) = \check{\beta}$. For otherwise, taking a generic H with $p \in H$, we would get that $\tau[H]$ applied to α gives two distinct values, contradicting the fact that $\tau[H]$ is a function.

Define $\beta_{p,\alpha}$ to be this unique ordinal when it exists, and otherwise leave $\beta_{p,\alpha}$ undefined. Since the relation $p \Vdash_{\mathbb{P}}^{M} \tau(\check{\alpha}) = \check{\beta}$ is definable over M, the assignment $p, \alpha \mapsto \beta_{p,\alpha}$ ($p \in \mathbb{P}$, $\alpha \in \lambda$) is definable over M, and by the Replacement axiom it belongs to M. Let A be the range of the assignment. Then A belongs to M and the cardinality of A in M is at most $\mathrm{Card}^M(\mathbb{P}\times\lambda)$. Since $\lambda<\kappa$ and $\mathbb{P}$ is countable in M, this cardinality is smaller than κ. Hence, $M \models \mathrm{Card}(A)<\kappa$.

But since f is a surjection onto κ, for every $\beta<\kappa$ there is some α so that $f(\alpha)=\beta$. Using (1), there is $p\in G$ which forces $\tau(\check{\alpha})=\check{\beta}$. Since both p and p_0 belong to G they are compatible in G, and so strengthening p we may assume $p<p_0$. But then p and α are such that $\beta_{p,\alpha}=\beta$, and so $\beta\in A$.

We have shown on the one hand that $M\models \mathrm{Card}(A)<\kappa$, and on the other hand that $\kappa\subseteq A$. This contradiction completes the proof of the claim, showing that $f\in M[G]$ cannot embed $\lambda<\kappa$ onto κ. The driving force in the proof is our ability to estimate f inside M using (1) and (2) for a relation definable over M. The estimate is sufficiently close to derive a contradiction, because $\mathbb{P}$ is countable. □

Continuing to work with the definitions of the last proof, it is easy to see that if $p<q$ and $\beta_{q,\alpha}$ is defined, then $\beta_{p,\alpha}$ is defined and equal to $\beta_{q,\alpha}$. It follows that the size of $\{\beta_{p,\alpha}\mid p\in\mathbb{P}\}$ is the size of a maximal antichain in $\mathbb{P}$. Thus we could replace the assumption in the last claim that $\mathbb{P}$ is countable in M, with the assumption that, in M, all antichains of $\mathbb{P}$ are countable.

Definition 2.9. A poset $\mathbb{P}$ has the *countable chain condition* if all antichains of $\mathbb{P}$ are countable.

Claim 2.10. *Suppose $\mathbb{P}$ has the countable chain condition in M, and G is an M-generic filter on $\mathbb{P}$. Then all cardinals of M remain cardinals in $M[G]$. Moreover cofinalities are preserved between M and $M[G]$: If $M\models \mathrm{Cof}(\kappa)=\delta$ then $M[G]\models \mathrm{Cof}(\kappa)=\delta$ too.*

Proof. Similar to the proof of Claim 2.8. In the second part, adapt the argument to work with cofinal $f\colon \lambda\to\kappa$ in $M[G]$ for $\lambda<\delta$, to obtain an approximation in M which on the one hand is cofinal in κ and on the other has size λ, contradicting the fact that $\mathrm{Cof}(\kappa)>\lambda$ in M. □

Claim 2.11. *Suppose M is a model of* ZFC, *and G is M-generic for $\mathbb{P}$. Then $M[G]$ is also a model of* ZFC

Proof sketch. Use the forcing relation in M and conditions (1) and (2) to reduce ZFC axioms in $M[G]$ to statements in M that are provable from ZFC. Let us explicitly handle only one axiom here. We show that Pairing holds in $M[G]$.

Fix $x,y\in M[G]$. We show that $\{x,y\}\in M[G]$. Let τ and σ be such that $x=\tau[G]$ and $y=\sigma[G]$. Let $\rho=\{\langle\tau,p\rangle\mid p\in\mathbb{P}\}\cup\{\langle\sigma,p\rangle\mid p\in\mathbb{P}\}$. Then using the axioms of ZFC in M (for example Union, Pairing, Comprehension), ρ is a set that belongs to M. It is clear that $\rho[G]=\{\tau[G],\sigma[G]\}=\{x,y\}$, so $\{x,y\}\in M[G]$.

For some other axioms the arguments are more involved. For example in the case of Comprehension, one is given a set $x=\dot{x}[G]\in M[G]$, and a formula φ, and must show that the set $\{u\in x\mid M[G]\models\varphi(u)\}$ belongs to $M[G]$. The proof uses the forcing relation to define in M a name $\dot{z}$ for this set. Briefly, $\dot{z}=\{\langle\dot{u},p\rangle\in V_\alpha^M\mid p\Vdash_{\mathbb{P}}^M \dot{u}\in\dot{x}\wedge\varphi(\dot{u})\}$ (for $\alpha\in M$ large enough that $\dot{x}\in V_\alpha^M$). $\dot{z}$ belongs to M since the forcing relation is definable in M, and M satisfies Comprehension. Using (1) and (2) one can show that $\dot{z}[G]=z$, so $z\in M[G]$ as needed. We refer the reader to Kunen [**20**] for a full proof, and proofs of the other axioms of ZFC in $M[G]$. □

Forcing was first used by Cohen to obtain a model where the continuum hypothesis fails, that is a model where there are more than $\aleph_1$ subsets of $\aleph_0=\omega$. This is done by using a poset G which "adds" subsets of ω.

DEFINITION 2.12. $\mathrm{Add}(\omega,\kappa)$ is the poset $\mathbb{P}$ of finite partial functions from $\kappa\times\omega$ into $\{0,1\}$, ordered by reverse inclusion, meaning $p<q$ if $p\supseteq q$.

FACT 2.13. $\mathrm{Add}(\omega,\kappa)$ has the countable chain condition. This is a combinatorial claim. We omit the proof, and refer the reader to [**20**, Chapter 2 exercise 20].

CLAIM 2.14. *Let $\kappa=\aleph_2^M$ and let $\mathbb{P}=\mathrm{Add}(\omega,\kappa)$. Let G be M-generic on $\mathbb{P}$. Then $M[G]$ satisfies "there are at least κ subsets of ω" and "κ is $\aleph_2$". In particular $M[G]$ is a model of the failure of the continuum hypothesis.*

PROOF. Each element of $G\subseteq\mathbb{P}$ is a finite partial function from $\kappa\times\omega$ into $\{0,1\}$. Since all elements of G are compatible, these finite partial functions agree with each other on any common part of their domain. So $f=\bigcup\{p\mid p\in G\}$ is itself a function from $\kappa\times\omega$ into $\{0,1\}$. Using genericity it is easy to see that f is total. (For any $\langle\alpha,n\rangle\in\kappa\times\omega$, the set $\{p\mid\langle\alpha,n\rangle\in\mathrm{Dom}(p)\}$ is dense and belongs to M. The fact that G meets this set implies that $\langle\alpha,n\rangle\in\mathrm{Dom}(f)$.)

Again using genericity, it is easy to see that for every $\alpha\neq\beta$, there is n so that $f(\alpha,n)\neq f(\beta,n)$. (Since all elements of $\mathbb{P}$ are finite, the set $\{p\mid(\exists n)\ p(\alpha,n)$ and $p(\beta,n)$ are both defined and take different values$\}$ is dense.) Letting $x_\alpha=\{n\mid f(\alpha,n)=1\}$ it follows that $x_\alpha\neq x_\beta$ for $\alpha\neq\beta$.

The sequence $\langle x_\alpha\mid\alpha<\kappa\rangle$ is built from G and it is not hard to show that it belongs to $M[G]$. Since the x_αs are distinct subsets of ω it follows that in $M[G]$ there are at least κ subsets of ω.

It remains to show that κ is the second uncountable cardinal of $M[G]$. But this is immediate since M and $M[G]$ have exactly the same cardinals, by Claim 2.10. □

More generally forcing can be used to increase the powerset of any regular cardinal δ.

A poset $\mathbb{P}$ has the δ^+ *chain condition* if every antichain of $\mathbb{P}$ has size at most δ. For $\delta<\kappa$ let $\mathrm{Add}(\delta,\kappa)$ be the poset of partial functions of size $<\delta$, from $\kappa\times\delta$ into $\{0,1\}$. The arguments above generalize to give the following facts:

- $\mathrm{Add}(\delta,\kappa)$ has the $(2^{<\delta})^+$ chain condition. ($2^{<\delta}$ is $\sup\{2^\lambda\mid\lambda<\delta\}$.) In particular, if $2^\lambda=\lambda^+$ for all $\lambda<\delta$, then the poset has the δ^+ chain condition.
- If $\mathbb{P}$ has the δ^+ chain condition in M, and G is M-generic on $\mathbb{P}$, then M and $M[G]$ have the same cardinals above δ. They also agree on cofinalities of cardinals whose cofinality in M is at least δ^+.
- If $\mathbb{P}=\mathrm{Add}(\delta,\kappa)^M$ and G is M-generic for $\mathbb{P}$ then in $M[G]$ there are at least κ subsets of δ.

It follows from the above facts that if we start from a model M that satisfies the generalized continuum hypothesis ($2^\lambda=\lambda^+$ for all λ), and pass to a generic extension $M[G]$ using the poset $\mathbb{P}=\mathrm{Add}(\delta,\kappa)^M$ for $\kappa=(\delta^{++})^M$, then $M[G]$ satisfies "$2^\delta\geq\delta^{++}$". However this by itself does not mean that we managed to increase the powerset of a cardinal greater than $\aleph_0$, since we have not yet ruled out that $M[G]\models$"δ is countable". The next two claims rule this out.

DEFINITION 2.15. A poset $\mathbb{P}$ is $<\delta$-closed if for any $\lambda<\delta$ and any decreasing sequence $\langle p_\xi\mid\xi<\lambda\rangle$ of elements of $\mathbb{P}$, there is $q\in\mathbb{P}$ below all p_ξ.

CLAIM 2.16. *If δ is regular then* $\mathrm{Add}(\delta, \kappa)$ *is* $<\delta$ *closed.*

PROOF. Fix a decreasing sequence $\langle p_\xi \mid \xi < \lambda \rangle$, with $\lambda < \delta$, and let $q = \bigcup_{\xi<\lambda} p_\xi$. Since the sequence is decreasing, all functions p_ξ agree with each other, and q is therefore also a function. q is a union of λ sets each of size $< \delta$. Since δ is regular, q too has size $< \delta$. So q is an element of $\mathbb{P}$. It is clear that $q < p_\xi$ for each ξ. □

CLAIM 2.17. *Let $\mathbb{P}$ be a poset which is $<\delta$ closed in M. Let G be M-generic on $\mathbb{P}$. Then M and $M[G]$ agree on cardinals below δ, and on cofinalities of cardinals whose cofinality in $M[G]$ is $< \delta$.*

PROOF. We show that any sequence of ordinals of length $< \delta$ that belongs to $M[G]$ in fact belongs to M. This implies the statement in the claim.

Let $f \in M[G]$ be a sequence of ordinals of length $\lambda < \delta$. Let τ be such that $f = \tau[G]$ and let $p \in G$ force that τ is a function with domain $\check{\lambda}$, taking ordinal values.

Working in M, and using the closure of $\mathbb{P}$, construct sequences $\langle p_\xi \mid \xi \leq \lambda \rangle$ and $\langle \beta_\xi \mid \xi < \lambda \rangle$ so that:

- $p_0 = p$.
- For each ξ, $p_{\xi+1} < p_\xi$ and $p_{\xi+1}$ forces $\tau(\check{\xi}) = \check{\beta}_\xi$.
- For limit α, p_α is a lower bound in $\mathbb{P}$ for the sequence $\langle p_\xi \mid \xi < \alpha \rangle$.

For the second item, $p_{\xi+1}$ and β_ξ can be chosen using Remark 2.7. The third item uses the closure of $\mathbb{P}$.

Since $p_\lambda < p_{\xi+1}$, p_λ forces $\tau(\check{\xi}) = \check{\beta}_\xi$ for each ξ. Let h be the function $\xi \mapsto \beta_\xi$. Since the entire construction was done in M, we have $h \in M$. We also have $p_\lambda \Vdash^M_{\mathbb{P}} \tau = \check{h}$.

Working in M, let $D = \{q \mid (\exists h) q \Vdash^M_{\mathbb{P}} \tau = \check{h}\}$. We showed that D is non-empty. The same argument, starting with an arbitrary $p_0 < p$, produces an element of D below p_0. So $D \in M$ is dense below p. Since $p \in G$, this together with M-genericity implies that G meets D. Fix then a condition $q \in D \cap G$. From the definition of D and the fact that $q \in G$ it now follows that $f = \tau[G]$ is equal to $h = \check{h}[G]$ for some $h \in M$, and therefore $f \in M$. □

Forcing can thus be used to violate the continuum hypothesis at any regular cardinal. For example, if M is a model of the generalized continuum hypothesis, and α is such that in M, $\aleph_\alpha$ is regular, then there is a generic extension $M[G]$ of M so that the two models agree on all cardinals, yet in $M[G]$, $2^{\aleph_\alpha} \geq \aleph_{\alpha+2}$. Furthermore, with posets $\mathrm{Add}(\delta, \kappa)$ for $\kappa = (\aleph_\beta)^M$, possibly much bigger than $(\aleph_{\alpha+2})^M$, we could make $2^{\aleph_\alpha} \geq \aleph_\beta$ hold in $M[G]$ for our favorite $\beta \in M$, no matter how large.

To increase the powerset of $\aleph_\alpha$ we had to assume it is regular; the assumption was used in Claim 2.16. It had been expected initially after Cohen's introduction of forcing that more sophisticated forcing constructions would allow similar manipulations of the powerset of singular cardinals. But it turned out that manipulating the powerset function at singular cardinals is much harder, and indeed in many cases it is outright impossible.

The *singular cardinals hypothesis* (SCH) asserts that for every singular κ, if $2^{<\kappa} = \kappa$, then $2^\kappa = \kappa^+$. One example of a theorem showing that some manipulations of the powerset function at singular cardinals are outright impossible is a result of Silver [**40**], that the smallest failure of the singular cardinals hypothesis cannot occur at a cardinal of uncountable cofinality. Another example is Shelah's famous result [**37**] that if $2^{\aleph_n} < \aleph_\omega$ for all n, then $2^{\aleph_\omega} < \aleph_{\omega_4}$. This is in stark contrast to the situation at a regular $\aleph_\alpha$, where $2^{\aleph_\alpha}$ can be made arbitrarily large.

Even in cases where the singular cardinals hypothesis can be violated, matters are complicated, and in fact the forcing notions involved make use of large cardinal ultrafilters.

3. Forcing with large cardinal ultrafilters

The basic idea for forcing a failure of the singular cardinals hypothesis is simple: start with a regular cardinal κ and force to increase its powerset, then force to make κ singular. The end result is a cardinal κ where $2^\kappa > \kappa^+$ through the first stage, and κ is singular through the second.

It is essential of course that κ remains a cardinal through both stages. But in general there is no guarantee that κ can be singularized without being collapsed. This is where large cardinals come in. We shall see that if κ is measurable, then there is a forcing notion that singularizes it without collapsing any cardinals. This will take care of the second stage, provided we can ensure that κ is measurable following the first stage.

The first stage involves iterated forcing. One could attempt to simply use the methods of the previous section, to construct a model $M[G]$ where $2^\kappa \geq \kappa^{++}$. But then κ may fail to be measurable in $M[G]$. Indeed, if κ is to be measurable, then it cannot be the *first* cardinal δ of $M[G]$ so that $2^\delta \geq \delta^{++}$. (The proof of this is similar to the arguments in Section 1 that a measurable cardinal cannot be the first α with a property whose definition involves quantifiers only over subsets of α.) But if M satisfies $2^\delta = \delta^+$ for all $\delta < \kappa$, then the preservation theorems in Section 2 show that $M[G]$ satisfies the same. It follows in this case that κ cannot be measurable in $M[G]$.

To get to a situation where in $M[G]$, κ is measurable and $2^\kappa = \kappa^{++}$, we must force to increase not only the powerset of κ, but also the powerset of many $\delta < \kappa$. This is where iterated forcing comes in. Through an iteration whose basic step is the forcing $\mathrm{Add}(\delta, \delta^{++})$, one creates an extension $M[G]$ satisfying $2^\delta = \delta^{++}$ for all inaccessible $\delta \leq \kappa$ in M.

One then argues that κ is measurable in $M[G]$, using a large cardinal assumption on κ in M. It is not enough to assume that κ is measurable in M; greater large cardinal strength is needed in M to ensure that measurability survives to $M[G]$. The first theorem along these lines was due to Silver, and the assumption needed for the proof was that κ is supercompact in M. The assumption of supercompactness was reduced by Woodin to the existence of an elementary embedding j with critical point κ so that $j(\kappa) > \kappa^{++}$ and the target model of j is closed under sequences of length κ. The existence of a model with such an embedding was proved by Gitik [**10**] to follow from the existence of many measures on κ, and all components combined led to Theorem 3.2 below:

DEFINITION 3.1. The Mitchell relation $\lhd$ on ultrafilters is given by $U \lhd U^*$ iff U belongs to the ultrapower of V by U^*. This relation is wellfounded. The *Mitchell order* of U, denoted $o(U)$, is its rank in the relation $\lhd$. The Mitchell order of a cardinal κ is $\sup\{o(U) \mid U \text{ a measure on } \kappa\}$.

THEOREM 3.2. *Let M be a countable model of* ZFC. *Suppose that in M, κ has Mitchell order κ^{++}. Then there is a countable model M^* where κ remains* *measurable, and* $2^\kappa = \kappa^{++}$. *All cardinals of* M *remain cardinals of* M^*. *(Indeed,* M^* *is a cardinals and cofinalities preserving generic extension of an inner model of* M.*)*

Let us now consider the second stage, that is singularizing a measurable cardinal κ. Fix a measurable cardinal κ in a model M. We attempt to find a forcing notion $\mathbb{P}$ so that in the resulting generic extension $M[G]$, $\mathrm{Cof}(\kappa) = \omega$.

Consider first the poset of finite partial increasing functions from ω to κ. If G is a generic filter on this poset, then $\bigcup G$ is a total increasing function from ω to κ. Using genericity it is easy to see that the function is cofinal in κ. So in $M[G]$ the cofinality of κ is ω. Alas, this forcing notion has no property that implies preservation of cardinals $\leq \kappa$. Indeed it is easy to see that it collapses κ (and all cardinals below it) to ω, meaning that in $M[G]$, κ is countable. So in $M[G]$, κ is not a cardinal at all, let alone a singular cardinal. We need a different poset, one that adds a cofinal function from ω into κ, but does not collapse cardinals below κ.

DEFINITION 3.3. Let κ be measurable, and let U be a normal, $<\kappa$ complete non-principal ultrafilter on $\mathcal{P}(\kappa)$. *Prikry forcing* using U is the poset whose elements are pairs $\langle s, A\rangle$, where s is a finite increasing sequence of ordinals in κ, and A is a subset of κ that belongs to U. s is called the *stem* of $\langle s, A\rangle$. The ordering of $\mathbb{P}$ is the relation $\langle t, B\rangle < \langle s, A\rangle$ iff t extends s, and all elements of $t - s$ belong to A.

Prikry forcing is similar to the simple forcing of increasing finite functions, but it includes restrictions on the possible growths of the function: an element $\langle s, A\rangle$ of the poset not only specifies an initial segment s of a function from ω into κ, but also a set A from which all elements in extensions of s must be taken. The set A is required to be large, in the sense that it must belong to the ultrafilter U.

LEMMA 3.4. *Let* κ *be measurable in* M *and let* U *be a normal,* $<\kappa$ *complete non-principal ultrafilter on* $\mathcal{P}(\kappa)$ *in* M. *Let* $\mathbb{P}$ *be Prikry forcing defined from* U *in* M. *Let* G *be* M*-generic on* $\mathbb{P}$.

Then $M[G] \models \mathrm{Cof}(\kappa) = \omega$, *and every cardinal of* M *is a cardinal in* $M[G]$. *Moreover if* $\mathrm{Cof}^M(\lambda) \neq \kappa$, *then* $\mathrm{Cof}^{M[G]}(\lambda) = \mathrm{Cof}^M(\lambda)$.

PROOF. Let $f = \bigcup\{s \mid (\exists A)\langle s, A\rangle \in G\}$. Since all conditions in G are compatible, f is the increasing union of finite increasing sequences of ordinals in κ. Using genericity it is easy to see that there are conditions $\langle s, A\rangle \in G$ with s of arbitrarily large finite length. (For every $n < \omega$, the set $D = \{\langle s, A\rangle \mid \mathrm{lh}(s) \geq n\}$ is dense.) Hence f is an increasing length ω sequence of ordinals in κ. Again using genericity it is easy to see that f is cofinal in κ. (For every $\alpha < \kappa$, The set $D = \{\langle s, A\rangle \mid s$ has elements above $\alpha\}$ is dense. This is because all sets $A \in U$ are unbounded in κ.) Thus, in $M[G]$, f is a sequence of length ω cofinal in κ. In particular $\mathrm{Cof}^{M[G]}(\kappa) = \omega$.

It remains to show that cardinals are not collapsed in the move from M to $M[G]$, and indeed no cofinalities are changed other than the fact that cardinals of cofinality κ in M have cofinality ω in the extension. For cardinals and cofinalities above κ the next claim, and the results of the previous section, suffice.

CLAIM 3.5. $\mathbb{P}$ *has the* κ^+ *chain condition in* M.

PROOF. It is enough to show that any two elements of $\mathbb{P}$ with the same stem are compatible. Since there are only $\kappa^{<\omega} = \kappa$ possible stem it follows that antichains cannot have size greater than κ.

Fix two elements $\langle s, A\rangle$ and $\langle s, B\rangle$ of $\mathbb{P}$ with the same stem s. To see that the elements are compatible, simply note that $\langle s, A\cap B\rangle$ is an element of $\mathbb{P}$ ($A\cap B\in U$ since U is closed under intersections), and it is below both $\langle s, A\rangle$ and $\langle s, B\rangle$. $\square$

Given the last claim, we know that M and $M[G]$ have the same cardinals above κ, and agree on cofinalities of cardinals whose cofinalities in M are greater than κ. It is for cardinals below κ that we really need the ultrafilter U. The methods of the previous section would allow us to prove preservation of cardinals below κ if $\mathbb{P}$ were $<\kappa$ closed. Of course Prikry forcing is not $<\kappa$ closed, as the stems are required to be finite. But for any stem s, the restriction of $\mathbb{P}$ to elements with stem equal to s *is* $<\kappa$ closed, because of the completeness of U. We intend to mirror a preservation proof from the previous section, using this restriction. The following claim provides the first step. It mirrors Claim 2.6.

CLAIM 3.6. *For any finite increasing sequence s of ordinals in κ, and any $\varphi(\tau_1,\dots,\tau_n)$, there is $A\in U$ so that either $\langle s,A\rangle \Vdash^M_{\mathbb{P}} \varphi(\tau_1,\dots,\tau_n)$ or $\langle s,A\rangle \Vdash^M_{\mathbb{P}} \neg\varphi(\tau_1,\dots,\tau_n)$.*

PROOF. Fix s and $\varphi(\tau_1,\dots,\tau_n)$. Working in M using the relation $\Vdash^M_{\mathbb{P}}$, define a coloring c on finite $a\subseteq\kappa$ as follows. If a has elements below $\sup(s)$, then $c(a)=$"small". Otherwise let t be the extension of s obtained by appending the elements of a to s in increasing order. If for some $Y\in U$, $\langle t,Y\rangle$ forces either $\varphi(\tau_1,\dots,\tau_n)$ or $\neg\varphi(\tau_1,\dots,\tau_n)$, then set $c(a)=$"true" or $c(a)=$"false" depending on whether the formula forced is $\varphi(\tau_1,\dots,\tau_n)$ or its negation. (They cannot both be forced, since any two conditions with the same stem t are compatible.) Otherwise set $c(a)=$"undefined".

c is a 4-coloring of finite subsets of κ. Using Claim 1.5 in M, there is a set $A\in U$ so that c is homogeneous on A. We claim that $\langle s,A\rangle$ provides the conclusion of the claim.

Suppose not. Then by Remark 2.5 there are two extensions $\langle t_1,B_1\rangle$ and $\langle t_2,B_2\rangle$ of $\langle s,A\rangle$ forcing $\varphi(\tau_1,\dots,\tau_n)$ and $\neg\varphi(\tau_1,\dots,\tau_n)$ respectively. Extending the conditions if needed we may assume that their stems have the same length l. Since both conditions are below $\langle s,A\rangle$, all elements of t_1-s and t_2-s belong to A. So $c(t_1-s)=c(t_2-s)$. But $c(t_1-s)=$"true" since $\langle t_1,B_1\rangle$ forces $\varphi(\tau_1,\dots,\tau_n)$, while $c(t_2-s)=$"false" since $\langle t_2,B_2\rangle$ forces $\neg\varphi(\tau_1,\dots,\tau_n)$. This contradiction completes the proof of the claim. $\square$

CLAIM 3.7. *Let $f\colon\lambda\to\delta$ belong to $M[G]$, where both λ and δ are smaller than κ. Then f belongs to M.*

PROOF. Let τ be such that $f=\tau[G]$. Let $p=\langle s,A\rangle$ force "τ is a function from $\check\lambda$ into $\check\delta$". We prove that there is a function $h\in M$ and $q<p$ so that $q\Vdash^M_{\mathbb{P}}\tau=\check h$. It then follows, as in Claim 2.17, that $f\in M$.

Let $\langle\alpha_\xi,\beta_\xi\rangle$ for $\xi<\mathrm{Card}(\lambda\times\delta)$ enumerate $\lambda\times\delta$. Working in M, fix for each ξ a set $A_\xi\in U$ so that $\langle s,A_\xi\rangle$ either forces $\tau(\check\alpha)=\check\beta$, or forces $\tau(\check\alpha)\neq\check\beta$. This is possible using the previous claim. Let A^* be the intersection of all the sets A_ξ, and the set A. Since $\mathrm{Card}(\lambda\times\delta)<\kappa$, and since U is $<\kappa$ complete, A^* belongs to U.

$\langle s,A^*\rangle$ is stronger than each $\langle s,A_\xi\rangle$, so for every $\alpha\in\lambda$ and $\beta\in\delta$, either $\langle s,A^*\rangle$ forces $\tau(\check\alpha)=\check\beta$, or else it forces $\tau(\check\alpha)\neq\check\beta$. Holding α fixed, the latter cannot hold for all β, since $\langle s,A^*\rangle$ is also stronger than $\langle s,A\rangle$ which forces $(\exists\beta\in\check\delta)\tau(\check\alpha)=\beta$.

So for each $\alpha \in \lambda$ there exists $\beta \in \delta$ so that $\langle s, A^* \rangle$ forces $\tau(\check{\alpha}) = \check{\beta}$. This β must be unique, since $\langle s, A^* \rangle < \langle s, A \rangle$ forces "τ is a function".

Continuing to work in M, define $h(\alpha)$ to be the unique β so that $\langle s, A^* \rangle \Vdash^M_{\mathbb{P}} \tau(\check{\alpha}) = \check{\beta}$. Then $h\colon \lambda \to \delta$ belongs to M, and it is clear that $\langle s, A^* \rangle \Vdash^M_{\mathbb{P}} \tau = \check{h}$. □

Recall that Claim 3.5 implies that M and $M[G]$ have the same cardinals above κ, and agree on cofinalities of cardinals whose cofinalities in M are greater than κ. Our latest claim, 3.7, implies that M and $M[G]$ have the same cardinals below κ, and agree on cofinalities of cardinals whose cofinalities in M are smaller than κ. This completes the proof of Lemma 3.4. □

REMARK 3.8. Note that we did not parallel Remark 2.7 for the restriction of $\mathbb{P}$ to conditions with a fixed stem s. We only paralleled Claim 2.6. It is because of this difference that the argument above is not an exact parallel of Claim 2.17. It establishes preservation of cofinalities for cardinals whose cofinality is smaller than κ in M; Claim 2.17 applied more generally to cardinals whose cofinality is smaller than κ in $M[G]$. The more general preservation fails for Prikry forcing, since the cofinality of κ is not preserved.

COROLLARY 3.9. *Suppose M is a countable model of* ZFC *satisfying "there exists κ of Mitchell order κ^{++}". Then there is a countable model of* ZFC *where the singular cardinals hypothesis fails.*

PROOF. Using Theorem 3.2 there is a model M^* in which κ is measurable and $2^\kappa = \kappa^{++}$. Now using Prikry forcing over M^* one obtains an extension $M^*[G]$ where κ is singular of cofinality ω, and such that all cardinals of M^* remain cardinals in $M^*[G]$.

Since κ is measurable in M^*, $2^\delta < \kappa$ for $\delta < \kappa$, and using preservation of cardinals the same holds in $M^*[G]$. Since $2^\kappa = \kappa^{++}$ in M^*, and κ^{++} is the same in M^* and $M^*[G]$, $2^\kappa = \kappa^{++}$ also in $M^*[G]$. Thus, in $M^*[G]$, the singular cardinals hypothesis fails at κ. □

4. Further results

Corollary 3.9 is one of the most spectacular uses of large cardinal ultrafilters in forcing. There are many later results that build on similar techniques. For example Magidor [**21, 22**] develops techniques that combine Prikry forcing at a cardinal κ, with collapses between the points of the cofinal sequence of length ω added by the generic. Enough cardinals are collapsed below κ, that κ becomes $\aleph_\omega$ in the extension. The result is a model in which the singular cardinals hypothesis fails at $\aleph_\omega$:

THEOREM 4.1. *Suppose there is a countable model of* ZFC *satisfying "there exists a cardinal κ of Mitchell order κ^{++}". Then there is a countable model of* ZFC *satisfying failure of the singular cardinals hypothesis fails at $\aleph_\omega$.*

Additional methods to reach such a failure were discovered later, for example in Gitik–Magidor [**12, 13**].

In a different direction, Prikry forcing can be generalized to yield an arbitrary cofinality for κ in the generic extension. The method for doing this was developed by Magidor [**23**]. It has seen many applications since, including some that are not immediately related to singularizing cardinals. Among them for example is a recent theorem of Gitik–Neeman–Sinapova [**14**] that (assuming large cardinals) it is possible to construct models $M \subseteq M[G]$ which have the same reals and cardinals, yet the class of ordinals of countable cofinality in M is so thin in $M[G]$, that it is not stationary at any regular cardinal λ of M beyond a starting cardinal κ.

In yet another direction, Prikry forcing can be modified to use supercompactness measures. Some of the most recent applications include constructions by Gitik–Sharon [**15**], obtaining a model where failure of the singular cardinals hypothesis is combined with failures of various combinatorial properties. The model was analyzed further by Cummings–Foreman [**7**], and the construction was combined with Magidor forcing by Sinapova [**41**] to obtain similar results on singular cardinals of arbitrary cofinality. Later on it was combined in Neeman [**29**] with a result of Magidor–Shelah [**24**] that the tree property holds at the cardinal successor of a limit of supercompact cardinals, to yield the following theorem:

THEOREM 4.2. *Suppose M is a countable model with ω supercompact cardinals. Let κ be the first one. Then there exists a generic extension $M[G]$ of M in which* $\mathrm{Cof}(\kappa) = \omega$, $2^{<\kappa} = \kappa$, $2^\kappa = \kappa^{++}$, *and the tree property holds at κ^+.*

The theorem answers a question of Woodin and others on whether failure of the singular cardinals hypothesis at κ implies failures of the tree property at κ^+. More specifically Woodin asked whether this is the case for $\kappa = \aleph_\omega$, and this question remains open.

Note that the failure of the singular cardinals hypothesis given by the theorem is qualitatively different from that given by Corollary 3.9. The Corollary is proved by singularizing a measurable cardinal κ without collapsing any cardinals. The tree property *fails* at successors of measurable cardinals, and the manner of failure is sufficiently absolute that it continues to fail in cardinal preserving generic extensions. The theorem in contrast reaches a situation where the tree property holds at the successor of κ.

Despite this qualitative difference, the theorem too is proved using Prikry forcing. But this time the Prikry forcing is relative to supercompactness measures, and many cardinals are collapsed. Indeed, letting τ be the supremum of the ω supercompact cardinals of M, all cardinals between κ and τ are collapsed, and κ^+, κ^{++} of $M[G]$ are the cardinals τ^+, τ^{++} of M.

Among the many other applications of forcing with ultrafilters there are some that are done under the axiom of determinacy, rather than the axiom of choice. By $\mathbb{R}$ below we mean $\mathbb{N}^\omega$ with the product topology of the discrete topology on $\mathbb{N}$. (The space is isomorphic to the irrational numbers with the usual topology.) In a *game* with payoff $A \subseteq \mathbb{R}$, two players I and II alternate turns playing $x(i) \in \mathbb{N}$, $i < \omega$. Player I wins a run $x = \langle x(i) \mid i < \omega\rangle$ if $x \in A$, and otherwise player II wins. The game is *determined* if one of the players has a winning strategy. The set A is called the *payoff* for the game. Determinacy for a pointclass $\Gamma \subseteq \mathcal{P}(\mathbb{R})$ is the statement that every such game with payoff in Γ is determined.

Perhaps surprisingly, determinacy turns out to provide the appropriate axiomatization for studying definable sets of reals, for example Borel sets, the sets generated by projections and complementations starting from Borel sets, all sets in $L(\mathbb{R})$, and more. Determinacy for all Borel sets is a theorem of ZFC, proved by Martin [**25, 26**]. For the larger pointclasses mentioned above, determinacy is provable using large cardinal axioms in the region of Woodin cardinals, see Martin–Steel [**27**] and Woodin [**45**] for the original proof, Neeman [**30, 31**] for a proof from optimal assumptions, and Neeman [**33**] for a self-contained proof.

The full axiom of determinacy (AD) asserts that all games of the kind described above, with any payoff set, are determined. Full determinacy contradicts the axiom of choice, and one can only expect it to hold in a submodel of V where choice fails. Under large cardinal axioms it holds in the model $L(\mathbb{R})$, consisting of all sets constructible from reals. Determinacy turns out to provide the correct axiomatization for studying this model.

One of the consequences of $\mathsf{AD}^{L(\mathbb{R})}$ is that there are many countably complete ultrafilters in $L(\mathbb{R})$. Even the first uncountable cardinal, $\aleph_1$, carries a measure. (Note that this does not mean there is an elementary embedding with critical point $\aleph_1$ definable over $L(\mathbb{R})$; such a map would contradict the discussion following Claim 1.2. It only means there is a countably complete, normal, non-principal ultrafilter on $\aleph_1$. The connection between these two statements breaks down without the axiom of choice.) These ultrafilters can be used for forcing over $L(\mathbb{R})$. We say more on this below.

A set A is $\mathbf{\Sigma}^1_1$ if it is the projection (for example from a plane to the line) of a Borel set. A set is $\mathbf{\Pi}^1_n$ if it is the complement of a $\mathbf{\Sigma}^1_n$ set, and $\mathbf{\Sigma}^1_{n+1}$ if it is the projection of a $\mathbf{\Pi}^1_n$ set. A set is $\mathbf{\Delta}^1_n$ if it is both $\mathbf{\Sigma}^1_n$ and $\mathbf{\Pi}^1_n$.

A *prewellorder* is a relation which is transitive, reflexive, linear, and wellfounded. Given a prewellorder R, define an equivalence relation $\sim$ by $x \sim y$ iff $x\, R\, y \wedge y\, R\, x$. Then R induces a wellorder on the equivalence classes of $\sim$. The *ordertype* of R is the unique ordinal isomorphic to this wellorder. We use prewellorders rather than wellorders below, since in contexts where the axiom of choice fails, for example in the context of definable sets of reals, wellorders are too restrictive. (Moving from prewellorders to wellorders requires choosing representatives for equivalence classes, and this cannot always be done in a definable way.)

For each n, $\boldsymbol{\delta}^1_n$ is the supremum of the ordertypes of $\mathbf{\Delta}^1_n$ prewellorders on $\mathbb{R}$. It is clear that every such ordertype is $< \mathrm{Card}(\mathbb{R})^+$, so $\boldsymbol{\delta}^1_n < (2^{\aleph_0})^+$. $\boldsymbol{\delta}^1_n$ is a measure of the size of the continuum in term of definable orders. We saw that the size of $2^{\aleph_0}$ can easily be increased beyond $\aleph_1$ by forcing. As one can imagine, it is much harder to change the size of the $\boldsymbol{\delta}^1_n$s. The ordinals are specified using definable orders, and the mere addition of reals will not change them. Nonetheless, it is consistent that some of these ordinals are cardinals greater than $\aleph_1$. For example by forcing over a model of determinacy, Steel–Van Wesep [**42**] showed it is consistent with ZFC that $\boldsymbol{\delta}^1_2 = \aleph_2$. They assumed more than determinacy, and forced over a model bigger than $L(\mathbb{R})$. But the assumption was reduced to AD for forcing over $L(\mathbb{R})$, by Woodin [**44**].

It is still open whether any $\boldsymbol{\delta}^1_n$ can be a cardinal greater than $\aleph_2$. However it is known that $\aleph_2$ can be realized by other $\boldsymbol{\delta}^1_n$s, not just $\boldsymbol{\delta}^1_2$.

THEOREM 4.3. *It is consistent with* ZFC *that* $\boldsymbol{\delta}^1_3 = \aleph_2$.

The theorem is due independently to Neeman and Woodin, see Neeman [**32**]. It generalizes to $n > 3$. The proof is done by forcing over $L(\mathbb{R})$ assuming $\mathsf{AD}^{L(\mathbb{R})}$. The forcing notion uses a supercompactness measure on $\mathcal{P}_{\omega_1}(\lambda)$, where λ is the cardinal predecessor of $\boldsymbol{\delta}^1_3$, in $L(\mathbb{R})$. It is very similar to Prikry forcing, but the stems are countable rather than finite, and the measure used is not the supercompactness measure itself, but a certain product of the measure, that applies to sequences of variable countable length. This product is analyzed in [**32**] using the close connections between determinacy and inner model theory.

We concentrated above on uses of large cardinals in forcing, that directly involve the large cardinal ultrafilters. There are many other uses that involve reflection properties of the large cardinals, rather than the resulting ultrafilters. Perhaps the most famous are the proofs of the consistency of the *proper and semi-proper forcing axioms* (PFA and SPFA), and *Martin's maximum* (MM) from supercompact cardinals. The axioms were developed through work of Baumgartner, Shelah, and Foreman–Magidor–Shelah , see [**3, 4, 36, 35, 9**]. They assert the existence, in the ground model, of partial generics for all forcing notions in specific classes of posets, called proper, semi-proper, and stationary set (in ω_1) preserving. The third axiom follows from the second. The consistency of each of the first two is obtained by forming an iteration, composed of a selection of posets in the respective class, reaching up to a supercompact cardinal. The reflection properties of the supercompact cardinal are such that, with a careful selection of the posets via a mechanism due to Laver, the resulting model has partial generics for *all* posets in the class. For a survey of this subject we refer the reader to Abraham [**1**].

Let us now return to our original application of ultrafilters in forcing, Corollary 3.9. Its initial form, using the existence of a supercompact cardinal in the hypothesis, is due to Silver. The large cardinal assumption was brought down to the existence of a cardinal κ of Mitchell order κ^{++} through work of Woodin and Gitik [**10**]. This lower assumption is *necessary*:

THEOREM 4.4. *Suppose it is consistent that the singular cardinals hypothesis fails. Then it is consistent that there exists a cardinal κ of Mitchell order κ^{++}.*

The theorem is due to Gitik [**11**]. It is proved using two very deep subjects in set theory: PCF theory which we did not discuss and which the reader can find in Shelah [**38**] and Abraham–Magidor [**2**]; and inner model theory, which we discussed briefly in Section 1.

Inner model theory is concerned with the construction of canonical models for large cardinal axioms, and is used to extract large cardinal strength from statements about the universe. The application of inner model theory in the proof of Theorem 4.4 builds heavily on the development of inner models at the level of measures of high order, due to Mitchell [**28**]. One of the key results is a *covering lemma* for these models, due to Mitchell, that allows approximating elements of the actual universe V using elements of the core models and ordinals generated from the measures in the core model. The first result of this kind is due to Jensen [**8**] and deals with the model L. It states that either (a) every uncountable set of ordinals in the universe can be covered by a set of the same size that belongs to L; or (b) there is a class of ordinals that is homogeneous with respect to truth in L (such classes are called classes of indiscernibles for L, and they capture the content of a measurable cardinal, restricted to L). Covering lemmas continue to hold with L replaced by inner models for large cardinals, but their forms become more restricted as one allows greater large cardinals. One of the most recent examples can be found in Jensen–Schimmerling–Schindler–Steel [**17**], where it is used to derive large cardinal strength from the proper forcing axiom. But the exact strength of the proper forcing axiom remains open. It may well be at the level of supercompact cardinals.

Bringing inner model theory to this level is one of the longest standing open projects in set theory.

References

[1] Uri Abraham. Proper forcing. In Aki Kanamori and Matthew Foreman, editors, *Handbook of Set Theory*, volume I, pages 333–394. Springer, 2010.
[2] Uri Abraham and Menachem Magidor. Cardinal arithmetic. In Aki Kanamori and Matthew Foreman, editors, *Handbook of Set Theory*, volume II, pages 1149–1228. Springer, 2010.
[3] James E. Baumgartner. Iterated forcing. In *Surveys in set theory*, volume 87 of *London Math. Soc. Lecture Note Ser.*, pages 1–59. Cambridge Univ. Press, Cambridge, 1983.
[4] James E. Baumgartner. Applications of the proper forcing axiom. In *Handbook of set-theoretic topology*, pages 913–959. North-Holland, Amsterdam, 1984.
[5] Paul Cohen. The independence of the continuum hypothesis. *Proc. Nat. Acad. Sci. U.S.A.*, 50:1143–1148, 1963.
[6] Paul J. Cohen. The independence of the continuum hypothesis. II. *Proc. Nat. Acad. Sci. U.S.A.*, 51:105–110, 1964.
[7] James Cummings and Matthew Foreman. Diagonal prikry extensions. *J. Symbolic Logic*. to appear.
[8] Keith I. Devlin and R. B. Jensen. Marginalia to a theorem of Silver. In $\models$*ISILC Logic Conference (Proc. Internat. Summer Inst. and Logic Colloq., Kiel, 1974)*, pages 115–142. Lecture Notes in Math., Vol. 499. Springer, Berlin, 1975.
[9] M. Foreman, M. Magidor, and S. Shelah. Martin's maximum, saturated ideals, and nonregular ultrafilters. I. *Ann. of Math. (2)*, 127(1):1–47, 1988.
[10] Moti Gitik. The negation of the singular cardinal hypothesis from $o(\kappa) = \kappa^{++}$. *Ann. Pure Appl. Logic*, 43(3):209–234, 1989.
[11] Moti Gitik. The strength of the failure of the singular cardinal hypothesis. *Ann. Pure Appl. Logic*, 51(3):215–240, 1991.
[12] Moti Gitik and Menachem Magidor. The singular cardinal hypothesis revisited. In *Set theory of the continuum (Berkeley, CA, 1989)*, volume 26 of *Math. Sci. Res. Inst. Publ.*, pages 243–279. Springer, New York, 1992.
[13] Moti Gitik and Menachem Magidor. Extender based forcings. *J. Symbolic Logic*, 59(2):445–460, 1994.
[14] Moti Gitik, Itay Neeman, and Dima Sinapova. A cardinal preserving extension making the set of points of countable V cofinality nonstationary. *Arch. Math. Logic*, 46(5-6):451–456, 2007.
[15] Moti Gitik and Assaf Sharon. On SCH and the approachability property. *Proc. Amer. Math. Soc.*, 136(1):311–320 (electronic), 2008.
[16] Thomas Jech. *Set theory*. Springer Monographs in Mathematics. Springer-Verlag, Berlin, 2003. The third millennium edition, revised and expanded.
[17] Ronald Jensen, Ernest Schimmerling, Ralf Schindler, and John Steel. Stacking mice. *J. Symbolic Logic*, 74(1):315–335, 2009.
[18] Akihiro Kanamori. *The higher infinite*. Springer Monographs in Mathematics. Springer-Verlag, Berlin, second edition, 2003. Large cardinals in set theory from their beginnings.
[19] Kenneth Kunen. Some applications of iterated ultrapowers in set theory. *Ann. Math. Logic*, 1:179–227, 1970.
[20] Kenneth Kunen. *Set theory*, volume 102 of *Studies in Logic and the Foundations of Mathematics*. North-Holland Publishing Co., Amsterdam, 1983. An introduction to independence proofs, Reprint of the 1980 original.
[21] Menachem Magidor. On the singular cardinals problem. I. *Israel J. Math.*, 28(1-2):1–31, 1977.
[22] Menachem Magidor. On the singular cardinals problem. II. *Ann. Math. (2)*, 106(3):517–547, 1977.
[23] Menachem Magidor. Changing cofinality of cardinals. *Fund. Math.*, 99(1):61–71, 1978.
[24] Menachem Magidor and Saharon Shelah. The tree property at successors of singular cardinals. *Arch. Math. Logic*, 35(5-6):385–404, 1996.
[25] Donald A. Martin. Borel determinacy. *Ann. of Math. (2)*, 102(2):363–371, 1975.

[26] Donald A. Martin. A purely inductive proof of Borel determinacy. In *Recursion theory (Ithaca, N.Y., 1982)*, volume 42 of *Proc. Sympos. Pure Math.*, pages 303–308. Amer. Math. Soc., Providence, RI, 1985.

[27] Donald A. Martin and John R. Steel. Projective determinacy. *Proc. Nat. Acad. Sci. U.S.A.*, 85(18):6582–6586, 1988.

[28] William J. Mitchell. The core model for sequences of measures. I. *Math. Proc. Cambridge Philos. Soc.*, 95(2):229–260, 1984.

[29] Itay Neeman. Aronszajn trees and failure of the singular cardinal hypothesis. *J. Mathematical Logic*. To appear.

[30] Itay Neeman. Optimal proofs of determinacy. *Bull. Symbolic Logic*, 1(3):327–339, 1995.

[31] Itay Neeman. Optimal proofs of determinacy. II. *J. Math. Log.*, 2(2):227–258, 2002.

[32] Itay Neeman. Inner models and ultrafilters in $L(\mathbb{R})$. *Bull. Symbolic Logic*, 13(1):31–53, 2007.

[33] Itay Neeman. Determinacy in $L(\mathbb{R})$. In Aki Kanamori and Matthew Foreman, editors, *Handbook of Set Theory*, volume III, pages 1877–1950. Springer, 2010.

[34] Dana Scott. Measurable cardinals and constructible sets. *Bull. Acad. Polon. Sci. Sér. Sci. Math. Astronom. Phys.*, 9:521–524, 1961.

[35] Saharon Shelah. Iterated forcing and changing cofinalities. *Israel J. Math.*, 40(1):1–32, 1981.

[36] Saharon Shelah. *Proper forcing*, volume 940 of *Lecture Notes in Mathematics*. Springer-Verlag, Berlin, 1982.

[37] Saharon Shelah. Cardinal arithmetic for skeptics. *Bull. Amer. Math. Soc. (N.S.)*, 26(2):197–210, 1992.

[38] Saharon Shelah. *Cardinal arithmetic*, volume 29 of *Oxford Logic Guides*. The Clarendon Press Oxford University Press, New York, 1994. Oxford Science Publications.

[39] J. R. Shoenfield. Unramified forcing. In *Axiomatic Set Theory (Proc. Sympos. Pure Math., Vol. XIII, Part I, Univ. California, Los Angeles, Calif., 1967)*, pages 357–381. Amer. Math. Soc., Providence, R.I., 1971.

[40] Jack Silver. On the singular cardinals problem. In *Proceedings of the International Congress of Mathematicians (Vancouver, B. C., 1974), Vol. 1*, pages 265–268. Canad. Math. Congress, Montreal, Que., 1975.

[41] Dima Sinapova. A model for a very good scale and a bad scale. *J. Symbolic Logic*, 73(4):1361–1372, 2008.

[42] John R. Steel and Robert Van Wesep. Two consequences of determinacy consistent with choice. *Trans. Amer. Math. Soc.*, 272(1):67–85, 1982.

[43] Stanisław Ulam. On measure theory in general set theory (doctoral dissertation). *Wiadom. Mat.*, 33:155–168, 1997. Reprint of the 1933 original.

[44] W. Hugh Woodin. Some consistency results in ZFC using AD. In *Cabal seminar 79–81*, volume 1019 of *Lecture Notes in Math.*, pages 172–198. Springer, Berlin, 1983.

[45] W. Hugh Woodin. Supercompact cardinals, sets of reals, and weakly homogeneous trees. *Proc. Nat. Acad. Sci. U.S.A.*, 85(18):6587–6591, 1988.

Department of Mathematics
University of California Los Angeles
Los Angeles, CA 90095-1555
ineeman@math.ucla.edu

Titles in This Series

530 **Vitaly Bergelson, Andreas Blass, Mauro Di Nasso, and Renling Jin, Editors,** Ultrafilters across mathematics, 2010

529 **Robert Sims and Daniel Ueltschi, Editors,** Entropy and the quantum, 2010

528 **Alberto Farina and Enrico Valdinoci, Editors,** Symmetry for elliptic PDEs, 2010

527 **Ricardo Castaño-Bernard, Yan Soibelman, and Ilia Zharkov, Editors,** Mirror symmetry and tropical geometry, 2010

526 **Helge Holden and Kenneth H. Karlsen, Editors,** Nonlinear partial differential equations and hyperbolic wave phenomena, 2010

525 **Manuel D. Contreras and Santiago Díaz-Madrigal, Editors,** Five lectures in complex analysis, 2010

524 **Mark L. Lewis, Gabriel Navarro, Donald S. Passman, and Thomas R. Wolf, Editors,** Character theory of finite groups, 2010

523 **Aiden A. Bruen and David L. Wehlau, Editors,** Error-correcting codes, finite geometries and cryptography, 2010

522 **Oscar García-Prada, Peter E. Newstead, Luis Álverez-Cónsul, Indranil Biswas, Steven B. Bradlow, and Tomás L. Gómez, Editors,** Vector bundles and complex geometry, 2010

521 **David Kohel and Robert Rolland, Editors,** Arithmetic, geometry, cryptography and coding theory 2009, 2010

520 **Manuel E. Lladser, Robert S. Maier, Marni Mishna, and Andrew Rechnitzer, Editors,** Algorithmic probability and combinatorics, 2010

519 **Yves Félix, Gregory Lupton, and Samuel B. Smith, Editors,** Homotopy theory of function spaces and related topics, 2010

518 **Gary McGuire, Gary L. Mullen, Daniel Panario, and Igor E. Shparlinski, Editors,** Finite fields: Theory and applications, 2010

517 **Tewodros Amdeberhan, Luis A. Medina, and Victor H. Moll, Editors,** Gems in experimental mathematics, 2010

516 **Marlos A.G. Viana and Henry P. Wynn, Editors,** Algebraic methods in statistics and probability II, 2010

515 **Santiago Carrillo Menéndez and José Luis Fernández Pérez, Editors,** Mathematics in finance, 2010

514 **Arie Leizarowitz, Boris S. Mordukhovich, Itai Shafrir, and Alexander J. Zaslavski, Editors,** Nonlinear analysis and optimization II, 2010

513 **Arie Leizarowitz, Boris S. Mordukhovich, Itai Shafrir, and Alexander J. Zaslavski, Editors,** Nonlinear analysis and optimization I, 2010

512 **Albert Fathi, Yong-Geun Oh, and Claude Viterbo, Editors,** Symplectic topology and measure preserving dynamical systems, 2010

511 **Luise-Charlotte Kappe, Arturo Magidin, and Robert Fitzgerald Morse, Editors,** Computational group theory and the theory of groups, II, 2010

510 **Mario Bonk, Jane Gilman, Howard Masur, Yair Minsky, and Michael Wolf, Editors,** In the Tradition of Ahlfors-Bers, V, 2010

509 **Primitivo B. Acosta-Humánez and Francisco Marcellán, Editors,** Differential algebra, complex analysis and orthogonal polynomials, 2010

508 **Martin Berz and Khodr Shamseddine, Editors,** Advances in p-Adic and non-archimedean analysis, 2010

507 **Jorge Arvesú, Francisco Marcellán, and Andrei Martínez-Finkelshtein, Editors,** Recent trends in orthogonal polynomials and approximation theory, 2010

506 **Yun Gao, Naihuan Jing, Michael Lau, and Kailash C. Misra, Editors,** Quantum affine algebras, extended affine Lie algebras, and their applications, 2010

505 **Patricio Cifuentes, José García-Cuerva, Gustavo Garrigós, Eugenio Hernández, José María Martell, Javier Parcet, Alberto Ruiz, Fernándo Soria, José Luis Torrea, and Ana Vargas, Editors,** Harmonic analysis and partial differential equations, 2010

For a complete list of titles in this series, visit the
AMS Bookstore at **www.ams.org/bookstore/**.